OPS NORMAL

The Authorised Operational History of the Army Air Corps, 1957–2017

Volume 1

John Greenacre & Mike Peters

Helion & Company

Helion & Company Limited
Unit 8 Amherst Business Centre
Budbrooke Road
Warwick
CV34 5WE
England
Tel. 01926 499 619
Email: info@helion.co.uk
Website: www.helion.co.uk
Twitter: @helionbooks
Visit our blog at blog.helion.co.uk

Published by Helion & Company 2024
Designed and typeset by Mach 3 Solutions (www.mach3solutions.co.uk)
Cover designed by Paul Hewitt, Battlefield Design (www.battlefield-design.co.uk)

ISBN 978-1-804515-34-1

British Library Cataloguing-in-Publication Data.
A catalogue record for this book is available from the British Library.

For details of other military history titles published by Helion & Company Limited contact the above address or visit our website: http://www.helion.co.uk.

We always welcome receipt of book proposals from prospective authors.

Ops Normal is the only complete and authorised account of British Army Aviation operations undertaken by the Army Air Corps. It is a fitting tribute to the men and women who wore the sky blue beret, especially those that made the ultimate sacrifice. This rich history is also a testimony to those that survive with the physical and mental scars of their service.

Contents

Lists of Illustrations

List of Maps

List of Contributors

Adams-Cairns, Ian
Ashton, Mick
Barnes, Micky
Baynes, Bob
Blaxland, Allan
Bonner, Sean
Brooke, Hugo
Burdett, Bill
Canterbury, David
Cash, Andy
Carpenter, David
Cole, Peter
Courtenay, Peter
Edwards, Paul
Fox, Douglas
Gater-Smith, Howard
Grist, Robin
Harrod, Bob
Hathaway, John
Hayhurst, Tony
Holtom, Spencer
Hyslop, Christopher
James, Ron
Kalinski, Dick
Keating, Martin
Kellock, David
Kirkcaldy, Kevin
Mallock, Ross
Marfleet, David
McGee, Dave
Morton-Race, John
Petrie, Bob
Ralls, David
Richards, Michael
Roberts, Dave
Scott, John
Sibun, Colin
Simkins, Andrew
Skinner, David
Small, Fred
Thursby, Nigel
Turner, Jim
Waldock, David
Walker, John
Walters, Stephen
Ward, Jason
Wellesley, Andy
Whidborne, Dick

Foreword

Over the past 30 years several books have been written that have covered the long, impressive history of Army flying. They have extensively recorded in words and pictures Army Aviation's people, aircraft types and operations from before the Second World War to Afghanistan but none have focused in detail solely on the operations of the Army Air Corps since its formation in 1957. This new book by former Army Air Corps officers Dr John Greenacre and Mike Peters has, in exquisite detail, been written to capture the authoritative history of those operations with extensive research enhanced with the personal testimonies of those that took part.

The Army Air Corps has come of age through the era of reconnaissance and attack aviation, with the Corps and what it can deliver now at the forefront of all Army deployments. The ingenuity, determination, and sheer bravery of putting aircraft in harm's way in the years leading up to today is an inspiration to us all, borne out by so many decorations for skill and courage. From fixed wing to rotary aircraft; from experimenting with the helicopter as a flying machine to hastily mounted machine guns; from wire guided to laser guided missiles, the Army Air Corps has always pushed the boundaries of innovation to provide the Army and Defence with this ubiquitous capability. As well as fighting machines, the utility and reconnaissance airplane and helicopter have provided the backbone to supporting those operations. The first volume of this book covers an unprecedented period of development starting at a time when the helicopter's uses in battle were clear but untested. The next will cover operations from the 1980s onwards, which will be familiar to many, and where we cemented our operational reputation in the eyes of the world.

As Colonel Commandant I am immensely proud of what the Army Air Corps has achieved over the years of its history; and what our people continue to deliver every day on current operations. Our heritage provides the inspiration for the present and the future; this book enshrines that in an authorised history that will become the reference standard for years to come.

Lieutenant General Sir Nicholas Borton, KCB DSO MBE
Colonel Commandant Army Air Corps

Preface to Volume 1

'Ops Normal' is a radio call routinely used by Army Air Corps aircrew during a flying sortie. Short for 'operations normal' the call denotes that the aircraft is safe and operating according to plan, in the absence of anything more significant to report. When the authors were initially approached by the AAC to consider writing this history it was the first idea we had for a title. We have found no reason to change our minds during the writing process, particularly as 'Ops Normal' could be considered to have a second meaning in the context of this history. For the first 60 years of its existence the soldiers and officers from the AAC and the wider Army aviation family were constantly on operations around the world. From Malaya to Borneo, Cyprus to Aden and British Guiana to Belize. In Hong Kong, Rhodesia and the Falkland Islands. Across Northern Ireland and the Balkans. From Kuwait and Iraq, through Afghanistan to Libya. Over 60 years, for generations of AAC soldiers and officers, being on operations was indeed normal.

Several authors have published works on aspects of the AAC over the past couple of decades, but they have resulted in the history of the Corps becoming fractured. The works tend to concentrate on individual squadrons, such as Guy Warners' *Auster to Apache* and *First in the Field*.[1] Others focus on particular operations, Steven Taylor's *Air War Northern Ireland* and Will Laidlaw's *Apache over Libya* for instance.[2] More complete histories of the AAC, Anthony Farrar-Hockley's *The Army in the Air*, published in 1994, for example do not include the Corps' more recent operational experience and tend to focus on the AAC's antecedents.[3] This work is an attempt to collate the historiography, pull the AAC's history together and bring it up-to-date, from the establishment of the Corps in 1957 to its Diamond Jubilee in 2017, and it has been deliberately researched, structured and written as an operational history. It does not focus on organisational or technological change, except where they have a direct effect on operations. It is not a social history and so many of the more sensational tales of misadventure that we have been sent are, alas, not incorporated in it. The authors debated the inclusion of the AAC's contribution to the British Army of the Rhine between 1957 and 1989. Many of the aspects of the AAC's presence in Germany undoubtedly qualify as operational experience. It is, however, such an immense and detailed subject, that we decided it deserves a separate history of its own. It is unquestionable that the Royal Electrical and Mechanical Engineers have made a crucial contribution to the success of AAC operations. The officers and soldiers of the REME make frequent appearances in this history, maintaining the AAC's equipment and keeping its aircraft in the air, often in difficult and dangerous circumstances. Members of the REME have contributed their experiences, but *Ops Normal* is not intended, nor can it be a full record of their Corps' engineering achievements.

This work has had a long gestation period. We had our first discussions with the AAC in 2013 but did not formally begin the process of research and writing until the autumn of 2019. Almost

1 G. Warner, *First in the Field; 651 Squadron AAC* (Barnsley: Pen & Sword, 2011), and G. Warner, *From Auster to Apache; The History of 656 Squadron RAF/AAC, 1942–2012* (Barnsley: Pen & Sword, 2013).
2 S. Taylor, *Air War Northern Ireland* (Barnsley: Pen & Sword, 2018) and W. Laidlaw, *Apache Over Libya* (Barnsley: Pen & Sword, 2016).
3 A. Farrar-Hockley, *The Army in the Air: The History of the Army Air Corps* (Stroud: Alan Sutton, 1994).

immediately we ran into two issues that rendered our estimated publication date obsolete. In the spring of 2020, the Covid pandemic closed archives and prevented us travelling to conduct interviews and research. When we were able to restart the process, we began to realise we had underestimated the task we had undertaken. The volume of material in the archive at the Army Flying Museum at Middle Wallop exceeded our expectations. More significantly, we were pleasantly surprised with the response we received from AAC veterans and serving soldiers, and their desire and willingness to contribute their own experiences to this history. We were nearly overwhelmed by the amount of material we had to work through. The overall result was twofold. First, it considerably delayed our projected publication date. Second, it meant that we needed to split the work over two volumes.

This first volume covers operations that began between 1957 and 1979. It includes operations in Malaya, Borneo, Hong Kong, Cyprus, Aden, British Guiana, Belize and Northern Ireland. Where those operations continued beyond 1979 or had a non-operational legacy that extended later than 1979 it has all been included in this volume. Borneo, Hong Kong, Cyprus and Belize all fit that category. The exception is Northern Ireland. The history of the AAC's contribution to Operation Banner up until 1979 is included in this volume. The second volume will continue and complete that experience, including the work of the fixed wing fraternity, along with all the other operations conducted by the AAC up to 2017.

We did not begin this process with any thesis in mind or any academic argument we sought to prove. Rather, we just wanted to unite and present the otherwise fractured history of the men, women and machines of the AAC during the big wars and small wars that took place across the globe between 1957 and 2017. Nevertheless, during the research and writing process themes and trends have become apparent. Difficulties with equipment are eternal; aircraft were delayed into service and did not always meet expectations; role equipment for specific tasks often had to be improvised and adapted. Aviation tactics, techniques and procedures (TTPs) are cyclical. Some TTPs that were developed for a specific operation were often found to have more enduring utility across multiple theatres. Aviation tactics would sometimes fade from the corporate consciousness at the end of one operation, only to resurface, perhaps in a modified form and sometimes many years later, in a subsequent operation.

Above all, however, the most enduring factor that has arisen from our research, has been the constant dedication, ingenuity, enthusiasm, perseverance and sacrifice of the generations of soldiers and officers who served with the AAC and the wider Army aviation family over the 60 years this history records. Where possible we have allowed those veterans and serving soldiers and officers to speak in their own words. Many of their accounts are contemporary to the events they are describing and therefore often reflect the language and attitudes of the time. We offer no apology but invite understanding.

John Greenacre and Mike Peters
Suffolk, 2024

Acknowledgements

First and foremost, the authors would like to acknowledge the contribution made by the former, long-serving Army Air Corps Regimental Secretary, Lieutenant Colonel (Retired) Chris Ions MBE AAC in preserving and promoting the history of the Corps. Without Chris's vision, enthusiasm, dedication and perseverance this history would not have been conceived or completed and the authors are grateful for his enduring support and backing.

On Chris's well-deserved retirement, Lieutenant Colonel Bernie Fowler AAC has stepped into the breach and continued to be a committed advocate for this project. We are extremely grateful for the support and sponsorship of the AAC Charity (Charity Commission No.1184333). We would also like to acknowledge the support of successive Colonels Army Air Corps including Colonels Murray Whiteside, Andy Cash, Richard Green, Jon Bryant, Julian Facer and Ed Butterworth. We would like to thank the Regimental Committee for its patience and for continuing to support us despite having considerably overshot our initial deadline.

The research, writing and publication of this book has been made possible by generous legacies left to the AAC. We hope therefore, that this history is, in part, a fitting memorial to Major (Retired) Neville 'Sam' Law AFC and his wife Margaret, and Staff Sergeant Fenton Sear AAC and his father Gordon.

We could not have tackled this project without the dedication and patient support of the staff at the Army Flying Museum at Middle Wallop. We would especially like to acknowledge the professional advice and support given freely by the museum's archivists Marjolijn Verbrugge, Kyle Thomason and Henry Whitington. We would also like to thank curator Susan Lindsay for her constant enthusiasm and interest. Andrew Simkins and Mark Meaton have also been generous with allowing access to the 656 Squadron Association archives and to quote from the Association publication, *The Chinthe*.

Many other individuals have made significant contributions to this volume. Jason Grover has accumulated a wealth of knowledge on honours and awards made to Army aviation personnel since 1957. He has freely and cheerfully shared his expertise and saved the authors many hours of research as a result. Our thanks also go to Samantha von Huetz who generously gave us access to the papers of her father, Francis Graham-Bell. We thank Duncan Rogers at Helion for guiding us through the publication process and George Anderson for his work producing the maps in this volume. Alex Reeve-Howarth at Regimental Headquarters AAC has been most helpful in publishing our requirements for information each month in *Hawkeye*. Richard Munro, a former AAC officer, has also made an important contribution as a professional indexer. Lieutenant Colonel Taniya Morris MBE AAC has reverted to adjutant, working as a thorough proof-reader and, more importantly, as a critical friend.

The authors would also like to acknowledge previous historians of the Corps on whose work we have been able to build. In particular we would like to single out the late Major Steve Elsey AAC who was great friend and inspiration to us both. We would also like to thank Guy Warner for his willingness to allow us to quote from his previous works.

As ever, both authors unreservedly thank their wives, Jo-anne and Karen, children, grandchildren and wider families for their patience and understanding in allowing us to pursue this time-consuming project.

Finally, this volume is built on the accumulated experience of generations of serving soldiers and veterans who have been members of the Army Air Corps or worked within the wider Army aviation family. We are grateful to all those who, in the past contributed to the *Army Air Corps Journal* and its forerunners, and the *Army Air Corps Newsletter*. The wealth of information they have bestowed on subsequent generations is invaluable. In particular we would like to thank those veterans and serving soldiers and officers, recorded in the List of Contributors, who reached out and made contributions specifically for this volume. Any misinterpretation of their words or errors of fact are the fault of the authors alone.

Abbreviations and Acronyms

2IC	Second in Command
AA	Anti-Aircraft
AAC	Army Air Corps
ABFAC	Airborne Forward Air Control
AFC	Air Force Cross
AGL	Above Ground Level
ANZUK	Australia, New Zealand and the United Kingdom
AOP	Air Observation Post
AQMS	Artificer Quarter Master Sergeant
ARF	Airborne Reaction Force
ATC	Air Traffic Control
AUW	All-Up Weight
BAOR	British Army of the Rhine
BASO	Brigade Air Staff Officer
BRITCON	British Contingent
CAA	Civil Aviation Authority
CAP	Combat Air Patrol
CAS	Close Air Support
Casevac	Casualty Evacuation
CBAS	Commando Brigade Air Squadron
Cdo	Commando
CO	Commanding Officer
CP	Command Post
CT	Communist Terrorist
DCM	Distinguished Conduct Medal
DFC	Distinguished Flying Cross
DFM	Distinguished Flying Medal
DZ	Drop Zone
EOKA	*Ethniki Organosis Kyprion Agoniston* (National Organisation of Cypriot Fighters)
ESBA	Eastern Sovereign Base Area
FAC	Forward Air Control
FARELF	Far East Land Forces
FGA	Fighter Ground Attack
FLOSY	Front for the Liberation of Occupied South Yemen
FOB	Forward Operating Base
FRA	Federal Regular Army
GPMG	General Purpose Machine Gun
GPR	Glider Pilot Regiment
GS	General Service
HLG	Helicopter Landing Ground

HLS	Helicopter Landing Site
HQ	Headquarters
HQNI	Headquarters Northern Ireland
IED	Improvised Explosive Device
IO	Intelligence Officer
IRA	Irish Republican Army
JEHU	Joint Experimental Helicopter Unit
JHU	Joint Helicopter Unit
LAD	Light Aid Detachment
LMG	Light Machine Gun
LZ	Landing Zone
MBE	Member of the British Empire
MC	Military Cross
MEC	Middle East Command
Medevac	Medical Evacuation
MM	Military Medal
MNLA	Malayan National Liberation Army
MOD	Ministry of Defence
NAAFI	Navy, Army and Air Force Institute
NGFS	Naval Gun Fire Support
NLF	National Liberation Front
OC	Officer Commanding
OP	Observation Post
Ops	Operations
Orbat	Order of Battle
PFF	Police Field Force
PLA	People's Liberation Army
POR	Post Operational Report
QCVSA	Queen's Commendation for Valuable Service in the Air
QHI	Qualified Helicopter Instructor
QRF	Quick Reaction Force
RA	Royal Artillery
RAAF	Royal Australian Air Force
RAC	Royal Armoured Corps
RAF	Royal Air Force
RAMC	Royal Army Medical Corps
RASC	Royal Army Service Corps
RCT	Royal Corps of Transport
Recce	Reconnaissance
REME	Royal Electrical and Mechanical Engineers
RFA	Royal Fleet Auxiliary
RHKAAF	Royal Hong Kong Auxiliary Air Force
RHKP	Royal Hong Kong Police
RM	Royal Marine
RTR	Royal Tank Regiment
RUC	Royal Ulster Constabulary
SAA	South Arabian Army
SAAF	South Arabian Air Force
SAP	South Arabian Police

SARBE	Search and Rescue Beacon
SAS	Special Air Service
SF	Special Forces
SITREP	Situation Report
SLR	Self-Loading Rifle
Tac HQ	Tactical Headquarters
UDR	Ulster Defence Regiment
UN	United Nations
UNBZ	United Nations Buffer Zone
UNFICYP	United Nations Force in Cyprus
VCAS	Vice Chief of the Air Staff
WO	War Office
WSBA	Western Sovereign Base Area

Chapter 1

Suez

> We in JEHU all knew how great an asset the helicopter could be to the soldier, but we had to prove it...[1]

The Army Council instigated the formation of the Joint Experimental Helicopter Unit (JEHU) on 1 April 1955. Those that were posted to the JEHU reported to what was then still very much an RAF station at Middle Wallop. JEHU was a collaborative partnership between the Army and the RAF, and this was reflected in both its structure and its manning. The Royal Navy had its own aircraft and therefore had little to do with this new venture between the other two services. Lieutenant Colonel J. F. T. 'Jock' Scott of the Royal Artillery was appointed to command JEHU. His second in command was Squadron Leader D. L. 'Danny' Kearns RAF.

Pilots and ground crew for JEHU were drawn equally from the two sponsoring services. The RAF providing a group of trained helicopter pilots, who had either accrued their flying hours while flying on search and rescue squadrons, or in the case of some, operational flying in Malaya. The Army pilots were, in the main experienced fixed wing pilots from the RA Air Observation Post (AOP) squadrons. The gunner pilots were joined by Royal Army Service Corps cap-badged pilots who had all completed fixed wing tours in the Glider Pilot Regiment before completing rotary conversion. Aircraft maintenance and servicing was entirely an RAF responsibility. All other supporting roles were carried out by Army personnel. These tasks included ground handling, driving, refuelling, and also acting as aircrew on an ad hoc basis. The initial aircraft strength of JEHU was a single flight of six Bristol Sycamore Mk14 helicopters. Less than a year later, in March 1956, a second flight of six Westland Whirlwind Mk2 helicopters was added to the JEHU establishment. Although JEHU was certainly a joint venture, the unit and the trial programme was financed solely from the Army budget.[2]

In 1955 Army aviation existed on a small scale. The capability that did exist was made up of the scattered remnants of a series of post-war cuts in defence spending. Critically, the RAF owned and controlled all the light aircraft operated and flown by the Army. Although light aircraft units were tasked by the Army and the aircraft flown by Army pilots, the squadrons and flights were, in effect, RAF units. The aircraft flown by the Army were procured by the RAF, serviced by the RAF and all Army pilots were trained by the RAF. A regular cause of friction was the fact that not only did the RAF select what aircraft types the Army flew, but also exactly how many were purchased. The RAF was however less and less interested in owning Army aviation and the Army's growing aspirations in its development. This was the jet age and the shadow of the hydrogen bomb loomed

1 A. D. Fitzgerald, 'An Outline History of the Joint Experimental Helicopter Unit 1955–1959', *Army Air Corps Journal*, 1980, p.76.
2 R. A. Norman-Walker, 'The Formation of the AAC – 1955 to 1957', *Army Air Corps Journal,* 1982, p.13.

large in RAF thinking. This and the chance to reduce expenditure by relinquishing responsibility for Army aviation was seen as a golden opportunity by both the RAF and the Air Ministry.[3]

A census of the Army's aviation capability in 1955 did not take long. Army pilot training took place at Middle Wallop; once trained, pilots were posted to one of a handful of units. There were three deployed AOP squadrons of varying size and composition, a total of 11 AOP Flights of four aircraft were shared out between the squadron headquarters. In addition, five light liaison flights operated outside the squadrons. At the same time that JEHU was activated, negotiations between the Air Ministry, the War Office, and the fledgling MOD were edging closer to agreement on the establishment of a small independent aviation capability for the Army. The strength of what would eventually become the Army Air Corps (AAC) was envisaged at 1,500 personnel, dispersed around the existing squadrons and flights.[4]

Elsewhere in the world however, just over a year after its establishment, an international crisis was about to derail JEHU's progress. On 26 July 1956, Egyptian president Gamal Abdel Nasser nationalised the Suez Canal Company, which prior to that was owned primarily by British and French shareholders. The British and French governments conspired with Israel to engineer events that would allow them to retake the Canal Zone by force and re-establish Anglo-French control. On 29 October Israel invaded the Egyptian Sinai. Britain and France issued a joint ultimatum calling for a cease fire. The demand was ignored by both sides which then gave the British and French governments the pretext they needed to launch a military operation to seize control of the Suez Canal and secure the vital waterway for international navigation.[5]

On 14 August 1956, HQ JEHU, was formally warned that trials were to cease and JEHU was to begin mobilisation for war. Scott and the JEHU command team faced the immediate challenge of reconfiguring from a peacetime trials unit to a fully operational and deployable field unit. The use of helicopters as part of what was soon to be known as Operation Musketeer was conceived by Rear Admiral Guy Sayer of the Royal Navy. Sayer would also deploy on Musketeer as Flag Officer, Helicopter Group. The events of the following weeks and months would do much to reinforce the case for an independent aviation capability within the Army and, as importantly, the viability of the helicopter as a battlefield capability.

Before any deployment could take place there were some fundamental manning and logistic problems to resolve. Not all of the RAF helicopter technicians were deployable and so were quickly posted out along with their collective experience on the aircraft. Trial instruments and equipment had to be stripped from each of the airframes. In addition, the unit equipment table was based on short term deployments to Salisbury Plain, not carrier-borne operations or desert flying. The quartermaster department faced significant challenges in the days leading up to embarkation. The sourcing of a thirty-day spares pack-up for both aircraft types was critical work. This was especially true of the Whirlwind, as there were no spares for that aircraft type held in Middle East Command. In spite of all of this, JEHU was at full war strength, with a 30-day stores pack by 25 August 1956.

The loss of 28 non-deployable RAF technicians coupled with the need to remove all non-essential trials equipment from aircraft had an immediate impact on flying training and JEHU's operational output. Nevertheless, this picked up again through September as replacement technicians were posted into Middle Wallop. Amid rumour and counter-rumour as to when, where, and how deployment would happen, some dry training took place on the airfield. Deck landing techniques

3 Fitzgerald, 'An Outline History of the Joint Experimental Helicopter Unit', pp.76–80.

4 Norman-Walker, 'The Formation of the AAC', p.14.

5 For a more detailed account of the Suez Crisis see S. Lucas, *Britain and Suez: The Lion's Last Roar* (Manchester: Manchester University Press, 1996).

JHU Bowsers were employed on board HMS Ocean to augment the limited AVGAS refuelling facilities for the Operation Musketeer Lifts. (Army Flying Museum)

were talked through using scale models and initial deck landing training took place with the dimensions of a carrier deck marked out on the grass.

Finally, on 1 October JEHU sailed from Devonport on board HMS *Theseus* for 12 days of flying training at sea. The lessons learned over this period proved to be vital later during Musketeer. The integration of Army and RAF ground crew with the Fleet Air Arm handlers into combined watches proved to be the most efficient way to balance JEHU's technical knowledge of the Sycamore and Whirlwind with the experience and skills of RN deck handlers. By the end of the training JEHU crews had completed 611 deck landings and a total of 163 flying hours. Refuelling on deck, and the speed at which it could take place was identified as a key planning constraint. The carrier's normal complement of aircraft did not burn AVGAS and only a few of the ship's storage tanks could be turned over for helicopter fuel. The plan for 45 Commando Royal Marines' assault hinged on each carrier being able to refuel six aircraft simultaneously. The solution was to run three refuel lines from the ship's tanks, and another three from JEHU's two Army fuel bowsers onboard. The bowsers were positioned forward of the island on specially constructed platforms close to the edge of the flight deck. The positioning was deliberate; in the event of a fire, the bowsers could quickly be jettisoned over the side by a damage control party.[6]

6 F. Graham-Bell, *The Other Side of the Coin, Suez 1956* (Unpublished manuscript, 1967).

The original requirement for the JEHU when it was warned for Musketeer was to operate in the casualty evacuation role only. Early in the planning however, it became clear there was an additional requirement to lift 45 Cdo RM from ship to shore. This heli-lift during the initial assault was to seize two defended bridges across a water gap which could not be outflanked by amphibious or airborne forces once they were on land. However, it was subsequently calculated that there would not be enough helicopters available to carry sufficient force in a single lift to seize and secure the bridges. Even though the area around the bridges was due to be 'softened up' before the assault, the casualty rate in terms of helicopters was expected to be 10 percent on each lift. That would mean the loss of up to three of the 22 aircraft during the first lift and possibly a further two on each subsequent lift. In three lifts a third of the JEHU's helicopters would be lost. That rate of loss was deemed unacceptable and as a result 45 Cdo RM was kept as a ready reserve afloat, ready to be flown ashore by helicopter in a reinforcing or exploitation role. It was considered that such risks might be acceptable during a *coup de main* type of operation where, 'provided the job was done, the loss was of no accord.'[7] More generally however, it appeared that for the risks to remain reasonable, helicopters should be used to rapidly fly in follow-up forces, onto ground already secured by the amphibious and airborne landings.

After a brief respite and reconstitution of stores the newly designated Joint Helicopter Unit embarked on HMS *Ocean* with 12 aircraft. The carrier sailed from Devonport for Malta on 27 October. Briefings and mission planning continued at sea, interrupted on 1 November for a day's flying training off Malta. The training successfully focused on the emplaning and deplaning drills of 45 Cdo RM. The launching of the required waves of six aircraft was also rehearsed. Planning was based on each fully equipped marine weighing in at 220lbs. Seats were to be removed and troops were to sit on the cabin floor and, in the case of the Sycamore, two marines were allowed to sit with their legs outboard during flight. Lieutenant Jack Smith RM commanded Z Troop, 45 Cdo RM and described the passengers' view of the Sycamore:

> Each man, as well as carrying his own ammunition, weapon, rations, water, respirator and spare clothing, had to carry some support weapon ammunition. All the Commando's vehicles had already sailed in the slow LST [Landing Ship Tank] and LCT [Landing Craft Tank] convoy a week before and were thus unlikely to join up with the unit until some hours after landing.
>
> A loaded Sycamore presented an extraordinary sight. The back seats, side panels and unessential fittings had been stripped to increase the lift. The three passengers sat on the floor, one hunched in the middle with six mortar bombs on his lap, the other two with their legs dangling over the side, each holding a 106mm anti-tank shell about three feet long. The man in the centre was responsible for the two outboard members not falling out. The Whirlwind was a little more orthodox, but there were no seats, doors or windows. The five passengers held onto any handhold available. On approaching the landing zone, the Bren gunner was ordered to put down suppressive fire out of the window if necessary, while the riflemen covered the area of the door. Communication between troops and pilot in both aircraft types was either by shouting or by tugging at the pilot's legs.[8]

The laden weight of the troops dictated the maximum combat loading for each aircraft type. The Whirlwind Mk22 could carry seven men, the Whirlwind Mk2 five men and the Sycamore

7 The National Archives (hereafter TNA), War Office papers (WO) 288/76: 'Report on the employment of Joint Helicopter Unit on Op Musketeer – "Operational Control"', 14 January 1957.

8 R. Jackson, *Suez, The Forgotten Invasion* (Shrewsbury: Airlife Publishing Ltd, 1996), p.62.

A fully laden Sycamore waits to lift from HMS Ocean as three Westland Whirlwinds head ashore to land assault troops. (Army Flying Museum)

Mk14 three men. The decision was also taken to create additional lift capacity on the Sycamore by flying the assault without a second pilot. With training completed, the fleet sailed from Malta on 3 November. The JHU and 845 Naval Air Squadron headquarters were briefed on their collective mission. The JHU orders for Musketeer listed tasks in order of priority, with the 45 Cdo RM landing top priority, followed by casualty evacuation (casevac) and finally reconnaissance and liaison sorties.[9]

The two JHU flights were to launch from HMS *Ocean*. The flights comprised, six Sycamore Mk14s and six Whirlwind HAR2. The 12 helicopters were flown and supported by 21 officers and 103 other ranks. The heliborne assault force that they were to carry into battle was made up of 17 officers and 103 other ranks of 45 Cdo RM. Similarly, aboard HMS *Theseus*, 845 NAS, waited with their 10 Whirlwinds, a mix of eight HAR 22, and two HAR 2. The Whirlwinds were flown and serviced by a total of 23 Royal Navy officers and 105 ratings. The balance of the assault force flying with 845 NAS was made up of 23 officers and 286 other ranks of 45 Cdo RM. As well as the usual pre-mission tensions there was an additional, unique pressure that Major Graham-Bell RASC, who was in command of JHU's Sycamore flight, later recounted:

> Our aim was to lift into Port Said the whole of 45 Commando in waves, in an assault landing. This was to be the first time helicopters were to be employed in an operational assault landing

9 TNA, WO 288/76: 'Report on the employment of Joint Helicopter Unit on Op Musketeer – 'Operational Control'', 14 January 1957.

> anywhere in the world, and the future of helicopters as a military vehicle depended to a large extent on its success.[10]

Two separate operation orders were issued. The first detailed a landing on the beaches and the second, a landing on the golf course; both these orders were issued at 0730 hours on 4 November (timings throughout were given as ZULU time, i.e. Greenwich Mean Time). There was a great deal of speculation as to which plan would be adopted on the day, but the final decision would be left until the last safe moment. Anticipating that precise moment was the trick and Captain Roger Harris RASC, a pilot with the JHU remembered the last hours before they started their rotors and lifted from *Ocean's* deck on their historic mission:

> Dawn found us all on the flight deck. A naval and air bombardment had started and oil fires on shore were well alight and we could see the initial waves of assault craft heading for the shore. Off to port we could just see the big French battleship Admiral Richelieu doing her stuff. Then the bell went, and we all assembled in front of Squadron Leader Danny Kearns… who was to lead our formation. He said, 'Sorry chaps, I still don't know where we are to land, but follow me, and when in doubt, lash out.'[11]

The helicopter lift was called forward once the airborne and amphibious landings had taken place. Situation reports coming in from ashore confirmed that the marines of 40 and 42 Cdo RM had secured their respective beaches and were pushing inland. The parachute landing by 3rd Battalion, Parachute Regiment on El Gamil Airfield had also been successful, and the paratroopers were clearing the airfield. At 0445 hours, all aircraft and troops were placed at 10 minutes notice to move and at 0541 hours a signal was received from HQ 3 Commando Brigade calling 45 Cdo RM forward from the carriers. X Hour (the time of first take off) was set for 0610 hours and loading of the first wave of aircraft was ordered.

While the final preparations for battle got underway a single Whirlwind from 845 NAS lifted from the deck of HMS *Theseus*. The second pilot was Captain John Shaw RASC, on detachment from JHU. The lone aircraft carried Lieutenant Colonel Norman Tailyour, the CO of 45 Cdo RM, accompanied by Major Ian De'ath RM, and three marines. The CO and his tactical headquarters (Tac HQ) were flying ahead of the main force to reconnoitre the landing zone that had been designated for the Commando. The proposed site was closer to brigade headquarters than an alternative site on the edge of docks close to the entrance to the canal where there was a statue of Ferdinand de Lesseps, the man responsible for the Suez Canal. However, as the Whirlwind flew over the landing zone they found it to be obscured by smoke and also identified high-tension cables that were a hazard to landing. The decision was taken to drop the CO and his Tac HQ in a nearby sports stadium.

> Almost as soon as they left the helicopter, the reconnaissance party came under heavy fire from an Egyptian position on the far side of the stadium. The five men took refuge in the players' entrance and prepared to fight it out. Fortunately, the Whirlwind crew, circling overhead, spotted their predicament and braved the enemy fire to land back in the stadium. The passengers dashed across two hundred yards of open ground and scrambled into the helicopter, which got away despite sustaining twenty-two hits. The pilot was slightly wounded but

10 TNA, WO 288/76: 'Report on the employment of Joint Helicopter Unit on Op – 'Landing Areas'', 14 January 1957.

11 R. Harris, *Memories of a Happy Soldier* (Publisher unknown, 2018), p.55.

> he managed to locate another landing zone, a rubble strewn patch of waste ground to the west of Lesseps' statue. This was judged to be suitable for a landing, and shortly before 0600 hours Tailyour sent a radio message ordering 45 Commando's mission to proceed.[12]

Unaware that the recce sortie from HMS *Theseus* were under fire ashore, the JHU crews on board HMS *Ocean* were waiting for confirmation of the LZ and the order to begin flying 45 Cdo RM ashore. Major Graham-Bell recalled the final minutes:

> Finally, at 0356 hrs on 6 November, both carriers dropped anchor some nine miles off Port Said and five cables apart. We had arrived, ready for anything, and sat in the swept channel, at immediate notice for slow speed and prepared to slip cables at a moment's notice. As dawn broke, we could see Port Said immediately in front of us, a great pall of black smoke rising into the sky above it. The helicopters were ranged on the flight deck and a signal was received on board to inform us that the first wave was required ashore at 0610hrs. We still did not know where we were supposed to land but readied ourselves by manning the helicopters with the pilots and the first load of marines to be ferried ashore.
>
> Danny Kearns, as Second in Command, was going in the lead Whirlwind with the Flight Commander, while I, as the Sycamore Flight Commander, would lead the second flight from Ocean. With everything ready to go, Danny and myself stood apart from the aircraft, looking at Port Said and quietly speculating whereabouts we would finish up. We did not have long to wait before the CO rushed up and said, 'You are to land in de Lesseps' Square.'
>
> We looked at each other and then at the colonel.
>
> 'Where is de Lesseps' Square? Is there a street map of Port Said?'
>
> 'I don't know' replied our colonel.
>
> I looked back at Danny and thought deeply.
>
> 'My mother sailed out to Malaya a few years ago and she sent me a picture postcard from Port Said, saying that as you enter the Suez Canal there is a significant statue of Ferdinand de Lesseps on the right-hand side. I wouldn't mind betting that is where we are supposed to land.'
>
> And so, with this priceless piece of information, we launched ourselves towards the first ever heliborne assault landing, guided by my mother's postcard.'[13]

The first wave of six JHU Whirlwinds lifted off from HMS *Ocean* as planned, the Sycamores were then moved forward on the flight deck. Within five minutes the pilots had run up their engines and the marines had emplaned ready for take off. Once both waves were airborne the 12 aircraft formed into four vics ('v' shaped formations) of three aircraft. The formation flew at low level, maintaining 70 knots airspeed. The leading wave was made up of six JHU Whirlwinds, followed by five Whirlwinds from 845 NAS. The six JHU Sycamores came next and finally, four more Whirlwinds from 845 NAS.

The pilots flying in the follow-on waves into Port Said had a dilemma to deal with. Once any element of surprise was compromised by the arrival of the leading aircraft, rapid reinforcement was absolutely essential. The marines needed to rally and establish themselves in force before the Egyptians reacted to the landing and counter attacked but the landing site could only accommodate six aircraft at any time. Careful judgement would be required if the follow-on waves were to arrive immediately after the preceding waves had disgorged their troops and lifted out of the landing site. If they arrived too early, they would have to hover or circle in vicinity of the landing

12 Jackson, *Suez, The Forgotten Invasion*, p.104.

13 Graham-Bell, *The Other Side of the Coin*.

site in full view of any defenders and vulnerable to ground fire. Timing between each of the waves was therefore critical to success. Smoke from fires in the area of the Lesseps statue and dust generated by the aircraft on the ground obscured any observation of the landing site.

See Map 1

In spite of the smoke and the dust, the landing of 45 Cdo RM went exceptionally well. The helicopter fly in was uninterrupted, with JHU flying seven consecutive waves of marines onto the landing site without loss to enemy fire. One fully laden Sycamore, XG502, suffered an engine failure during the second wave.[14] Graham-Bell recalled the incident:

> …our only setback was when a Sycamore had an engine failure on its second lift from the deck. Captain Bryan Shaw RASC, the pilot was unhurt but slightly shattered to discover that a large anti-tank missile had been resting just behind his head when they thumped back onto the deck. If the jolting of their enforced landing had set it off, Bryan's head plus a large section of the Sycamore would have been blasted away. Anyway, no damage was done, and the Sycamore was pushed out of the way and the operation continued short of one aircraft.[15]

A second aircraft was lost when a damaged 845 NAS Whirlwind ran out of fuel in sight of *Theseus*. Luckily, the pilot and his passengers were recovered from the water close to their ditched aircraft.

Once the assault troops were established, follow-on waves of underslung loads brought heavy equipment and ammunition ashore. After two and a half hours JHU had lifted 178 marines into Port Said and 12.5 tons of underslung loads. When added to the 845 NAS lift totals, the two helicopter squadrons had brought ashore 415 troops and 25 tons of stores. The combined helicopter force also evacuated 96 casualties from Port Said for medical treatment offshore. Casualty evacuation using helicopters on this scale had not been attempted before and it was not without challenges. The need to minimise weight on the aircraft had resulted in the decision not to fly stretchers into Port Said. Prone casualties therefore arrived at the aircraft on a mixture of wooden doors, blankets or sheets of corrugated metal. Many of these improvised stretchers could not be separated from the casualty and were taken back to the surgical teams waiting offshore.

The JHU Post Operational Report (POR) commented on the speed and efficiency of the medical evacuations by air and the positive effect on the morale of the troops fighting in Port Said. The POR also highlighted, for the first time, a simple design problem that would plague AAC medevac operations for decades – the compatibility of the Army's General Service (GS) stretcher with Army helicopters. The standard issue stretcher would not fit into the Sycamore and was also considered to be extremely heavy. JHU had not been able to carry stretchers to the wounded, nor had they been able to carry casualties on the GS stretcher that medical units were using during the operation. A local modification was therefore made after the initial landing operation:

> Our routine settled down to liaison flights, evacuation of casualties, and the reconnaissance of the surrounding area. On several occasions our helicopters came under fire in the El Cap area, but luckily, we escaped damage and casualties. We also managed to work out a modification of the Sycamore to enable it to carry the issue stretcher. I obtained some lengths of angle iron and these, fixed to the floor of the cabin made excellent runners for the stretcher legs. Netting over the open side of the cabin secured the stretcher in place and gave the patient some piece of mind.[16]

---

14 Sycamore XG502 became a gate guard at Middle Wallop and is currently on display at the Army Flying Museum at Middle Wallop, Hampshire.

15 Graham-Bell, *The Other Side of the Coin.*

16 Graham-Bell, *The Other Side of the Coin.*

Over the three weeks following the landings and occupation of Port Said, the 11 JHU aircraft were employed in support of patrol and policing activity. The JHU remained ashore, operating from El Gamil Airfield from 8 November. They were reinforced at El Gamil by the Austers of 1913 AOP Flight RAF which were flown in from Cyprus. On 10 November a Sycamore was fired on, but no damage was sustained. On 18 November a patrol of the 1st Battalion, The Argyll and Sutherland Highlanders was inserted by JHU to cut off a group of armed civilians that was approaching the airfield in a felucca boat. The felucca was successfully captured and 18 armed insurgents were captured with their personal weapons, a light machine gun, and 3,000 rounds of ammunition.

This enduring task that developed had not been anticipated when JHU landed in Egypt. Initially the JHU had planned to operate from Gamil Airfield for only two or three days during the early stages of the Suez operation before re-embarking. Instead, the unit's helicopters spent 10 days flying from the narrow strip of land west of Port Said leading to problems caused by the environmental conditions. By late November the conditions at Gamil were 'the very worst', surrounded by salt water on three sides the condition of the airframes and major components of the aircraft deteriorated rapidly. Gale force winds drove sand and spray across the airfield that caused serious corrosion to the Whirlwind's magnesium alloy skin on the rear fuselage. It was concluded that the aircraft should have returned to the carriers where the effects of the elements would be reduced and any corrosion monitored.[17]

The Whirlwinds were grounded at El Gamil on 19 November and the decision to withdraw JHU was taken two days later. The Whirlwinds were flown out to HMS *Eagle* on 22 November and were then taken by the carrier to Cyprus for inspection and maintenance. The Sycamores were embarked on HMS *Eagle* on 28 November, before sailing to Malta to meet HMS *Ocean* at sea. Once the transfer was complete, HMS *Ocean* turned for home with the Sycamores on board. Meanwhile, the remaining Anglo-French forces at Port Said were beginning their withdrawal. They were being replaced by a multinational UN peacekeeping force. The ground support element of JHU withdrew from El Gamil on 30 November, embarking on HMS *Ravager*, a very slow LST. After a rough and eventful voyage, the ground crew and their vehicles disembarked at Plymouth on 23 December.

Lieutenant Colonel John Scott RA compiled a unit POR for the JHU. In addition to thoughts on the risks involved in the use of helicopters in the direct assault role, or as a *coup de main* force, he reviewed the major lessons from Musketeer. These included the location of temporary operating bases, the limitations of the Sycamore, and command and control arrangements which had caused issues after the initial assault operations. Once the JHU had established itself on El Gamil Airfield a Joint Operations Centre (JOC) had been set up to control helicopter flying operations. The JOC immediately ordered the JHU not to undertake any sorties unless specifically ordered to do so. The JHU was forced to disregard those instructions and accept tasking direct from units when it became clear that the JOC could not react quickly enough to requests for support from HQ II Corps. The JOC subsequently reiterated its instructions and the JHU was forced to adhere to them. There were also tasks being duplicated and poor coordination between the JHU, the AOP flight and the liaison flight which were also operating around Port Said. It was clear that a single point of control for all helicopter operations was required on land, embedded in the headquarters requiring support, which in this case was II Corps.[18]

17 TNA, WO 288/76: 'Report on the employment of Joint Helicopter Unit on Op Musketeer – 'Landing Areas'', 14 January 1957.

18 TNA, WO 288/76: 'Report on the employment of Joint Helicopter Unit on Op Musketeer – 'Operational Control'', undated.

Scott's observations also concerned the shortcomings of the Sycamore helicopters that the JHU operated in Suez alongside the Whirlwinds. The Sycamore had its roots in the very earliest years of British helicopter development. The Bristol Aeroplane Company began to form a helicopter division in the autumn of 1944, with Austrian engineer and helicopter pioneer Raoul Hafner as its chief designer. In 1945 Hafner began work on the Bristol Type 171 which developed into the Bristol Sycamore. Between June 1951 and June 1952, the Army accepted four variants of the Sycamore HC11 for AOP and liaison duties. They were operated by 1906 Flight RAF at Middle Wallop. The aircraft operated by the JHU in Suez were Sycamore Mk14 variants, which had come into service in 1953.[19] Having operated the aircraft in the amphibious assault role, the JHU listed numerous faults in the Sycamore's design. These included main rotor blades that were too flexible, which made them impossible to start in winds over 30 knots; an ineffective rotor brake; a tail rotor that was too low and therefore caused a hazard to crew working on the deck of the carriers during take off and landing; a tail skid that fouled obstructions on the carrier's deck, and undercarriage tyres that were prone to deforming during vertical landings. The main issue with the Sycamore, however, was its size. The aircraft was longer than the Whirlwind and yet only carried four passengers, and therefore was considered 'quite useless' for casualty evacuation. It was used extensively for liaison and reconnaissance where its size was uneconomical and a disadvantage.[20] The problem was that the Sycamore fell between two requirements. In 1946 the War Office had defined three classes of helicopter that would be needed in the future to support the Army: a two-seat AOP, reconnaissance and liaison aircraft; an 8 to 10 seat general purpose helicopter; and a heavy-lift, logistic aircraft capable of lifting 22,400lb (10 tons).[21] The Sycamore fell between the first two of these requirements and fulfilled neither satisfactorily. It was recommended by Scott that the Saunders-Roe Skeeter represented a better option than the Sycamore to fulfil the AOP, reconnaissance and liaison role.

The Westland Whirlwind Mk2 was considered ideal in terms of its size as a utility helicopter but was found to be underpowered. Scott suggested the American Sikorsky S58 (which would be flown as the Westland Wessex from 1958) as an alternative, being the same size as the Whirlwind but capable of carrying 12 to 15 soldiers. He went on to comment that a larger helicopter, such as the Fairey Rotodyne, which might fulfil the War Office's heavy-lift requirement would be too large to be useful in operations such as those conducted during Musketeer. Its effectiveness as a casualty evacuation aircraft would be limited by the increased size of landing site required and as a troop carrier in the assault it would be too vulnerable to enemy fire.[22] As Musketeer was the first time helicopters had been used in the amphibious assault role the JHU's lessons and observations from its time in Suez were undoubtedly valuable.

The Suez Crisis was a political disaster for the British Government and the aftermath of Operation Musketeer had far reaching consequences for Imperial strategy and defence. At a tactical level however, the operation had established the utility of the helicopter in support of an assault, as well as being a useful platform for logistics and medical evacuation. The JHU had flown 340 sorties carrying approximately 300 troops from ship to shore and around the Port Said battlefield. The unit's Sycamores and Whirlwinds had also evacuated nearly 100 British and French casualties to the medical facilities afloat and ferried over 12 tons of equipment and stores into the bridgehead.[23]

19 J. Everett-Heath, *British Military Helicopters* (London: Arms and Armour Press, 1986), pp.41–48.

20 TNA, WO 288/76: 'Report on the employment of Joint Helicopter Unit on Op Musketeer – 'Capacity of Utility Helicopters'', 14 January 1957.

21 Everett-Heath, *British Military Helicopters,* p.41.

22 TNA, WO 288/76: 'Report on the employment of Joint Helicopter Unit on Op Musketeer – 'Capacity of Utility Helicopters'', 14 January 1957.

23 TNA, WO 288/136: 'JHU War Diary, Summary of Operations for 6–8 November 1956', undated.

In doing so, along with 845 NAS, the JHU had demonstrated that helicopters in assault operations were robust, survivable, and sustainable, and represented a valuable additional capability alongside amphibious and airborne forces.

Chapter 2

The Formation of the Army Air Corps

Birth Pangs

Although Operation Musketeer was a tactical success and proved that the helicopter could contribute to offensive operations, Suez was a strategic disaster for the British Government and was to have far reaching effects. The USA put pressure on the Prime Minister, Sir Anthony Eden, to end the conflict through an oil embargo and threats to devalue the pound. British and French forces were forced to make a full withdrawal from Egypt and the Sinai by 22 December 1956. Eden was accused of misleading Parliament and resigned on 9 January 1957, being succeeded by his Chancellor of the Exchequer, Harold Macmillan. It was clear that Britain had lost its imperial pre-eminence. Macmillan accelerated the withdrawal from empire and the process of decolonisation. The British economy remained fragile and it was evident that government policy needed revision in light of the new strategic and economic reality. This included policy within the MOD led by Duncan Sandys who had been promoted into the post by Macmillan. Sandys published his White Paper 'Statement on Defence, 1957' on 1 April that year.[1] Sandys' paper introduced root and branch reforms and rebalanced the relationship across the three services. It ended National Service in a bid to save money and manpower and led to amalgamation and disbandment across the infantry, cavalry and RA. It prioritised the development of missile technology, reduced manned aircraft projects and led to a restructuring of the British aircraft industry. Although not specifically directed within the White Paper, it also formalised the conditions for the British Army to establish a new dedicated aviation branch on 1 September 1957.

The Army had lost its previous internal air and aviation capability, the Royal Flying Corps when it was amalgamated with the Royal Naval Air Service to form the Royal Air Force on 1 April 1918. In the inter-war years the nascent RAF jealously guarded its fiefdom and resisted any attempts from the other two services to reclaim any degree of control. The RN did reclaim some autonomy with the return of the Fleet Air Arm to Admiralty command on 24 May 1939. The Army however, continued to have to rely on the RAF for its air and aviation requirements. Prior to 1957 there was at least one line of thought within the Army that saw no issue with the RAF retaining control. 'Army aviation took form as a new Corps on 1st September 1957. The lateness of its appearance must be ascribed to economics sought by the Treasury. This tardiness was tolerated by the Army itself partly on account of the prevalent attitude that the operation of aircraft is the prerogative of air forces.'[2]

1 TNA, Cabinet Papers (CAB) 129/86/C57/84/34: 'Statement on Defence, 1957', 1 April 1957.
2 R.J. Parker, 'The Third Flank', *Army Air Corps Association Journal*, 1959, p.16.

The establishment of an Army Aviation Organisation had indeed been a long and tortuous affair and Sandys' White Paper only put a cap on extended and often acrimonious negotiations. Almost immediately after the end of the Second World War the Chief of the Imperial General Staff, Field Marshal Alan Brooke, requested responsibility for 'its own aircraft for certain specifically Army functions.' In response, the Chief of the Air Staff, Air Chief Marshal Charles Portal, explained that it would be uneconomic and inefficient for the Army to try to operate such a small fleet of aircraft and that responsibility for providing them should remain with the RAF. The same argument in various forms rumbled on with no resolution for another decade; however it was the increasing military utility of the helicopter that eventually forced the issue. The RAF was wary of the Army's enthusiasm for rotary flight and was concerned that any increasing use of helicopters on the battlefield for liaison, observation, evacuation, transport and even armed action could pose a financial strain on the Air Ministry's budgets to the detriment of other, more iconic air projects. In June 1956 the Air Ministry offered a compromise that it believed would demonstrate it was not being unnecessarily obstructive while not giving too much away. It proposed that:

i) The Army should assume full responsibility for purchasing, manning, maintaining and operating unarmed aircraft used solely for its own purposes in the AOP and light liaison roles. The maximum all up weight – the weight of the aircraft plus whatever load was to be carried in flight – would not exceed 4,000 pounds.
ii) The Air Ministry in conjunction with the War Office would remain responsible for formulating specifications.
iii) In the field, Army aircraft would be subject to the overall direction of the air force commander.[3]

It was a clever ploy that gave the Army its much sought after responsibility and control while the all-up weight limit and requirement that aircraft should remain unarmed would prevent it from encroaching too far into the RAF's bailiwick. The proposal was agreed in principle between the Secretary of State for War, John Hare, and the Secretary of State for Air, Nigel Birch, in November 1956 and it was rubber stamped by Sandys early in his tenure as Minister for Defence. The decision did have critics and there were members of the new aviation branch that believed the Army had conceded too much in its creation.

> It was in my view a sorry day for Army flying in general and the liaison branch in particular. The introduction of the 4,500 pounds [*sic*] all up weight as the top limit to Army flying has introduced an artificial barrier to further development. We have moved out of a large professional organization whose job is the provision and manning of flying machines and have set ourselves up in competition in a small way by ourselves.[4]

In fact, the agreement in principle between the two Secretaries of State had provided some pliancy over the question of the AUW limit with Birch writing to Hare that, '…it is only reasonable that you should suggest some flexibility [in all-up weight]. The time may come when there will be no suitable aircraft with a weight under 4,000lbs.'[5] As will be seen, events soon conspired to render both the AUW limit, and the specification that Army aircraft should remain unarmed, obsolete.

3 Farrar-Hockley, *The Army in the Air*, p.178.
4 Army Flying Museum (AFM) 179: 'Newsletter for 1957, 10 Indep Recce Flight Army Air Corps', 23 December 1957.
5 Farrar-Hockley, *The Army in the Air*, p.179.

Symbols of Unity

With the new organisation now given ministerial approval there were many subsequent decisions to be made, some operational and others domestic, albeit no less important at the time. The first task was to agree a title for the new organisation for which the following criteria were suggested:

a) The initials of the title should not be either rude or already in use by some other Corps.
b) The title should cover the whole organisation, both Officers and Other Ranks, and all roles of the organisation. For instance, the "Air Reconnaissance Corps" does not cover the air OP or liaison roles, while the "Pilot Regiment" does not cover the ground crew.
c) The title must sound good.[6]

Dozens of titles were suggested including Army Aviation Organisation, Light Aircraft Corps, Army Air Regiment, Army Flying Regiment and the Army Flying Corps. There was even the idea of returning to the Royal Flying Corps. Some of the potential names elicited violent objections. 'Army Aviation Organisation – Horrible! It includes both "Aviation" and "Organisation" and comes straight from America.' Major Maurice Sutcliffe, the Commanding Officer of the GPR agreed. 'Don't put the word "Aviation" in. Let's be British!… I strongly suggest Army Air Corps. It sounds well, is to the point and no one from the most horse loving to the REME mech could object.'[7] The majority of those canvased agreed with Major Sutcliffe, and Army Air Corps was adopted and in general use by March 1957. The title harked back to its original incarnation formed in February 1942 as the overarching Corps for the Parachute Regiment and the GPR but it was also felt that it was forward looking and inclusive.

That inclusivity would be important as the initial members of the AAC came from a disparate collection of corps and regiments. Early estimates of the manpower establishment of the AAC for operational and training purposes settled on a figure of 400 officers and 1,410 non-commissioned officers and soldiers. That was later reduced to around 1,500 personnel in total. Most of that manpower came from the RA AOP organisation and the RAF's liaison flights. The balance would come from the remaining personnel of the GPR, resulting in the disbandment of that Regiment on 31 August 1957, and finally secondments from other regiments and corps. The principle of continuing to employ NCOs as pilots was also established. The wide range of backgrounds of the original members of the AAC when it was formally established on 1 September 1957 needed a focus in order to forge a unified identity.

> Now that the insignia of the Glider Pilot Regiment is no longer worn, a startling number of Regiments are physically apparent in the Flight. Though all personnel owe allegiance to their Regiments, it will be better for the unit when a decision has been made on the mode of dress of the Army Air Corps, producing some semblance of uniformity.[8]

The accoutrements of the new Corps, particularly the question of a cap badge and beret, were vital in creating that uniformity. The colour of the beret seemed at first to be a relatively straight forward choice. The original members of the Second World War AAC, both Parachute Regiment

6 TNA, WO 32/17164: Letter, 'Major R. A. Norman-Walker to officers commanding all light aircraft units', 10 January 1957.
7 TNA, WO 32/17164: Letter, 'M. Sutcliffe', 16 January 1957 and letter, 'Lieutenant Colonel J.F.T. Scott to Major R. A. Norman-Walker', 24 January 1957.
8 AFM 182: '13 Liaison Flight AAC, Quarterly Newsletter for July, August and September 1957', 24 October 1957.

and GPR, wore the maroon beret and that seemed the logical choice for the new incarnation and attracted wide support. There was however, entrenched opposition to the idea as voiced by General Richard Gale:

> I am NOT happy about the Red Beret for the Army Air Corps...Boy Browning, I and [Major General] Hopkinson together fixed the Red Beret in the Dungeons at Home Forces during the War. In the [Parachute] Regiment we regard the Red Beret as something for which we fight. I myself NEVER wore a tin hat in battle – only my Red Beret. The Glider Pilot Regiment was part of us; we loved and respected them. There was NO question about their wearing it.
>
> But now the Glider Pilot Regiment is no more. That they or those who are now in the Air Corps should be permitted to wear it is reasonable and would have the support of every member of Airborne Forces – but that this should be extended 'ad lib' to all members, pilots and ground crews of all aircraft flown by the Army Air Corps is a very different matter.
>
> I fully appreciate that this, as with everything else, is a WO [War Office] decision. My views, and those held by the Regimental Committee, are as above.[9]

With the maroon beret not an option another distinctive colour was sought. The process was later recalled by Colonel Bob Begbie AAC, who had been involved in the decision as a young major:

> When we looked around for a suitable colour which was distinctive yet practical, we found it difficult to find one that some other regiment or service did not already wear. Green berets were worn by the Royal Marine Commandos; black by tank regiments; dark blue by the gunners and infantry; khaki of varying shades by the SAS and Guards regiments; brown by the 11th Hussars; grey by the Scots Greys and so on. It seemed that whatever colour we considered might be suitable, it had already been adopted and jealously guarded by someone else.
>
> So we looked at aspects of our antecedents other than the maroon beret and decided that the colours that predominated were Cambridge Blue (the facings and shoulder titles of the Glider Pilot Regiment and of the Army Pilot Wings worn by the Air OP and the Glider Pilots) and the Slate Blue of the RAF which would symbolise the RAF Air OP Squadrons. So two berets were produced, one light blue, the other slate blue for the inspection and decision of an ad hoc committee consisting of, among others, our Colonel Commandant, General Stockwell, the Brigadier AAC, Brigadier Pat Weston, myself and the Chief Aircraft Engineer...
>
> When the two berets were produced to the Committee there was a general sucking of teeth. The slate blue beret was rejected mainly on the grounds that it might make us indistinguishable from the RAF Regiment which if I remember at that time wore khaki battledress and the slate blue beret; it was too RAF in fact. These comments were mild compared to those uttered when the light blue beret was modelled. It should be remembered that in 1957 the Army was very conservative: the dreary war years were not all that long gone and somehow the light blue seemed then more than garish than with years of acceptance it does now. However most of the comments, (once Brig Pat Weston had convinced us that the punch-ups we would have defending our light blue beret against the ribald remarks of the other soldiery would make us more feared than any 'red devil' in a maroon beret – and if we lost out on that, then the girls would certainly fall for anyone wearing it) reduced our objections to doubts on its workaday practicality. Nothing daunted, Pat Weston, ready for this, produced a bucket of oil and a bucket of petrol, plunged the light blue beret into the oil

9 TNA, WO 32/17903: Letter, 'R. Gale', 11 August 1957.

> where it emerged as we expected, in a horrible mess. He then sloshed it around in the bucket of petrol where surprisingly it emerged almost glowing clean! So we were persuaded that the light blue beret was practical (besides being our only choice) and that it should become the distinctive headgear of the AAC.[10]

One of the rejected designs for the AAC cap badge. (The National Archives)

With the beret settled the choice of cap badge was the next major regimental decision to require attention. Various submissions were made most of which included an eagle or similar bird of prey. A bald eagle with its wings held high as if about to land with the motto *Vide A Supra* (Observe from Above) on a scroll beneath its outstretched talons. An eagle with its wings folded as if in a stoop, one talon grasping a sword, the other a bolt of lightning. Both were rejected, possibly because they resembled badges of the Second World War *Wehrmacht*, although the former did appear to live on in the AAC collar badges.[11]

Towards the end of 1957 the AAC submitted a potential badge to the Army Dress Committee and to J. R. Gaunt and Son Ltd, badge and button makers to the British military. The badge was based on that of the GPR with the surrounding title replaced by a laurel wreath and the Queen's Crown instead of the King's. The AAC admitted there was probably room for improvement and Gaunt duly obliged, returning a refined illustration of the badge on 16 December noting that, 'we think it is a great improvement with regards to the sweep of the body and head etc.'[12] The modified design was agreed and went into production. By the end of 1958 all questions of AAC regimental identity, including beret, cap badge, collar badge, buttons, arm badge, mess kit and stable belt had been settled and authorised by the Army Dress Committee and by the Queen.[13]

10 B. Begbie, 'Why the Light Blue Beret?', *Army Air Corps Journal*, 1982, pp.47–49.

11 TNA, WO 32/17164: 'Drawings of potential AAC cap badges', undated.

12 TNA, WO 32/17903: Letter 'J.R. Gaunt & Son Ltd', 16 December 1957.

13 TNA, WO 32/17903: Letter 'Director Land/Air Warfare', 7 November 1958.

The Equipment Challenge

Although the AAC would be taking over existing units and sub-units on 1 September 1957 they did not all come from the same background, so a degree of reorganisation was required in order to bring some unity to the new Corps. The AAC was given authorisation to use the RAF squadron designation system and was granted squadron numbers 651 to 699 for Army use. On formation the AAC established three squadron HQs: 651, 652 and 656. It was envisaged that each squadron would support an army division and be made up of three flights plus a light aircraft workshop. The flights would either be reconnaissance or liaison flights comprising, in theory, one section of three fixed wing aircraft and one section of three helicopters. The original RAF designation of the flights was abbreviated and altered from 1905 AOP Flight to 5 Reconnaissance (Recce) Flight for example. On establishment the AAC was made up of the following flights:

1 Recce Flight
2 Recce Flight
3 Recce Flight
4 Recce Flight
5 Recce Flight
6 Independent Liaison Flight
7 Liaison Flight
8 Independent Recce Flight
9 Recce Flight
10 Independent Recce Flight
11 Liaison Flight
12 Liaison Flight
13 Liaison Flight
14 Recce Flight
15 Independent Liaison Flight
20 Independent Recce Flight

The independent flights were not part of the three squadrons and were based in Middle Wallop (6 Flight), Malta (8 Flight), Cyprus (10 and 15 Flights) and Hong Kong (20 Flight). The Light Aircraft School RAF at Middle Wallop became the Army Air Corps Centre, the AAC's HQ and spiritual home.[14]

The early decisions on dress and organisation were important steps in establishing AAC unity and uniformity but there were also immediate operational questions that required attention. In particular, the types of aircraft that the AAC required going forward was at the forefront of many minds. The Army inherited mainly Auster V and AOP6 fixed wing light aircraft, although those would be steadily relegated to training roles as the Auster AOP9 came into service during the late 1950s. The envisioned rotary contribution to the liaison and recce flights was however, causing more difficulty. The Army already had a history with the helicopter. During the Second World War a few Army pilots, including Major Jack Nicolson and Major Bill Nelson, were given helicopter instruction on the Sikorsky R4 in the USA. Unfortunately, they were demobilised at the end of the war and the experience was lost. In 1946 C Flight (later redesignated 1901 Flight) of 657 AOP Squadron RAF was equipped with the Sikorsky R6 Hoverfly which were flown by Army pilots. The Flight was disbanded after a year, but it had piqued the Army's interest in the utility

14 B. Robertson, *The Army and Aviation* (London: Robert Hale, 1981), pp.186–187.

Hiller UH-12B and Westland Dragonfly of Helicopter Flight, 657 AOP Squadron RAF at Middle Wallop in 1954. (Army Flying Museum)

of the helicopter. In May 1950, 1906 Helicopter Flight was established as part of 657 Squadron at Middle Wallop and equipped with the Hoverfly. A year later, the Flight received Bristol Sycamore aircraft. The helicopters were being used to support exercises in Germany and the UK and the Army was becoming convinced that it provided a capability that it could not do without. The RAF however, remained unconvinced that the helicopter was the right sort of support for the Army and brought forward objections based on cost versus payload and mechanical inefficiency. As a compromise it was decided that a joint Army and RAF experimental unit should be set-up and on 1 April 1955 the JEHU was established at Middle Wallop.[15]

Following Operation Musketeer and the establishment of the AAC the provision of a capable helicopter to the recce and liaison flights continued to pose problems. The idea of mixed fixed wing and rotary flights was proving a challenge:

> On the formation of the Army Air Corps in September 1957, 6 Independent Liaison Flight, to give the full title, was formed from 1906 which had been the War Office V.I.P. Flight equipped with Sycamores. Commanded by Major Sutcliffe 6 Flight went to Germany in the autumn of 1957 to take part in the Army manoeuvres in order to put into action the War Office conception of the mixed fixed-rotary wing recce and liaison Flight. The aircraft used were three Sycamores, two Skeeters and two Auster AOP9's supplied with pilots by the School at Middle Wallop.
>
> In both roles the disadvantages of a mixed Flight from the servicing side became apparent. The Flight put itself at the disposal of various formations in the recce and liaison role, carrying

15 F. Graham-Bell, 'The Army Goes Rotary', *Journal of Army Aviation*, 1967, pp.6–12.

> out a withdrawal with [3rd Royal Tank Regiment] and an advance to contact and recce of a contaminated area under nuclear conditions with the Royal Dragoons. The helicopter was still a novelty to some formation commanders and great use was made by them in this introduction to rotary wings .... The Flight undertook the role of liaison Flight to Corps H.Q. for the final exercise "Cross Swords" during which the five seater Sycamores proved invaluable.[16]

Following the exercises in Germany the helicopters were withdrawn from the Flight. The Sycamore was too large and ungainly for the Army's needs and in any case, it had an AUW of 5,600lb and therefore broached the arbitrary limit. The Saunders-Roe Skeeter was the only realistic option to meet the Army's immediate rotary requirements but in 1957 it suffered from being underpowered, had ground resonance issues and required further development. Admittedly, the Army's requirements were demanding:

> The essential characteristic for any Army aircraft is the ability to land and operate anywhere without the need for airstrips. Helicopters and other V.T.O.L. [vertical take off and landing] devices can do this but certain technical difficulties remain to be overcome. Available cruising speeds are at the moment adequate for the combat zone. Army machines must however be able to fly very low in bad visibility, and at night, to avoid enemy aircraft, radar, and homing missiles. Simplicity of maintenance and small size are essential, as is worldwide operating ability and tolerance to normal land vehicle fuels. Most of these requirements have already been fulfilled but the ideal machine is yet to be designed.[17]

Having lost its helicopters after the 1957 exercises 6 Flight was re-equipped with fixed wing de Havilland Beavers and the Sud Aviation Alouette II helicopter on a trial basis. The latter was found to be very capable. 'The Alouette jet turbine helicopter with its rugged characteristics and unrivalled performance showed how much the Army Air Corps were in need of an aircraft of a similar type.'[18] Although it would later prove a useful stop-gap, the Alouette was only a trial aircraft in 1957 and not a solution to the immediate helicopter problem. The Beaver, like the Sycamore, broached the artificial 4,000lb AUW limit. It did, however, appear that the limit was not going to prove as serious a constraint as originally feared. Shortly after the formation of the AAC an informal discussion between John Hare and Vice Chief of the Air Staff, Air Marshal Geoffrey Tuttle, indicated that it was not a rigid limit. After the meeting Hare's Permanent Under Secretary noted that:

> ...if [the AAC] wanted a particular helicopter for use in the role which was already agreed, which weighed more than 4,000lbs, this would be perfectly negotiable with the Air Ministry. [VCAS] agreed that it would be quite wrong for us to buy a less satisfactory helicopter or lighter craft because it was just within the 4,000lb limit.[19]

16 Anon., '6 Liaison/Depot Flight', *Army Air Corps Association Journal,* 1960, p.39.
17 Parker, 'The Third Flank', pp.17–18.
18 Parker, 'The Third Flank', p.40.
19 TNA, WO 32/17164: Note, 'E.W. Playfair, PUS Defence', 9 August 1957.

The Operational Challenge

The helicopter question contributed to the main challenge facing the AAC on its establishment. Several of the flights and squadrons the Corps inherited were already committed to operations across the world on 1 September 1957 and the AAC had to ensure they provided continuity of support. The Malayan Emergency that had been declared in 1948 continued despite the country having gained its independence the previous day, 31 August 1957. 656 (Air Observation/Light Liaison) Squadron RAF became 656 (Light Aircraft) Squadron AAC based at Sembawang. Its four flights, 2 and 14 Reconnaissance Flights and 7 and 11 Liaison Flights, were spread across the Malayan Peninsula and engaged in counterinsurgency operations to combat the Malayan National Liberation Army. In Cyprus a state of emergency had been in existence for nearly two years when the AAC was established. 10 Recce Flight and 15 Liaison Flight AAC were assisting ground troops in countering the insurgency being conducted by the National Organisation of Cypriot Fighters (*Ethniki Organosis Kyprion Agoniston*, EOKA). Finally, closer to home, in Northern Ireland the Irish Republican Army had been conducting its border campaign, Operation Harvest, since December 1956. On 1 September 1957 13 Liaison Flight AAC was in the air assisting the Army and police on the ground in Ulster. In all cases those flights engaged in operations were equipped with the Auster. In Malaya and Cyprus the RAF was making good use of helicopters and gaining valuable experience in their operation. In both theatres AAC officers urged the Army to expedite the fielding of its own rotary aircraft or the new Corps might find itself obsolete before it had a proper chance to prove its utility. The challenge for the AAC on 1 September 1957 could not be overstated.

> There is a threefold task now facing the Corps. Firstly, it must establish itself as a *corps d'elite;* having had some of the uses of air power properly demonstrated the Army must now press for the full development and control of its own aviation. Secondly, this small corps must do everything possible to stimulate commanders and their staff airwise; unit and brigade commanders should be given indoctrination courses and the capabilities of Army Air Corps units should receive greater emphasis at the Staff College. Thirdly, within its limited budget it must continue to develop new techniques and equipment. The Army Air Corps has adopted the eagle as its emblem and the soldier must now, like the eagle, take the air to find, strike and destroy.[20]

20 Parker, 'The Third Flank', p.18.

Chapter 3

Malaya

The Malayan Emergency

The creation of the Army Air Corps was not marked by any great public fanfare. The British Government, Army and public were far more interested in the implications of an announcement of international importance made the preceding day. On 31 August 1957 the independence of Malaya, known nationally as Merdeka, was formally announced marking a turning point in the counterinsurgency that had been fought there since 1948. Following the Second World War the Malayan Communist Party (MCP) began attacking symbols of British rule in the country, particularly the economically important rubber plantations. In response Britain declared a state of emergency in Malaya on 17 June 1948, beginning what became known as the Malayan Emergency. The MCP leadership under Chin Peng moved into the jungles and formed the Malayan National Liberation Army (MNLA). Many MNLA fighters were veterans of the Malayan Peoples' Anti-Japanese Army, a communist guerrilla army previously trained, armed and funded by the British to fight against Japan during the Second World War. The MNLA found support in the countryside, particularly from the Malayan Chinese community who provided members of the MNLA with shelter and sustenance.[1]

The initial British response was clumsy and relied heavily on the military's experience of fighting the Japanese in the jungles of Burma during the Second World War. The character of the Malayan conflict was, however, entirely different and required a new approach. The MNLA did not confront British forces directly, they relied on stealth and ambush, and they blended into the local population. Large British formations expended huge amounts of effort crashing through the jungle only to find the MNLA had disappeared before they reached them, sheltered by its civilian branch, the Min Yuen. The British needed to develop an effective counterinsurgency strategy. The situation began to improve when General Sir Harold Briggs, the British Army's Director of Operations, was appointed to Malaya in April 1950. Real improvement, however, was brought in with Briggs' replacement General Gerald Templer in January 1952. During the previous October the British High Commissioner in Malaya, Sir Henry Gurney had been assassinated by the MNLA. When Templer arrived he took over both Briggs' and Gurney's positions, thus unifying the political and the military response to the insurgency. Templer's tenure saw a dramatic reduction in MNLA numbers and incidents as he strove to separate the insurgents from their civilian support.

Templer also recognised the advantages of air support. Offensive air support, reconnaissance and transport from the RAF, the RAAF and the RMAF all made significant contributions during

1 For more information on the Malayan Emergency see R. Neillands, *A Fighting Retreat; The British Empire, 1947–97* (London: Hodder and Stoughton, 1997), pp.149–254 and G. van Tonder, *Malayan Emergency; Triumph of the Running Dogs* (Barnsley: Pen & Sword, 2017).

the Emergency under Operation Firedog.[2] In early 1949 the issue of casualty evacuation in the jungle prompted an urgent request for helicopters to be sent to Malaya. On 1 May 1950 the Far East Casualty Evacuation Flight was established by the RAF with three Westland Dragonfly HC2 helicopters. Despite teething problems, the experiment was a success and the establishment was expanded. In 1954 the more capable Bristol Sycamore began appearing in RAF squadrons in Malaya. Although casevac was a valuable and welcome contribution to operations Templer was interested in the potential for using helicopters for the tactical movement of troops, to carry soldiers quickly to points in the jungle where they could surprise and confront the MNLA. In response the RN deployed 848 Naval Air Squadron with 10 American-built Sikorsky S-55s that were soon replaced by RAF Westland Whirlwinds (the same aircraft built under licence), which began arriving in theatre in late 1954.[3]

See Map 2

656 (Air Observation/Light Liaison) Squadron RAF had been reformed from 1914 Flight at Sembawang in Singapore in June 1948 and equipped with the Auster. On 1 September 1957, the first full day of Malayan independence, the squadron became 656 (Light Aircraft) Squadron AAC with its headquarters at Noble Field on the outskirts of Kuala Lumpur. Its commander was Lieutenant Colonel Brian Storey RA. Its four existing flights, No.2 and No.14 Reconnaissance Flights and No.7 and No.11 Liaison Flights, were spread across the 500 mile long Malayan Peninsula from Sembawang in the south to Taiping in the north. Each flight was allocated under the operational control of a brigade which in turn had an area of responsibility of approximately 10,000 square miles. A fifth flight, No.16 Reconnaissance Flight, was added to the Squadron's Orbat and co-located with the headquarters in Kuala Lumpur.[4] 656 Squadron planned to initially mark its transition to the AAC with a disbandment parade for those remaining members of the GPR. Unfortunately, it had to be cancelled when 2 Flight were tasked to locate the crash site of RAF Vickers Valetta VX491 which had disappeared near Tanjong Malim, 40 miles north of Kuala Lumpur. The aircraft was found by the Squadron along with four surviving air despatchers. Despite this setback, the establishment of the AAC did not unduly affect the rhythm of flying operations.

> The transference from RAF to Army command on 1st Sept went very smoothly. The Royal Air Force as usual helped enormously with their advice and practical assistance. We can see no insuperable difficulties either, in the change over from RAF to REME servicing. There is an ample hand-over period for each man and the REME artisans who have arrived have impressed all with their enthusiasm, skill and friendly cooperation.[5]

The AAC inherited a formidable task in Malaya. The geography, topography and climate alone created immense challenges. The Malaysian Peninsular is generally mountainous throughout its 500 mile length south from the border with Thailand and is covered mostly by dense jungle. There are narrow coastal plains along both the east and west of the country that converge on the island of Singapore in the south. On those plains coconut palms, paddy fields, tin mines and rubber plantations were prevalent. The central mountain range, known as the Cameron Highlands, rise to 7,000 feet with narrow valleys, and winding streams and rivers. A main trunk road ran from Singapore up the western side of the country northwards to beyond the Thai border, as did the one

2 For a full examination see M. R. Postgate, *Operation Firedog; Air Support in the Malayan Emergency, 1948–1960* (London: Stationery Office Books, 1992).
3 For more information see D. Taylor, *The First Helicopter Boys* (Barnsley: Air World, 2019).
4 Postgate, *Operation Firedog*, pp.93–94 and Warner, *From Auster to Apache*, pp.106–107.
5 AFM 209: '656 Light Aircraft Squadron, Army Air Corps, Newsletter for the Period July – December 1957', undated, p.4.

Captain George Richey of 7 Flight AAC completes an entry in his log book following a sortie in 1958. Note the AAC cap badge on his flying helmet.

main single-track railway following the general line of the road. There was also a main road from the central capital, Kuala Lumpur, east to the coast road which ran from Johore Bahru in the south to Kota Bahru in the north and beyond.

The weather was a factor all year round in Malaya. The daily pattern, which could nearly always be relied upon, started off clear and windless but with some low cloud and fog in the valleys. When the sun rose cumulus cloud formed, which quickly developed into cumulonimbus. Widespread and substantial thunderstorms with torrential rain would follow in nearly all areas in the afternoons, often extending into the evening. Turbulence could be violent, especially when operating in the mountains where thunderstorms were generally to be avoided. More persistent rain became an issue during the monsoon period at the latter end of each year.

Malaya boasted a number of provincial airfields with concrete runways, as well as smaller airfields, operating internal Malayan Airways Douglas DC-3 aircraft, including Ipoh and Penang Island on the west coast. Across from Penang on the mainland was Butterworth, a military airfield operated by the RAAF and home to two squadrons of North American F-86 Sabres. Across the country, often co-located with military camps in remote areas, there were dozens of smaller strips with grass or laterite (compacted sandy clay) surfaces. These smaller strips were used extensively for liaison with infantry units, as jumping off points for remote operations, and for refuelling. Ground units generally kept these strips in good order and although there were no radio facilities they could provide fuel on demand from gallon tins known as 'flimsies' using funnels and chamois leathers for filtration.

Captain Malcolm Fleming of the Royal Highland Fusiliers flew with 2 Flight and was later the OC of 7 Flight. He described the organisation and day-to-day operations of an AAC flight in Malaya in the late 1950s:

> There were approximately 35 all ranks in each recce Flight; six pilots, a dozen aircraft technicians, the remainder consisting of radio operators, drivers and general duties men and one cook. The pilots were either officers, warrant officers or sergeants seconded from the cavalry, Royal Artillery or infantry regiments. Aircraft Technicians were REME NCOs led by a staff sergeant REME (Artificer). Transport was either radio or GS Land Rovers, and 3 ton trucks. All stores on charge, radio equipment and aircraft spares enabled Flights to operate independently.
>
> All personnel did a three year tour in Theatre and would have arrived by troopship at Singapore, thence proceeding to Squadron HQ in K-L [Kuala Lumpur] by train. Technicians would receive some indoctrination at the aircraft workshops before being allocated to Flights. Pilots were required to carry out six weeks theatre familiarisation at Squadron HQ under the Qualified Flying Instructor, and a fourteen day jungle survival course run by the RAF in Singapore before being authorised to carry passengers, and then they joined their Flights.
>
> Everyone was issued with jungle equipment, green shorts, jackets and slacks, and jungle hat and a pair of jungle boots. Everyday dress at work was invariably shorts, normal boots with socks rolled down and nothing else. The heat and high humidity demanded so. Pilots were issued with an Australian lightweight flying suit which was very popular. Jungle boots were worn when flying. Pilots were also issued with an RAF aertex inner helmet and an outer bone-dome shell together with an oxygen mask with microphone insert. White leather kid gloves were also issued. These became very sweaty in no time but were very much appreciated.
>
> The Auster AOP6s and T7s having been phased out, the workhorse of 656 Squadron in the late 50s and early 60s was the AOP9. It was a popular aircraft with pilots. It was reliable, agile and had a good endurance though power was somewhat reduced due to the high humidity and heat prevailing. This had an effect on its ability to cope with some of the short strips used

> and loads had to be reduced in consequence. However, the Mk 9's low stalling speed, balloon tyres and disc brakes were great assets indeed. It was a fine observation aircraft.
>
> The comprehensive radio fit of the AOP9 consisted of a 10 channel VHF (very high frequency) radio for air traffic, a B47 VHF standard Army radio and an Army 62 HF (high frequency) set (crystal controlled) with a trailing aerial facility, the hand operated winch for which was located in the cabin roof. These radios enabled pilots to communicate with virtually anyone on the ground and in the air. The HF 62 set was particularly vital when operating in remote areas when long-range communications and position reports could be made on the Squadron net on which all Flights listened out during daylight hours.
>
> Each aircraft always carried a survival kit containing medical supplies, ration packs with small, related items; a rifle was included together with a search and rescue beacon, (SARBE) which, if switched on, certain types of RAF rescue aircraft could home in on.
>
> ...The serviceability of the Auster AOP9 in both 2 and 7 Flights, which incidentally did by far the most flying in the Squadron, was first class. There were no total engine failures in either Flight, but there were one or two instances of rough running, none of which resulted in forced landings, and there was a tendency to gulp oil on full throttle climbs (quickly cured). The generally high standard of aircraft servicing was entirely due to the enormous dedication and skill of our REME technicians who invariably toiled in the heat and the rain and often at night. The traditional aim of rectifying unserviceability at the end of each day, ready for the next was rigidly maintained. Routine servicing was always carried out at Flight locations whilst a major servicing was done at K-L in Squadron Workshops.[6]

Although Templer's strategy and the declaration of independence had taken the sting out of the Emergency, the MNLA was still active and operations continued beyond September 1957. There was a stated aim to bring the Emergency to an end by 31 August 1958. As a result, there was an increase in counterinsurgency operations and a subsequent requirement for the continued use of aviation post-independence. All the Squadron's flights were fully employed on a number of tasks in support of Malayan Federal infantry brigades and British and Commonwealth infantry brigades.

2 Flight, based in Ipoh in Perak 170km north of Kuala Lumpur, were engaged in target marking for air strikes by RAF de Havilland Venoms and Avro Lincolns, and observing fire from 95 Field Battery RA as part of Operation Chieftain. Its Austers also recced landing zones for follow-up troops brought in by the RAF Whirlwinds.

A further 30 miles to the northwest at Taiping 7 Flight conducted a number of operations along the Thai border, found two considerable insurgent encampments, called in artillery and air support, and worked closely with 2nd Battalion, Royal Australian Regiment.

In the south of the peninsula, flying from Sembawang on Singapore Island, 11 Flight marked targets for airstrikes and conducted search sorties for a crashed RAF Venom and missing English Electric Canberras. During September 1957 alone the Flight's Austers delivered 4,775lbs of supplies to infantry patrols over 47 sorties. During Operation Huckster the Flight was engaged on extensive night flying operations, using flares to illuminate the jungle, and dropping 'battle simulators' (devices that replicated the noise of gunfire and grenade detonations) to flush out insurgents.

Meanwhile, 14 Flight based at Seremban, 30 miles south-east of Kuala Lumpur, was engaged in the recce of LZs and air photography. One of the Flight's photographs led directly to an ambush on an insurgent position which killed five members of the MNLA, also known as CTs in the language of the day.

6 M. Fleming, 'With the Auster Mk9 in Malaya 1959–62', *The Chinthe*, Spring 2009, pp.26–37.

16 Flight conducted liaison flying, air observation, recce and resupply, and provided a detachment for anti-piracy operations in Borneo. Propaganda leaflet dropping sorties increased. Between the declaration of independence and the end of 1957 656 Squadron dropped 28.5 million leaflets over the jungle representing four percent of the squadron's total flying effort.[7]

The dispersed Flight system meant that aircrew became intimately familiar with the area of responsibility of the brigade they were supporting and the features of the ground they flew over. This was not only critical for successful counterinsurgency operations it was also crucial for survival when sorties were flown in poor weather. Captain Fleming described sorties during which intimate knowledge of the ground helped him out of difficult situations:

> I recall having finished a task in the Kota Bahru area trying to find a way through the mountains to return to Ipoh one afternoon. After looking at several valleys that were all in cloud I eventually, after proceeding south for a while, found one which led to the Cameron Highlands area which at 5,000 feet or so at the top was just in the clear. I knew there was the rather twisty road from there down to the west so I decided to fly down into the tunnel between the road, halfway down the side of the valley, and the cloud base close above. I eventually arrived at the bottom of the valley after quite an exciting ride on to the coastal plain near Chemor, a small town on the north/south railway and then back to base.
>
> I had flown this route in the clear several times before and was familiar with the road and the Cameron Highlands, having been on leave there where there was a fine golf course. A precautionary landing on the course would probably have been an option if things had become impossible but dodgy. If completely baulked, then Kuantan or Trengganu would be alternatives but would result in an uncomfortable night!
>
> Another example of the detailed knowledge of the terrain being used to advantage was on occasions when in the early mornings a task would be called for at Grik, north of Ipoh. It was not unusual for Grik, and all the valleys running north to it, to be under low stratus cloud until being burned off by the sun later. In order to get there on time the trick was to descend under the overcast at low level at Chenderoh Lake north of Ipoh and proceed up the Perak river valley wide at this point which ran a few miles east of Grik after about 30 or so miles. The further north you were the narrower and twistier the river became and there was usually not a lot of room below the cloud and so you flew close to the water. You knew exactly the point at which a 90 degree turn was to be made in order to fly down west into Grik, a few miles on.
>
> This procedure was shown to me by a predecessor in the Flight when we flew the route in the clear previously. It worked extremely well but was demanding and it is of course well known that flying rivers needs to be approached with caution anywhere, even in Malaya where telephone wires across rivers had a nasty habit of appearing unannounced![8]

With the AAC and 656 Squadron established in Malaya the operational priority was to clear districts of their remaining CT groups so their designation could be changed from 'black' to 'white' on the map. Once the entire national map was 'white' the Emergency could be declared over. All means were sought to reduce support to the insurgents so they could not continue their operations. It was common for MNLA groups to operate illegal stills in the jungle producing local liquor known as 'samsu'. The samsu would either be consumed by the CTs or sold for profit. In

7 AFM 209: '656 Light Aircraft Squadron, Army Air Corps, Newsletter for the Period July – December 1957', undated, pp.2–3.

8 Fleming, 'With the Auster Mk9 in Malaya', pp.26–37.

August 1958, 11 Flight were commended by the comptroller of customs in Singapore for finding five illegal distilleries in a four-day period in the Seletar and Senoko river areas. Thanks to the Flight's careful reconnaissance the stills were broken up; 48 gallons of samsu and 1,648 gallons of fermented mash were seized. Auster pilot Captain David Crouch RTR was singled out for his contribution and was subsequently Mentioned in Despatches for his work in Malaya in September 1958 and April 1959.[9]

With CT support being steadily eroded the insurgents were forced into more remote areas where they had to be pursued and then captured, killed or persuaded to give up their cause. Support to these operations often involved 656 Squadron detachments deploying forward from their usual flight operating bases. This entailed the Austers temporarily operating from jungle strips with austere conditions for aircrew, ground crew and technicians alike. Craftsman Derek Walker REME recalled one such detachment with 14 Flight to a rubber plantation at Sungai Tinggi, 40km northwest of Kuala Lumpur:

> Sleeping in trucks and trailers or under them was a new experience for me. The application of anti mosquito oil was a must to avoid being heavily bitten by mosquitoes, however this used to result in excessive perspiration and generally people avoiding you. When you coupled this together with the rubber jungle boots we wore all day in the heat, the combination could be overpowering to say the least in the back of a vehicle. To avoid foot rot it was necessary to bathe your feet daily and powder them profusely with 'Army Foot Powder', the crotch too if you wished to combat prickly heat, sweat rash or tinea. The nightly guard stags in the jungle – with all the weird noises and movements of insects in the darkness – the patrolling of the airstrip to ensure the safety of the aircraft and equipment, one felt very lonely and most relieved to get the stag over and go to sleep in the safety of others, by dawn it got very cold…
>
> The aircrew flew early morning and dusk reconnaissance sorties to locate CT fires or smoke from encampments or movement and vegetation plots during the day. They also did target marking for 'bomb drop' areas. One early morning I carried out pre-flight inspections by torchlight on the aircraft, including a scavenge filter bowl check, on one of the aircraft. That aircraft then took off and I followed in the next one, with Captain Bell. We noticed the first aircraft had returned and landed, but we paid no heed. We returned to Sungai Tinggi an hour or so later from our sortie, having monitored a recent bomb drop area, to find ground runs being carried out, engine cowlings off. Apparently the scavenge filter bowl had come off on take-off and, but for the spray of oil being seen in the sun's rays by the passenger through the aircraft's open doorway, the flight would have continued on, possibly resulting in an engine failure and a jungle crash. Anyway, my neck was on the line even though I knew I had thoroughly checked the engine oil filter bowl for security on the pre-flight inspection. The flight artificer had convinced the OC that the filter could be checked after fitting, found secure, but later dislodge under oil pressure. After three failed attempts, he was successful [in demonstrating the issue] on the fourth, thereby allaying a technical charge against myself and a possible court-martial had an accident occurred. My standing was a little low for a while, but Willie Wastell, the pilot, lived to see another day….
>
> I recall the all night 'light up' of a specific section of railway track by 4-inch Parachute Flares. A task requiring rearming, refuelling and servicing checks for the aircraft to maintain a ten minute turn around between aircraft, this on a jungle airstrip lit only by 'gooseneck' lamps, all night! The completion at dawn was very satisfying. Next, we had three days of

9 AFM ALB 55: Letter, 'Comptroller of Customs, Singapore to the Officer Commanding Royal Naval Air Station Sembawang', 19 September 1958.

> torrential rain during our stay which turned the area into a quagmire of mud which, when dry, smelt awful, then more rain. At this time 'Rum' was issued, very warm and satisfying.[10]

The risks of flying over the jungle were ever-present. 656 Squadron suffered its first fatality on 14 February 1958 when Auster AOP9, XK378 crashed after loss of control in a stall turn on approach to Kroh strip, 80 miles north of Kuala Lumpur. The pilot, Captain Peter Dalley RA was killed in the accident.[11] The dangers were multiplied during remote detachments where aircrew and technicians had to trust each other implicitly. The majority of those junior technicians were National Service soldiers drawn from the aircraft industry and then pushed through a six to eight-week course at Middle Wallop to learn the intricacies of the Auster AOP9. A detachment was often just a single pilot accompanied by a single technician who had to be able to cope with faults in the engine, airframe, electrics, hydraulics, instruments or radio. This could strain the relationship at times as described by Captain Bob Coles RASC:

> The mechanic would look at the pilot and hope he wasn't rough with the aircraft and pray like mad that nothing major would go wrong while, on the other hand, the pilot would look at the mechanic and think "God! Your eight weeks training had better be good."
>
> It is fitting to mention at this juncture that no accidents occurred due to mechanic inefficiency and also very few due to pilot error. This is quite amazing if one could witness some of the places that pilots were expected to put the aircraft down, i.e. on sand beaches, on the bed of a partially dried up river, marshy fields, airstrips cut in the jungle complete with protruding tree stumps and even on a few occasions, using roads just wider than the undercarriage track, hoping that the other road users would notice what was happening and 'hop-it' quick. Those big balloon tyres were marvellous.[12]

In fact, Captain Coles should probably have had more confidence in his National Service technicians. Craftsman Glyn Chambers REME had just completed an apprenticeship with the de Havilland Aircraft Company before undergoing trade training at Middle Wallop and being posted to 7 Flight AAC at Taiping. Once there he found himself working alongside fellow de Havilland apprentices and others who had completed their training with Rolls Royce:

> The mechanics had a lot of work to do. The AOP Mk 9 was fitted with the Blackburn Bombardier engine. Although more powerful, it required very much more maintenance than the de Havilland engines I had worked on in civilian life. There always seemed to be at least one aircraft having a 'top overhaul' mainly required because lead built up on the valve stems and caused the valves to stick open. Crashes had occurred just before my arrival at other Flights because of this. Flight Sergeant Toogood RAF and Staff Sergeant Boam REME kept a close eye on us mechanics and we were assisted on the technical side by three 'Airy Fairies' Air Arm lads.[13]

Captain Coles estimated that 85 percent of accidents were caused by major mechanical failures beyond the influence of the local technicians. Approximately 10 percent of accidents were caused by turbulence or other weather conditions and the remaining 5 percent fell into the 'very few due

10 D. Walker, '656 Squadron Tour, Malaya, 1958–1961', *The Chinthe*, 2013, Iss.2, pp.15–17.
11 Anon., 'Army Air Corps Aircraft Losses Since 1 September 1957', HQ AAC, 13 June 2019.
12 AFM 208: B. Coles, 'Austering over Malaya', *Air Classics*, undated.
13 G. Chambers, 'Narrative of a National Serviceman', *The Chinthe*, Summer 2009, p.34.

REME aircraft mechanics working on a 7 Flight AAC Auster AOP9 at Taiping in 1960. (Army Flying Museum)

to pilot error.' One of the 10 percent of accidents caused by turbulence occurred in September 1958 and drew considerable attention from the press in Malaya, Singapore and Britain. On 4 September Auster WZ696 was being flown by Captain Dickie Jones AAC on a resupply mission in the area of Kota Tinggi, approximately 30 miles north of Singapore. On board with Jones was Lieutenant Mike Gibson of the 1st Battalion, 10th Princess Mary's Own Gurkha Rifles. Captain Jones recounted the incident for the press soon after the event:

> It was about 3.30pm. We were dropping supplies to a Gurkha patrol about 22 miles north-west of Kota Tinggi. Air conditions were turbulent. We made one drop. We tried to climb. Instead, the Auster was carried down into the trees. It went through with a crunching sound. Something dealt me an enormous clout on the back of my head. My helmet saved me.
>
> One wing was shorn off by a tree then we were down. I smelt petrol and heard it pouring from the tank. We scrambled out of the plane and got clear as fast as we could. I was scared stiff lest the plane go up in flames.

Lieutenant Gibson continued the story:

> We had crashed on the side of a valley. I was cut by jagged Perspex as I hauled myself through the shattered cockpit canopy. We ran from the wreck and stumbled across an old terrorist camp, about 100 yards away. It was about six months old and contained a natural pool of water. We had a wash, lit a fire, put up smoke and sent up a couple of Verey lights.

> We looked in the supply pack that we had not been able to drop. It contained clothing, cigarettes and rum. I sat down and had a swig. We got the seat cushion and an emergency survival pack out of the wreck.
>
> Then we found our only protection against terrorists was a Verey light pistol and a rifle whose barrel had been hopelessly bent in the crash. Dickie got on with the job of opening the survival kit, while I started to make a basha in the camp. We knew rescue was on its way but we also realised there was no hope of being brought out before the next day.
>
> We tried to settle down for the night but neither of us got much sleep. Our bruises and cuts made us too uncomfortable. We had arranged to take two-hour watches. Instead we talked for most of the night.
>
> At dawn we ate a bar of chocolate. Then a helicopter appeared overhead and four men from the jungle rescue team parachuted down. Shortly afterwards we heard shouting in the distance. We shouted and whistled back. Then we heard a nearby cry of "Hey Mambo!" the call-sign of the RAF rescue team.[14]

As WZ696 went down Jones had just enough time to send a mayday call which was heard by another Auster. A search was then conducted before night fell. Having located the crash site, the following morning at 0600 hours a four-man RAF parachute rescue team jumped from a Whirlwind to make contact with the two stricken officers. They found them and sent a report on their condition back to Singapore via radio while the team doctor, Squadron Leader Glen-Thomson, attended to their injuries. Despite neither Jones nor Gibson being seriously injured the rescue team found it could not move them to a helicopter landing zone that had been cut into the jungle some distance away by men of 1/10 GR. The rescue team therefore set about creating a new clearing for a landing zone closer to the crash site. Further helicopters flew in to drop axes and chainsaws to the team to aid their efforts. In the afternoon a storm rolled in that stopped all flying and so Jones, Gibson and the rescue team were forced to spend another night in the abandoned CT camp. The following day a Whirlwind piloted by Flight Lieutenant Bennet made an attempt to lift Captain Jones and Lieutenant Gibson out of the jungle. However, as the helicopter's main rotor blades sliced into the tree tops it became obvious the clearing was not big enough. The landing zone was enlarged with the assistance of men from 1/10 GR and later on 6 September the two officers and the rescue team were lifted out to safety.[15]

Although always accompanied by a certain amount of risk, 656 Squadron's routine tasking over the jungle was vital to the safety and security of the troops operating on the ground. Often, the positive effect on morale produced by knowing there was 'top cover' above the jungle canopy was just as valuable as the physical effect of resupply or the direction of firepower.

> It's the monotonous buzz of a light aircraft engine – the sound of which invariably means that help and support of one kind or another is on the way. The tiny Auster aircraft, 200 feet above the canopy of the jungle treetops maybe bringing food for a patrol, new boots and clothing, urgently needed medical supplies, fresh batteries for a wireless.
>
> Perhaps the pilot and the aircraft are bringing reassurance to a tired patrol that they are in the right place, on course and still in contact. But no matter what the material cargo, the plane and its pilot always bring an intangible, but invaluable boost to the infantryman's morale.

14 AFM ALB 10: N. Rutherfurd, 'Auster Men tell of Rescue Drama', unattributed newspaper cutting, undated.

15 AFM ALB 10: N. Rutherfurd, 'Auster Rescue Drama', *The Straits Times*, 6 September 1958; AFM ALB 55: unattributed newspaper cutting, undated.

> The men who fly the Auster liaison planes and the ground crew who keep them flying are the British Army's own airmen, members of the Army Air Corps.[16]

Nevertheless, the AAC, and 656 Squadron in particular, benefitted from the 10 years the RAF had spent developing specialist roles and techniques for the Auster during the Malayan Emergency up until 1957. The Austers had an important role to play in the field of psychological warfare. Aircraft known as 'voice planes' were fitted with powerful, external loudspeakers. Combined with accurate intelligence messages these could be targeted at individual insurgents and their leaders. 2 Flight at Ipoh flew day and night broadcasting messages across South Perak aimed at the local MCP Committee Secretary Choy Fong. The effort was boosted by the air drop of over a million propaganda leaflets and the 'Thunderer' vehicle-mounted loudspeaker system operated by the Emergency Information Service. Combined with focused offensive ground operations, the psychological warfare effort produced dramatic results. At the beginning of 1958 it was estimated there were approximately 165 CT in South Perak. By the end of the year there were just seven still unaccounted for and the district was on the verge of turning 'white'.[17]

Visual reconnaissance from the Auster was a difficult and time-consuming business. It was estimated that it took at least nine hours of flying to produce any sort of information or intelligence from visual recce. Given the huge areas to be covered, earlier reconnaissance had been conducted from 3,000 feet using binoculars in order to give the pilot the optimum combination of breadth and detail of view. The CTs became acutely conscious of aerial reconnaissance and the fact that it was often followed by an artillery or air strike. Insurgents were therefore likely to move location if they believed they had been spotted. Pilots had to employ various techniques and try to disguise reconnaissance sorties as routine liaison flights so as not to arouse suspicion on the ground. By 1958 the MNLA that were still active had withdrawn deep into the jungle and, as support for the MCP dwindled, insurgent groups were forced to grow their own food. The small, brown plots of cultivation were a tell-tale sign against the verdant background, but it was necessary for pilots to fly below 1,000 feet to spot them. The Auster's work as part of the food denial campaign was so effective that MCP issued instructions to engage the small aircraft with small arms and machinegun fire whenever possible. Nevertheless, it was extremely rare for an Auster pilot to receive incoming fire.[18] The work required great perseverance and dedication to reap results but when it did, it could be decisive. Captain John Stenson RA was employed over South Pahang during Operation Seladang searching for the remaining insurgents who were proving elusive.

> As a result of Captain Stenson's initial appreciation, and constant air reconnaissance over the area, a pattern of jungle camps and cultivations emerged, as a direct result of which, ground patrols were directed onto them. The effect of this was to harass the communist terrorists to such an extent that the deep jungle was denied to them which up to that point had been a safe refuge for them. The communist terrorists were forced out to the periphery and immediate results were obtained in the form of seven eliminations. Not content with constant flying during the hours of daylight in unusually adverse weather conditions – a characteristic of this part of the federation – Captain Stenson on his own initiative has also flown many hours at night. By dropping flares and battle simulators he kept up the pressure on the enemy around the clock. The success of these tactics has become all too apparent in the complete demoralisation of the enemy as described by surrendering communist terrorists.

16 AFMALB 10: 'Flying Soldiers Back Men in the Jungle', unattributed newspaper cutting, undated.
17 AFMALB 10: 'Night Flight Voice Planes Urge Seven Reds to Quit', unattributed newspaper cutting, undated.
18 Postgate, *Operation Firedog*, pp.129–130.

> Captain Stenson by his complete disregard for his own safety, outstanding ability as a pilot, endurance and above all dedication to his work, has impressed all ranks who have had the privilege of working with him and has contributed in no small measure to the high morale of the soldier in the jungle. It is not too much to say that without his gallantry and outstanding work in the field of air reconnaissance, supply dropping, harassing the enemy, marking for air strikes and selfless devotion to duty, the operation as a whole would never have reached a stage whereby complete success is now in site.[19]

For his efforts during Operation Seladang Captain Stenson was awarded the DFC. Due to the time required for success to be achieved 656 Squadron expended over half its effort, in terms of hours flown, on visual reconnaissance. There were, however, alternatives available. Visual reconnaissance could be complemented by photographic reconnaissance as described by Captain Coles:

> A specially adapted camera was mounted on the wing life-strut. Sighting of the camera was achieved by a datum line scratched on the aircraft's window – not the canopy proper – and when scratch, camera and subject were all in line, "click". If the operation was going to plan, undetected, we would at all times try to remain concealed. Surveillance from the air took the form of over-flying the doubtful area in one direction only once on each sortie at an altitude of approximately 2,000 feet and a distance of ¼ mile. In this way we hoped that it wasn't apparent that we were spying and that the CT would think it was just another aeroplane going from A to B. A large map in the crew briefing room indicated that nearly all the CT camps were known, including the types of armaments carried, in fact it was surprising what was known and what could be gleaned from the photographs by experts…
>
> Surveillance of a suspect area could take weeks and only concluded when completely mapped. Not only was it necessary to have full coverage of the camp site but also to establish the source of food and equipment supply, armament supply, transportation routes and hosts of other incidentals. Once all this information was accurately gathered the operation was finished. In the event of trouble the ground troops could move in without delay and ambush the supply routes, the supply source and patrol rivers. The sudden cessation of the terrorists' equipment and the general confusion allows enough time for the goodies to get the upper hand. The CT camps were usually arranged in patterns. The terrorist would inhabit a camp, eat the food that had grown, plant more and move on – at the same time recruiting more candidates to their cause. Hence our job was born – keeping a subtle eye on them until things looked like getting out of hand.[20]

Once a camp had been located, whether by visual or photographic reconnaissance, it could be attacked by ground troops. Any assault, however, could be preceded or supported by artillery or close air support. For the latter to be effective the target had to marked or indicated on the ground. This was not an easy task in dense jungle and several techniques were developed which often required the assistance of an AAC Auster. An early drill involved the Auster pilot making contact with the close air support aircraft at a rendezvous approximately 20 miles from the target. The marker aircraft would then proceed to a 'gate' position close to the target but far enough way so it could not be audibly detected by the insurgents on the ground. The strike leader would then count down the minutes to weapon release. The Auster left the gate in order to be over the target one and a half minutes before the appointed attack time. Marking was carried out by phosphorous

19 TNA, WO 373/136/32: 'Recommendation for Honours and Awards', Captain JMG Stenson, 19 January 1960.
20 AFM 208: B. Coles, 'Austering over Malaya', *Air Classics*, undated.

grenades or reconnaissance flares and, as soon as the Auster was clear, the strike force bombed visually at the aiming marker. The disadvantage of this technique was the minute or more lag between the marker being dropped and the attack, enough time for the insurgents to get under cover or vacate the target area.[21] Captain Fleming, who flew with 2 Flight and 7 Flight, described the intricacies and adaptation of the procedures in more detail when working with RAAF F-86 Sabres.

> Once a target strike was ordered the Auster pilot and the Sabre leader would brief each other before take-off, probably on the telephone, as to the precise location of the target, the nature of it, the IP (initial point) and the attack heading for the Sabres and, most important, the timings. All this would be agreed on the 50,000 map.
>
> It was essential of course that the Auster pilot ensured his markers were precise and so he would arrange to arrive in the area, usually in remote jungle, before the Sabres, so as to provide time for a good look round at height and at some distance away from the target. The pilot would probably carry a passenger or dispatcher from the Flight if the marker flares were to be carried internally. The markers were 4-inch parachute illuminating flares which would have their static lines attached to the passenger's seat leg so that the parachute would deploy as the flare was thrown out. These flares would burn with a bright light for 2 minutes or so and would exude white smoke as soon as its parachute opened.
>
> It was obviously necessary to despatch the flare from as low level as possible in order to achieve accuracy and a hang up on the tree tops where both the white parachute and the source of the white smoke would provide maximum visibility for as long as possible. The weather needed to be good for this type of mission. Once the pilot was quite certain of the target he would stand off and await the Sabre Leader's call that he was at his initial point. This would usually be about 10 miles from the target where his formation would orbit for a short time.
>
> The Auster pilot would how position himself for his 'run in' to mark the target having given a final brief over the R/T about any useful recognition features of the target area, confirm the attack heading, and finish off by saying "marking in 30 seconds", or whatever, "Target is 400 yards north of number 2 flare." The Sabre Leader would then report "leaving the IP". The first flare would be dropped well short of the target, the second approximately half a mile beyond it on an exact line to the target on the attack heading. The second flare, as already inferred, would not be dropped over the target but 300–400 yards short so as to try not to alert any occupants of the CT camp unduly.
>
> The trick was to drop the second flare as the attacking aircraft were coming into the area so that they could line up to attack straight away and not have to orbit. It was also important for the Auster pilot, as soon as he had dropped the second flare, to peel off at 90 degrees to the attack heading and make himself scarce. Timings were of the essence!
>
> As an alternative to flares carried internally they could be hung on a bomb rack beneath the Auster's fuselage. We found this method a touch unreliable as hang-ups sometimes occurred, besides the whole contraption tended to cause too much drag. Once the mission was completed the Auster pilot would make a close inspection of the target area and report.[22]

656 Squadron pilots could also use 'Lodestone', a small radio transmitter that had been developed by the Operational Research Staff in 1957. Lodestone could be put in position in the jungle by

21 Postgate, *Operation Firedog*, pp.56–57.
22 Fleming, 'With the Auster Mk9 in Malaya 1959–62', pp.18–19.

an infantry patrol and then switched on before they retired. It then emitted a continuous signal on a pre-set frequency for up to 24 hours. The Auster could detect the Lodestone signal with an accuracy of 100 yards at a distance of 5,000 yards from the transmitter. This allowed the aircraft to remain beyond the visual and audible range of the insurgents. The infantry patrol would then pass the location of the enemy camp to the Auster pilot in relation to the position of the Lodestone transmitter. That information could then be passed from the Auster to the strike aircraft allowing the target to be attacked with a degree of accuracy.[23] Lodestone sorties were flown to good effect by Captain Ian Bell RA at the beginning of 1958.

> At the end of November 1957 his flight was given the duty of fixing the jungle headquarters of the Central Committee Member of the Communist Terrorist organisation – Hor Lang. This was assumed to be in mountainous country near the River Rompin [in south Pahang] and fixation was by means of a secret radio directional device requiring skilful and very accurate flying. Only two pilots in the flight were trained to use it.
>
> In January 1958 it was decided that fixation by night would provide the only hope of success and it was of such importance that it was worth sacrificing a pilot and aircraft in the attempt. The monsoon was heavy in January and February particularly in the River Rompin area.
>
> In these treacherous and dangerous conditions, with complete disregard for personal safety, he repeatedly volunteered against advice and often after a heavy day's flying to carry out the sortie which he did once or three times a week until April 1958 when Hor Lang surrendered. It was a sortie often involving the maximum endurance of the aircraft and at this time the aircraft were not wholly reliable.
>
> Only his exceptional skill, his outstanding coolness in the face of danger and undoubted courage and tenacity of purpose enabled him to overcome repeatedly the hazards of flying a light aircraft in extremely turbulent conditions by night over treacherous mountainous country, inhabited by communist terrorists often in the cloud and storm conditions associated with a monsoon.[24]

If an air strike was followed up by a ground assault then the Austers could still be employed to good effect by dropping battle simulators. The simulators replicated the sound of battle and were designed to confuse the insurgents as to the direction and timing of the actual attack. The use of the battle simulator was revived by 656 Squadron's Captain Michael Badger RE who was subsequently commended for his efforts.

> Captain Badger .... has constantly maintained a very intense rate of flying for light aircraft on operations in Malaya. His outstanding energy, cheerfulness and willingness to undertake hazardous sorties and to exploit success beyond the bounds of duty has been an inspiration to all.
>
> As Flight Commander in particular his imagination and enthusiasm to give the utmost light aircraft support possible to the emergency operations has resulted in the development of new techniques. He is largely responsible for the recent revival in the use of Battle Noise Simulators and has evolved a much improved method of dropping them. To this can be allied the marking of air strikes by night, and the exploitation of the use of flares by night for illumination purposes and also other techniques which are still in the development stage…

23 Postgate, *Operation Firedog*, pp.56–57.

24 TNA, WO 373/136/27: 'Recommendation for Honours and Awards', Captain IE Bell, 26 January 1959.

> The techniques he has evolved and revised, by introducing an element of offensive action against the Communist Terrorists, have had a very real effect on operations. Captain Badger's ability and fearlessness are in the highest traditions of the already famous squadron to which he belongs.[25]

Sustaining the fighting troops on the ground was another task for the adaptable Auster. Security forces patrols and observation posts had to reach out into the most impenetrable areas of the Malayan mountains and jungle. To reach them by road, track or river could take days even if it was possible. Although the payload of the Auster was generally small it could drop critical supplies, equipment and spares accurately in the most difficult terrain. Captain Fleming recalled two incidents that demonstrated the Auster's capability and flexibility on this task. In the first he was required to drop a large jar of rum to an isolated infantry observation post on the 7,000 feet peak of Mount Korbu, 15 miles east of Ipoh:

> The rum was duly wrapped in a cardboard box with sawdust and was delivered in person by the Company Commander who wished to come along. This sortie had been arranged for a very early take-off so that there would not be any cloud build-up on top of the mountain and we duly took off and headed for the summit which could be seen for miles around. It soon became evident that there was a very strong wind close into the mountain and, contrary to what we all know about Katabatic winds in the early morning, strangely the wind was blowing up the mountain instead of down! The former proved to be of great assistance to our climb by flying close in so the top was reached in almost record time, but in having a look round the summit it soon became exceedingly turbulent. The top was quite small in area, about 40m square and one could see some figures huddled by their plastic shelters waving madly. They all looked very wet and cold as we tried out a few dummy runs.
>
> It was obviously important to fly into wind and as slowly as possible to get the free bundle into the right place. To miss would ensure an almost free fall some thousands of feet down, and much disappointment from the recipients. We eventually settled on a very low approach, into wind, slowed right down and on a line to miss the soldiers. All went well. The Company Commander hurled the package out on the word and was able to report that it landed smack on target, much to my relief. He did a fine job. We flew a final run by and noted that the boys were waving and looking pleased so we peeled off making a fast descent back down to base having been airborne for exactly one hour. We subsequently heard that the jar of rum had arrived intact and was much appreciated.
>
> On the second occasion the author was chatting in the Flight office at Taiping one morning with the AAC Examiner who, out from UK, was about to check him out for his annual standardisation trip, the plan for which was about to change rapidly! The door opened and in walked a Lieutenant of the local Royal Australian Battalion saying, "could you drop this?" He was brandishing an 8 foot double-handled saw! The Examiner, who we all knew was an experienced 656 Squadron Aviator in past times in Malaya, without a moment's hesitation said, "Let's go!"
>
> So, after a brief exchange of details of the problem, a grid reference of the D.Z was produced, the saw was firmly wrapped in hessian, whereupon we both walked to the aircraft. The starboard door was removed, the Examiner installed himself in the right-hand seat and told the crewman to put one end of the saw down by his right knee with the other end down the outside of the aircraft holding it into the side with his right arm and off we went. The story

25 TNA, WO 373/136/24: 'Recommendations for Honours and Awards', Captain M. G. Badger, 27 July 1958.

> was that a patrol was in trouble in an area northeast of Grik trying to fashion a helicopter landing zone (LZ) in order to lift out a casualty. It would be about 40 minutes to get there and from the map the drop zone (DZ) looked as if it might be a touch tricky as it was at the bottom of a steep valley by a sharp right turn half-way down a sizeable mountain. The flight up was fine, the weather was good and the slip-stream kept the saw hard up against the side of the aircraft, making no noticeable change to the trim.
>
> It was only when we flew over the grid reference given that we spotted the DZ and, as envisaged, it was right down by the river at the bottom of a steep valley next to a sharp right turn. There was a vestige of a clearing visible but obviously not fit for a helicopter yet. On looking around it was obvious the only approach possible would be down the valley with an immediate turn right after the drop, so a dummy run was carried out. Meanwhile we agreed that the Examiner would judge the release point himself as the view out of his side was much better. The first dummy run was too fast and not low enough so we climbed up again to try once more. This time the throttle was closed and half flap applied as we entered the top of the valley and flew on down. Again, this time we were a bit too fast over the DZ with half flap down so we went round again, deciding to use full flap, no power and keeping a sharp watch on the speed.
>
> This time all settled nicely in a steep descent and the decision was taken to drop. As we neared the DZ my passenger had manoeuvred the saw almost out of the aircraft, and at the precise moment quickly launched it down and out. Meanwhile it was full throttle, full flap reduced to half as rapidly as possible, then into a steep turn to starboard on down the valley, amid sounds of jubilation in the cockpit with the dispatcher enthusiastically shouting that he had seen it "going into the DZ like a dirty dart!" It sounded like a success, and we later heard that it was. That was my standardisation trip for the year![26]

When infantry patrols were moving on foot through the jungle they could also relay on 656 Squadron's Austers for support and assistance when they ran into problems:

> Whenever patrols were operating in the jungle it was normal for Flights to react quickly to calls for assistance. Patrols sometimes became unsure of their exact location on the ground, had become overdue, or had sustained a casualty, for example. In answering the call the pilot would first get a brief from Battalion or Company HQ on the last known position of its patrol, radio frequency and call sign and provide a time over the area. The pilot would then proceed into the area at an altitude and position affording the greatest view over the terrain. If the aim was to provide a position report, the pilot would keep well away from the patrol's last known position so as not to give the game away and then establish contact on the radio. This having been done the patrol would be instructed to initiate a white phosphorous grenade which would produce a distinctive puff of white smoke on the ground. These would take a minute or so to show above the jungle canopy. All patrols carried these grenades. The pilot exercising his skills at stand-off map reading would then produce an accurate grid reference over the radio to the patrol. A pair of binoculars was a useful aid on these occasions.
>
> If a patrol in similar circumstances needed to keep its own location secure – WP [white phosphorous] grenades made a discernible explosion – an orange carbide balloon was sometimes used. This device would be inflated by the resultant gas from mixing the carbide with water and when the balloon was fully developed it would be released on the end of a nylon cord up through the canopy so that it could be seen from a distance. This procedure took a

26 Fleming, 'With the Auster Mk9 in Malaya 1959–62', pp.18–19. Postgate, *Operation Firedog*, pp.16–17.

good deal longer than that using a WP grenade and of course the balloon kit was heavier to carry than the grenade, but it was a silent procedure when, for example, a patrol was trying to position itself relative to a known C.T. camp in order to launch an attack. Again, it was vital that the aircraft kept well away from the area the signal was expected to come from.

In the event that a patrol had not been heard of for some time and was well overdue, a call for an aircraft to try and locate and contact it would be sent. Again, the pilot approaching the area from height would over-fly the last known location of the patrol and give a call on its frequency just in case. Very often on hearing an aircraft near at hand patrols would let off a WP grenade in order to attract attention, its radio being unserviceable.

In such a case the pilot would make a close search of the location of the white smoke to try to see whether the patrol was showing a fluorescent panel. All patrols carried these 6' x 2' brightly coloured panels which would be laid out on the ground in a certain pattern in order to convey a message. All pilots carried a note of standard panel messages which could be acknowledged by message drop from the aircraft or by waggling wings.

This procedure was not altogether straightforward as panels would not be very easily read if laid out under the jungle canopy, nor could a message drop have much hope of reaching the ground. So patrols would normally try to find some sort of clearing of shorter undergrowth or, for example, a sandbank on a river on which to display their signal. All this would invariably take some time as these features could be some distance from the patrol. Once the state of the patrol was established, which might vary between "all OK – send radio spares" or "serious casualty – require assistance", the pilot would then call the Flight to pass on the information to the patrol's HQ. If all was fine but for a u/s radio the pilot would be able to deliver the parts in due course without too much delay. If on the other hand there was an emergency, a helicopter might have to be called for to lift a casualty out. If the patrol was in deep jungle and a long way from base a helicopter rescue would be the only solution. Helicopters were in very short supply and were based a long way down country at KL. This operation would be quite a lengthy one, especially if an LZ had to be cut or blown using explosives. All security would obviously be lost on these occasions.

Whatever transpired the Auster pilot would remain on hand to help ease any situation until resolved. This would involve remaining at a forward strip whilst top cover sorties, relaying messages, dropping supplies, guiding a helicopter in and so on, could be flown.[27]

Towards the end of 1958 656 Squadron was engaged in supporting an operation that would have sweeping consequences for the south of Malaya. In October the South Johore (Johor) Regional Committee (SJRC) of the MCP were identified by 'surrendered enemy personnel' operating east of Kluang, 55 miles north-west of Singapore. Operation Tiger was launched to kill or capture the insurgents.

On the evening of the 4th October, whilst giving a patrol of 2/7 GR [2nd Battalion, 7th Duke of Edinburgh's Own Gurkha Rifles] a contact recce, the pilot of an Auster aircraft sighted smoke approx 3,000 yards from the patrol, who went to investigate the next day. A camp for 25 was found and tracks followed resulting in a contact on the 6th October. As a result of indications from the contact more troops were sent to the area and an Auster denial called for. The same pilot undertook this. On arrival in the area he again sighted smoke 9,000 yards to the south of the cordoned area. This again was SJRC who had made good their escape. The chase was again taken up by a further patrol who started from this camp and a succession of

27 Fleming, 'With the Auster Mk 9 in Malaya 1959–62', pp.26–37.

> patrols followed their tracks for a further 25,000 yards, when another position indication was given by the same pilot; by this time SJRC was back in the original area. At this stage due to continual SF pressure that party of 25 dwindled quickly down to nine due to surrenders and with the aid of the SEPs [surrendered enemy personnel] working with Special Branch resulted in the kill [*sic*] of Ah An, the RCM, and his wife.[28]

Captain Mark Legg RA was one of the 656 Squadron pilots largely responsible for the success of Operation Tiger and was later cited for his efforts:

> During the period 3rd October to 21st October 1958 he repeatedly flew dawn sorties in support of 99 Gurkha Infantry Brigade's operation against the South Johor Regional Committee. These sorties were carried out partly by night under adverse monsoon conditions of low cloud and cloud at ground level over uneven country. With great skill and disregarding his own safety he succeeded in fixing the position of the communist terrorist party on several occasions by spotting their campfires at dawn. It was mostly due to the information he thus provided that the brigade was able to reduce this party to an ineffective force in a very short time. His splendid example, exceptional skill and fearlessness under adverse conditions have been an inspiration to those who worked with him.[29]

Although the Emergency continued beyond the 31 August deadline, 1958 had been a successful first full year of operations for the AAC and 656 Squadron in Malaya. Captain Badger was awarded the DFC on 9 December 1958 along with Captain Humphrey Crutchley RA who added the DFC to his Mentioned in Despatches, awarded earlier the same year. Staff Sergeant Bob Bowles AAC and Staff Sergeant Kenneth Mead AAC were both awarded the DFM for their flying during that year in the Emergency.[30] The success of Operation Tiger led to 1959 beginning with the welcome news that Johore, the southernmost province in Malaya had been turned 'white' on the national map.[31] Johore joined Negri, Sembilan, Selangor and South Perak in being declared 'white'. This resulted in a reduction of tasking for 11, 14 and 16 Flights in the south. In the north however, 2 and 7 Flights remained fully committed supporting the Malayan Federation Brigade and 28th Commonwealth Brigade respectively. In February the entire squadron was distracted by a celebration to mark 656 Squadron achieving 150,000 operational flying hours. A parade and flying display at Noble Field were followed by the presentation of a ceremonial Malaysian dagger, a kris, by the Commander in Chief Federation Armed Forces on behalf of the Malaysian Government and a Gurkha kukri by the GOC 17th Gurkha Division.

28th Commonwealth Brigade, with its headquarters alongside 7 Flight in Taiping, consisted of three infantry battalions, one each from Australia, New Zealand and the UK. Lieutenant Bernard Redshaw of the Royal New Zealand Signals Regiment was the Regimental Signals Officer with 1st Battalion, Royal New Zealand Regiment working as part of the Brigade. Early in 1959 a large insurgent camp was located just north of the border in Thailand in the Betong salient. The size of the camp led intelligence to believe it could be the lair of the Secretary General of the MCP, Chin Peng. A major Thai Police Field Force operation was planned to assault the camp, with 28th Commonwealth Brigade providing extensive ambushes along the Malayan side of the border. Two

28 Anon., '656 Light Aircraft Squadron', *Army Air Corps Association Journal*, 1959, pp.23–24.

29 TNA, WO 373/136/28: 'Recommendations for Honours and Awards, Captain M. P. E. Legg, 26 January 1959.

30 J. Grover, 'Honours & Awards to the Army Air Corps/RA AOP/GPR – Malaya', undated.

31 AFM ALB 55: 'All White – Crisis is Over in Johore', unattributed newspaper cutting, 1 January 1959.

Austers moved up to an airstrip at Grik (Gerik), 60 miles east of Penang. During the build-up to the operation Lieutenant Redshaw was a frequent observer flying with 656 Squadron:

> The Auster 9 that we flew in most frequently (XK375) was suffering a number of minor problems, which should possibly have warned me of what was to come. On one occasion, as we were coming into land, the pawl securing the sprocket for the HF antenna winding gear came adrift, the antenna unwound violently and we lost it. On another occasion, just as we were taking off (on half flap) the hydraulic flap actuating pump failed and we lost our flaps, losing height and narrowly avoiding the tin-mining lake at the end of the strip. On yet another flight, whilst returning in failing light, I noticed that my right side was getting wet with what seemed to be oil (we flew with the doors off). On landing, we discovered that it was indeed oil – coming from a leaking engine gasket.
>
> Then, two days before the attack was due to go in, a CT courier, travelling north towards the camp, was ambushed and killed by a patrol of 3 Royal Australian Regiment. It was considered imperative that this courier should be identified without delay, so the body was carried back in relays to a 1 NZ Regt position that was considered to be sufficiently far enough away from the CT camp area so as not to compromise the operation.
>
> The 1 NZ Regt patrol cleared an LZ from which an RAF helicopter was to bring out the body. However, at this particular moment a suspected design fault had grounded all the Sycamores, and the one remaining Whirlwind in the area was declared unserviceable. It was therefore decided to parachute in a camera and fingerprinting equipment – a mission for the 656 Squadron detachment. The same Auster 9 was tasked, with Captain Nigel Budd, 10th Hussars, as pilot, and myself flying as dispatcher.
>
> We were to drop the equipment into the same DZ that had been prepared for the helicopter. This, we noted, was sited on a ridge. We swooped in and, on Nigel's signal, I dispatched our load. As we circled, the parachute developed and started to descend perfectly into the clearing. Then, to our dismay, an up-draft from side of the ridge lifted the parachute upwards and sideways, depositing it into the top of a tall tree on the edge of the DZ.
>
> It soon became obvious that the patrol could not recover the parachute, and it seemed very likely that another parachute would also suffer the same fate. So we returned to Kroh to collect another set of equipment prepared for a free drop on a streamer. So back we came, and into the clearing we dived. As I threw the bundle out, Nigel pulled the Auster tightly up and out. Whilst I was looking downwards and backwards, focusing on the streamer, I felt a bump and heard a loud thumping, ripping noise on Nigel's side of the aircraft. I then saw that we were now flying very low, almost skimming the jungle canopy. "Oh Christ!" came Nigel's voice over the intercom, "I've never done that before." When I looked out to his side, I saw that we had lost a piece of our port wing, loose canvas was flapping and that we were carrying a piece of branch about 3½ metres long. We had clipped the same tree that had claimed the parachute!
>
> We were still flying! As we recovered and gained height, Nigel turned to me and said that, since we had lost our pitot head, he had to do a stall check to be certain of his ASI before we could land...This frightened me much more than hitting the tree! Then, as we made our way back home to Kroh, we were surprised to be joined firstly by the "grounded" Sycamore, then by the "unserviceable" Whirlwind and finally by a Pioneer that had been in the vicinity. Apparently the 1 NZ Regt patrol on the ground were sure that we had gone in and had put up a Mayday. And so this mixed convoy slowly made its way above the ocean of the jungle canopy – rather like the final procession in Peter and the Wolf.
>
> We eventually made a perfect landing on Kroh strip, with a large audience of worried-looking military personnel looking on. When the aircraft was duly inspected, it was discovered

> that we had lost the end cap of our port wing (including the pitot head) and that the fuselage was twisted out of alignment. As a result, the airframe was written off – or so I was told.
>
> The attack on the CT camp duly went in two days later, with Captain John Chandler, Royal Irish Fusiliers, flying the other Auster, observing and giving a running commentary in the closing stages. Unfortunately, the CTs sophisticated warning system allowed the occupants of the camp to escape before the PFF company fought their way in, and some 7 PFF personnel were badly wounded in the action. But, despite this, and although evidence recovered from the camp showed that it was not that of Chin Peng, it was still a substantial HQ of the MCP, and taking it out was a major step forward in the Malayan Operation.[32]

Incidents such as that experienced by XK375 were among the operational hazards of flying in Malaya. They tested the REME to their limits and called for endurance and ingenuity:

> Some tricky recovery problems were set when an Auster made an emergency landing in a paddy field in Malaya. The occupants were unhurt but the aircraft undercarriage was torn off. No roads or tracks led to the field which was reached only by very narrow banks between the other fields. It was the wet season and all the ground apart from the banks was under water. The local Malay policeman kept curious children away from the aircraft, and cheerful little Gurkha soldiers who could become invisible in the jungle edge supplied the guard against the terrorists.
>
> Mechanics, standing knee-deep in water among the young rice plants, removed the engine, which was then tied to a sledge and dragged a mile to a waiting Land Rover. This had been driven as near as possible to the scene but was finally stopped by a wide stream. The engine was rafted across the stream.
>
> The mainplanes were detached and they were carefully manhandled beyond the stream to a waiting Bedford QL. On the next journey the fuselage was moved in the same way. Once in the vehicles, it was a much simpler task to get the A/C into its workshop.[33]

Although insurgent activity continued to decrease, 1959 ended with no formal declaration of the end of the Emergency. 656 Squadron had, however, continued to play its part in eroding the capabilities of the MCP and the MNLA. In recognition of their efforts Captain Bell and Captain Legg were awarded the DFC while Sergeant Bernard Horsey was awarded the DFM. Four members of the Squadron were Mentioned in Despatches.[34]

1960 began with a tragic accident: on 20 January Sergeant William McCammont of the Cameronians from 11 Flight was flying Auster WZ728 from Seremban to Gemas with Private J. Finnerty of 1st Battalion, Cheshire Regiment on board as a passenger. Seven minutes after take off all trace of the aircraft was lost. 656 Squadron immediately instigated a large-scale search operation and was assisted by the RAF's airplanes and helicopters. Ten Austers from the Squadron flew over 500 hours of visual and photographic reconnaissance on the task burning some 3,600 gallons of fuel in the process. Nevertheless, despite the effort WZ728 and its crew remained hidden in the jungle. Nearly four months later, on 11 April, the aircraft wreckage was discovered by Chinese loggers. A party from 656 Squadron was immediately sent to the scene. Private Finnerty's body was still in his seat. There were signs he had been made as comfortable as possible and a tourniquet had been applied to his upper right arm. The aircraft's radio and battery had been removed and set up outside. A bed

32 B. Redshaw, 'A Near Miss – Auster AOP 9 XK375', *The Chinthe*, 2010, Iss.2, pp.25–29.
33 Anon., 'Recovery in the Rain – Malaya', *Army Air Corps Association Journal*, 1959, p.21.
34 J. Grover, 'Honours & Awards to the Army Air Corps/RA AOP/GPR – Malaya', undated.

The wreckage of Auster WZ728 in which Sergeant William McCammont and Private J. Finnerty died following the crash near Seremban on 20 January 1960. (Army Flying Museum)

had been improvised from seat cushions under the remaining wing, an indication that Sergeant McCammont had stayed with the aircraft for at least 24 hours, however there was now no sign of him. Several days later his body was discovered 400 metres from the aircraft, a position to which he had crawled despite being severely injured. Before succumbing to his injuries, he had covered half the distance to a nearby logging road from where he would probably have been found and rescued.

The entire wreckage was brought out of the jungle by Gurkha engineers with the assistance of indigenous soldiers who carried the engine a considerable distance through primary jungle. A subsequent board of enquiry found that XZ728's engine had failed through the seizure of the exhaust valves. In an attempt to crash-land onto the roof of the jungle canopy (an established albeit desperate technique in Malaya) the starboard wing root had hit a tree trunk about 80 feet above the ground. Both wings were ripped off and the engine broke clear. The Auster then fell to the ground and continued to roll a further 60 feet down a hill before coming to rest. The wreckage, which had been flown over many times, had been 'swallowed up by the jungle except for half a wing which remained in the tree at the point of impact. No amount of air search or air photography would have discovered it.' As a result of the accident, it was recommended that fluorescent material be used on parts of the aircraft. Orange dayglo panels were applied to the upper and lower leading edges of the wings and the upper surfaces of the fuselage and tailplane. The word 'Army' on the fuselage was also painted white, having previously been black.[35]

35 AFM 208: B. Coles, 'Austering over Malaya', *Air Classics*, undated. AFM ALB 56: various unattributed newspaper cuttings dated 22 and 23 January and 11 April 1960.

Despite the hazardous flying conditions, the weather and terrain, accidents were relatively infrequent. When they did occur, fatalities were rare thanks to the robust nature of the Auster and the skill of the pilots, as described by Captain Coles:

> It is with horror that I remember a day in 1960 whilst engaged on a supply drop to a party of Sarawak Rangers – a jungle patrol – of food, batteries and radio spares. The clearing had been prepared down in a valley, a good spot, no high obstacles in the vicinity and clear enough to get down very low and so prevent any possibility of getting valuable supplies hooked up in the trees. Wind and turbulence appeared to be at a minimum and in conditions such as these it was usual for the observer to slacken his harness in order that one foot could be placed outside on the wing lift strut and then the body moved as far across as possible making observation of the dropping zone and subsequent fall of the supplies easier.
>
> This procedure also eliminated the necessity for any fancy manoeuvres by the aircraft in what was often a restricted space – the observer now being in a position to guide the pilot who usually had the DZ on his blind side. The aircraft was being flown without doors as on all supply drops.
>
> All was ready as we crossed the valley ridge at approximately 1,400 feet. The goods had been pulled forward ready to be unleashed and tipped out. The nose was pushed down quite hard and we did a tightish left-hand spiral turn into the valley. Returning the aircraft to a horizontal attitude we put out about 20° of flap and kept heading down. We then realised that what had outwardly looked like a valley had now turned into a hole – having quite high ends. The airspeed had risen much too high to be able to get all the supplies out with one pass but we were going to need some of this speed to get ourselves out. Only half of the supplies went out on the first run, as expected so we would do another, easing the power up to 2,200 rpm and holding straight and level for as long as possible before finally pulling the nose up and climbing fairly steeply. We had been holding 2,300 rpm for about 10 or 12 seconds and just beginning to turn out over the ridge when "bang", no power. The propeller was still turning, terrific vibration, windscreen completely covered in oil with more oil coming along the sides of the fuselage and in the door apertures – what a mess! Get the back end up quick, haul the IX round tight and maintain flying speed by heading back down into the valley.
>
> We had no choice but to try putting the lot down in the dropping zone clearing – plenty of help there! Whip the radio round to channel J. "Mayday, Mayday, Mayday" – the area location was easy, about the only thing we did know at this point was the map reference for the dropping zone. It was soon obvious that we wouldn't make the clearing – the IX could manage to glide about one mile per 1,000 feet of altitude with a little bit spare if all the conditions were right. We were now at 800 feet, gliding quite quickly but with two up, downwind, and the 70lb load still on board. It wasn't worth taking any chances to reach the dropping zone, especially as a further small clearing was seen to port about 200 yards over and half a mile ahead. The IX slid across well and we lined up nicely – still with plenty of knots and comfortable altitude. As we got nearer we could see something on the ground – logs!
>
> It turned out, as we were soon to learn, that Chinese loggers were cutting timber. We arrived at the clearing unable to go elsewhere, sat the IX down with a "thud" and were rolling fast – luckily fast enough to get airborne again and clear the first heap of logs. We came down again with a second "thud", a three pointer, and headed straight up a track used by the timber laden lorries – what luck! Finally stopping we evacuated to sit on a log and enjoy a cigarette.
>
> It was at this point I found a large lump on my forehead and realised that my harness could have been tighter – I hadn't pulled it up at all. At some stage of that erratic landing I had

probably collided with the wing support structure, I don't know, but I still have traces of that lump today.

When the nervous trembling subsided, we strolled back to the aircraft and took the cowlings off – the trouble was soon found. As we peered in our gaze was returned by a connecting rod peering out through the side of a cylinder. The Rangers were with us in no time and had arranged for the aircraft recovery. They took their remaining supplies and, with handshakes all round, left. The recovery wagon arrived next day and within hours the IX was dismantled, loaded, and we moved off. Two weeks later the aircraft was back at 14 Flight, Paroi with new engine, all cleaned up, heavy landing checks done and fully serviced – a tough little bird.[36]

Leading up to mid–1960, 656 Squadron's flying effort was concentrated in the north of Malaya. Intelligence suggested that the few remaining active CT were making their way north in order to rebase themselves in Thailand where they would be more secure, out of the reach of British and Malay forces. From there Chin Peng endeavoured to exercise his declining leadership over the remnants of the MNLA still operating in Malaya. Agreement was reached between the Thai and Malayan governments to allow the PFF to operate across the border into Thailand. 656 Squadron's aircraft were also permitted to fly north of the border to provide visual recce and resupply the troops on the ground. 2 and 7 Flights reported over 200 sightings. A period of intensive operations resulted with CT resupply being ambushed and several insurgents killed. There were so few active members of the MNLA still operating that each one eliminated represented a significant reduction in their capability. The north of Malaya was progressively turned 'white' until, on 31 July 1960 the Emergency was declared over. A victory parade was held in Kuala Lumpur to mark the event including a mass formation of Austers as part of the fly past.[37]

Maintaining Security

With the official cessation of the Emergency a border security area was instituted which extended 15 miles either side of the border. 2 and 7 Flights continued to fly in support of operations in the border security area and although operational flying had to continue, the end of the Emergency gave the Squadron a chance to exercise neglected skills and tactics in 'orthodox' warfare. A camp was established on Malaya's east coast where aircrew and ground crew reacquainted themselves with the procedures for the reconnaissance and establishment of advanced landing grounds. There was also an opportunity to conduct deck landings on the Centaur class aircraft carrier HMS *Albion*. 656 Squadron also examined the possibility of being deployed on future operations in response to problems elsewhere in Southeast Asia. Squadron personnel began to experiment and work through the issues associated with air portability, but it was not without teething problems as described by Lance Corporal Roger Mitchell REME:

An Auster aircraft was dismantled and loaded onto a trusty Bedford RL three tonner. Each Land Rover was allocated a trailer and so, dressed in our olive greens, jungle boots and our 'hats ridiculous' we set off for a secret destination. Full of hope and expectation to not only see, but to load all this onto a Bristol Beverly Aircraft, a real full sized, man sized aircraft. The excitement was so much that we took the camp dog 'Sputnik' with us.

36 AFM 208: B. Coles, 'Austering over Malaya', *Air Classics*, undated.
37 AFM 209: '656 Squadron AAC Report for 1960', undated.

> ...but on arriving at the secret destination we were all very disappointed and dismayed because the 'Beverley Aircraft' turned out to be a concrete block about twenty feet wide and about forty feet long and about four feet high with metal ramps resting against it. Perhaps this was some form of secret weapon. It would take some pretty powerful engines to get it off the ground. The idea being that the vehicles were to be reversed up the ramps and parked on the concrete block in air portable order.
>
> So, the Bedford, complete with Auster, was duly reversed up the ramps and parked on the concrete block. Now the fun starts. It appeared that most of the Land Rover drivers had little or no experience of reversing a trailer. After a lot of attempts and mishaps it was found that the best way to approach the problem was to uncouple the trailer and manhandle it to the bottom of the ramps then couple it up again. This course of action prevented trailers from falling over the side of the concrete block and other mishaps. But alas two of the Land Rovers could not get up the steep ramps because of slipping clutches. This caused some consternation as they had to be pushed up the ramp by manpower.
>
> When all the vehicles were aboard the concrete block it was time for lunch and we could all settle down to eat our ration packs and have a brew of tea with the help of the little tin stove and our mess tins. The dog 'Sputnik' had a special ration supplied by the cook, he also did eat quite a lot of 'hard tack' biscuits that were in the ration packs. He had a wonderful day out as he went to sleep in the afternoon, his tail was wagging, he was very happy. As for the rest of us, the only thing that you could say was that it did not rain and we now had a dismantled Auster aircraft to put back together.[38]

On 4 September 1960, Captain Chris Roberts of the Gloucestershire Regiment was flying his Auster in southern Thailand when engine trouble led to a forced landing. The only available strip was a short stretch of laterite road on the edge of paddy fields. The aircraft was undamaged, but it presented an interesting recovery problem. The recovery team had to travel over 750 miles to reach the Auster and the Thai authorities would only allow one vehicle across the border. A Bedford RL was fitted with the recovery rig to take an Auster fuselage and frames to carry the main plane and the undercarriage. The remaining space was filled with ropes, slings, ration packs, fuel and water. An advance party was flown forward by RAF helicopter while the truck began the long drive north. At the border the recovery team was joined by an armed escort from the Malayan PFF and Thai police. After nearly two days' continuous driving the team reached the village closest to the stricken Auster, met the advance party, prepared a meal, and settled down for the night:

> The next morning everyone was up early, had a light breakfast and with everyone carrying some item of recovery gear made tracks to the aircraft. Evidently word had got around the village and surrounding areas that we were about to work on the aircraft; eating and drinking stalls were set up, and for miles around parties of local folk were streaming in. This caused the recovery team to resemble a carnival with village headman, police, Army officers and inevitably the hundreds of children. They all sat, watching intently, whilst we proceeded to strip down the aircraft completely, remove the engine and mainplanes and by 0930 hours we were ready for the helicopters to airlift the components out and across the river to the school compound which we had previously marked out as a LZ.
>
> This type of operation was a new venture, so the Sycamore Squadron Commander and his senior technical officer plus our Squadron Commander, Lt Col [John] Cresswell [RA], and 2IC Major [Michael] Pritchard-Davies [RA], arrived to watch and suggest ideas on the

38 R. Mitchell, 'Air Portable', *The Chinthe*, 2014, Iss.2, pp.30

> most practical way of doing this "sky-recovery." The engine was no problem and with it slung under the helicopter it went away beautifully. Unfortunately, the mainplanes were a problem. After numerous abortive attempts to get them fixed under the helicopter it was decided to man-handle them out.
>
> After removing the complete undercarriage we got the fuselage away making a very imposing sight as it was lifted out. After several more airlifts across the river the first part of the recovery was completed. At the school compound we put the bogey wheels on the fuselage, replaced the engine and with local help, placed the whole airframe into the 3-ton vehicle. Next the mainplanes were placed into the frames, the whole tied down and we were ready for the road.
>
> By this time it was quite late, so we remained in the village for a further night. Early next morning we departed, making Kroh five hours later. After a quick meal we pressed on making Kuala Lumpur the next day. The OC Workshops and myself left the party at Kroh to fly back to Kuala Lumpur and had a further experience when due to very bad weather we had to land at a palm oil estate and spend another night away from base. A great deal was learnt from this, our first long distance recovery exercise with the assistance of helicopters. The construction of a lightweight cradle to fix the mainplanes under a helicopter is being considered for future recovery operations.[39]

The move away from operations and towards training continued through 1961. It was estimated that there were only 350 'hard-core' terrorists still remaining in Malaya and so Commonwealth forces were withdrawn from the northern border areas, leaving the fight with Malayan forces. 7 Flight AAC also withdrew from the north, co-locating with 16 Reconnaissance Flight and Squadron Headquarters at Noble Field before moving on to a new home at Kluang. 11 Liaison Flight remained stationed at Sembawang in support of 99th Gurkha Infantry Brigade Group. 14 Liaison Flight was stationed at Paroi, Seremban in support of 17th Gurkha Division but altered its equipment scales taking on three Beavers and reducing to three Austers.[40] The year ended with tragedy. On 22 November 1961 Captain Peter Hills RA was flying a visual reconnaissance sortie in Auster XK374 along with Captain D. J. O'Grady Royal Army Dental Corps. The aircraft disappeared along with its crew in the north of Malaya. An intensive air search was conducted lasting seven weeks and including aircraft from 656 Squadron, RAF and the Malaysian, Australian and New Zealand air forces but to no avail; the fate of XK374 and Captains Hills and O'Grady remained unknown for over five years.[41]

Four months later, another accident resulted in a more positive outcome. On 13 March 1962 Captain John Tadman Royal Lincolnshire Regiment was flying Auster WZ674 around Kroh in Upper Perak. While flying along a ridge the aircraft was caught in a downdraught, a wing struck a tree and the Auster crashed in flames. Captain Tadman later recounted, 'It was just a routine reconnaissance flight until I got into trouble at the end of the sortie. I got caught in a sudden downdraught of air in a valley .... Had the plane hit a tree and not a bamboo clump it would have broken into pieces.' Captain Tadman recounted that as the left fuel tank caught fire, he flung himself from the wreckage. Once the fire had burnt out he sheltered under the tail of the Auster, using an empty ration tin to collect rain water until assistance arrived. A RAF Vickers Valetta located the wreckage 20 miles east of Grik later the same day. Later a

39 F. Penfold, 'Auster Mk9 Recovery in Thailand', *Army Air Corps Association Journal*, 1961, pp.38–40.

40 Anon., '656 Light Aircraft Squadron, Army Air Corps 1961', *Army Air Corps Association Journal*, 1962, p.29.

41 Anon., '656 Light Aircraft Squadron, Army Air Corps 1961', p.32; Anon., 'Army Air Corps Aircraft Losses Since 1 September 1957', HQ AAC, 13 June 2019.

The injured Captain John Tadman having a welcome meal during his stretcher move to an emergency HLS from which he was finally evacuated on 20 March 1962. (Army Flying Museum)

helicopter flew over the crash site and spotted Tadman who pointed to his legs to indicate he was immobile. It would, however, take some time before an effort could be mounted to rescue Captain Tadman. There was no useable helicopter landing site close to the position, so a different plan was needed. On 16 March an eight-man SAS team including a doctor, Flight Lieutenant J. M. Brown RAF, parachuted into the jungle close to the crash site and spent the rest of the day hacking a path to Captain Tadman. On arrival the rescue team found that Tadman was suffering from burns down his left side and left leg, and a dislocated right hip. The following day a stretcher and splints were dropped to the rescue team. At the same time the SAS team began constructing an emergency landing site 1,500 yards from the crash site using explosives. With the jungle cleared a helipad approximately eight metres square was constructed, but it took time. On 20 March the landing site was finally complete and Captain Tadman had been stretchered to it through dense jungle – a full week after the crash. At 1600 hours a RAF Sycamore flew into the site and evacuated Captain Tadman to the British military hospital in Taiping where he spent his recovery and was reunited with his pregnant wife Corinna.[42]

Aside from Captain Tadman's rescue the main event for the AAC in Malaya in 1962 was a move of squadron headquarters and all flights, concentrating to a single location at Kluang, 60 miles north of Singapore. 2 Flight left Malaya and reformed in Northern Ireland and other 656

42 AFM ALB 10: and AFM ALB 56: various press clippings, untitled and undated; Anon., 'Army Air Corps Aircraft Losses Since 1 September 1957', HQ AAC, 13 June 2019.

Squadron aircraft were detached to deal with the deteriorating situation in Brunei. Flights began a rotation between Kluang and Brunei in an attempt to quell a situation which would deteriorate into what became known as the Borneo Confrontation. By 1963 the tempo of operations in Borneo forced 656 Squadron to base two of its flights in Borneo and establish a system whereby they alternated at regular intervals with the two flights remaining in Kluang.[43]

The requirements of the counterinsurgency operation in Borneo were just one contributor to a period of upheaval for the AAC in Malaya which reached a peak during 1964. In March that year the first Westland Scout helicopters began to arrive in Malaya, requiring a flurry of activity including acceptance checks, pilot training and the adjustments needed to shift to rotary operations. The success of helicopters in general across several operational theatres had, by the early 1960s, led to calls for the control of the aircraft to be devolved more widely. In 1962 the Unit Light Aircraft concept had been trialled. The 1st The Queen's Dragoon Guards (QDG) undertook the trial and were allocated a flight of Saunders-Roe Skeeters. The pilots and crewmen were drawn from the QDG, trained at Middle Wallop and then operated their regiment's own aircraft in support of that regiment. The trial was deemed a success and led to the Integrated Flight Scheme being introduced across the Army. This led rapidly to a plethora of air troops, air platoons and air squadrons being created, often with different establishments and operating procedures depending on the whims of the commanding officer. The AAC however, remained responsible for the maintenance of the aircraft in the integrated flights, considerably increasing the workload of small units such as 656 Squadron.

In May 1964 the Integrated Flight Scheme was rolled out in the Far East. The first to be established was 30 Flight RASC, equipped with Beavers and under the command of Major John Riggall RASC. It immediately faced the complex task of providing four separate detachments spread across Malaya covering more than 1,250 miles. 10 Flight arrived in theatre to join 656 Squadron in March while 16 Flight was disbanded and reformed as 4th Royal Tank Regiment's (4 RTR) Air Squadron in Seremban. 656 Squadron HQ also took 20 Flight under command despite it being based 1,500 miles away at Kai Tak in Hong Kong.[44] The integrated flights also brought the Bell Sioux with them as well as the Auster, adding another layer of complexity to aviation operations in Malaya as recounted by Sergeant Howard Gater-Smith RTR serving with 4 RTR Air Squadron:

> I received training as an aircraft observer, both on the Sioux and Auster. The Auster cockpit was quite cramped in comparison to the Sioux which boasted a large Plexiglas bowl-like canopy. It was akin to sitting inside an enormous gold fish bowl. Elementary navigation procedures were practised and the art of aerial map-reading learnt. Most often or not journeys in Malaya took place over primary jungle. This necessitates accurate flight on a magnetic compass bearing for a set time and distance, whilst constant reference to the map looking for landmarks to confirm one's position. These might be a small hamlet, bend in a river, track junction or high ground…
>
> At some stage C Squadron [4 RTR] deployed on exercise to an area between Seremban and Kuala Lumpur. We were tasked to fly out to the area and drop bread and mail at various troop locations, these being clearly marked on the map. Each troop sack was labelled and stored securely inside the back of the aircraft in order of the drop. The front doors of the Auster were removed and we took off, transiting to the area in about thirty minutes. Approaching each location, the boss [Major Lindsay Smith RTR] made a slow pass overhead, flew a circuit and let down to tree-top height. At the right moment I would hoick the appropriate sack from

43 Anon., '656 Squadron, A.A.C.', *Army Air Corps Association Journal,* 1963, pp.61–62.

44 Anon., '656 Squadron, A.A.C. Headquarters and Workshop', *Army Air Corps Association Journal*, 1965, p.46.

A Sioux of the Scots Guards Air Platoon on a hilltop HLS at Tampilat in 1966. (Army Flying Museum)

> behind my seat and toss it earthwards. All went well until the last drop when the map flew out of cockpit along with the sack. I made a fruitless attempt to snatch it back, alas too late. Some unprintable expletive followed the downward path of this vital piece of paper and I said something like, "Oh dear! What do we do now?" The boss replied, "Well, we can't land. You got us into this mess, now get us out of it." I thought for a moment and replied that we had flown for thirty minutes on a heading of 320 degrees to reach this area so I would suggest a reciprocal heading for the same amount of time and that this would put us back in the right parish. He said, "OK," climbed to our transit height, trimmed the aircraft to a seventy knot cruise and handed over the controls to me. "Keep the nose just below the horizon and aim between those two hills in the distance." It was my first attempt at flying an aircraft – what a joy, and what a reward for losing the map![45]

The introduction of the Scout was plagued by technical issues, particularly with its Rolls Royce Nimbus engine which had a spectacularly low life-expectancy of less than 10 hours during early operations. There was also a significant ergonomic problem in the cockpit with the fuel shut-off control and the cabin heating control levers being situated side by side, looking similar and working in the same aspect. On 15 July 1964 Scout XR596 was being flown by Captain Daniel de Valencourt Jacot de Boinod of the Coldstream Guards along with 656 Squadron's QHI WO2 William 'Hutch' Hutchings AAC, when the aircraft disappeared close to its base in Kluang. The wreckage was found three days later with the crew dead in the cockpit. A subsequent investigation found that a fuel control unit had failed causing the crash. As a result, the Scout fleet world-wide was grounded for a month while further investigations were conducted. Captain Jacot de Boinod had only been married for six weeks and was described by the Household Brigade as 'a fine officer, destined for high position.' WO2 Hutchings had joined the RASC in 1945 and transferred to the GPR in 1951, becoming a QFI in 1956 and an AAC QHI in 1960. According to the AAC he was 'greatly liked and respected wherever he went' and was one of the Corps' 'finest instructors'.[46]

1965 brought yet more reorganisation and turbulence to the units operating in Malaya. Given the continued focus on supporting operations in the Borneo Confrontation, HQ 656 Squadron left Kluang and moved to Kuching in Sarawak to control aviation operations on the island. The move ended 17 years continuous service on the Malayan Peninsula by 656 Squadron. To fill the void in command and control created by the squadron's departure HQ 4 Wing AAC was created, co-located with HQ FARELF in Singapore, to oversee operations and training across Malaysia. HQ 4 Wing controlled the flying training elements, 75 Aircraft Workshop REME and the remaining AAC flights at Kluang as well as the increasing number of integrated flights arriving in theatre. By the end of 1965 these included Air Platoon 40 Commando RM, Air Troop QDG, and Air OP Troop 45 Light Regiment RA. 20 Flight AAC and the integrated flights in Hong Kong also fell under control of HQ 4 Wing AAC.[47]

The peace agreement signed on 11 August 1966 between Malaysia and Indonesia brought an end to the Borneo Confrontation and threw the future of FARELF, including the AAC, into doubt. It was clear from defence statements that there would be a dramatic drawdown of the British military presence in the Far East but at first at least, it seemed to have very little effect on Army aviation. Although 10 Flight left Kluang and moved to Netheravon in January 1967,

45 H. Gater-Smith, unpublished account, passed to the authors 20 November 2020.

46 Anon., 'Obituaries' and Anon, '656 Squadron, A.A.C. Headquarters and Workshop', *Army Air Corps Association Journal,* 1965, pp.11 & 46.

47 Anon., 'Headquarters 4 Wing Army Air Corps', *Journal of Army Aviation*, 1967, pp.65–66.

Sergeant R. J. Moreton of 2 Flight AAC checks his map before a Scout sortie in the late 1960s. (Army Flying Museum)

2 Flight, 7 Flight, 11 Flight and 14 Flight were all based back on the Malayan Peninsula by the end of the year, at Kluang, Seremban and Sembawang. Integrated flights continued to proliferate with Air Squadron Life Guards, Air Platoon 7th Duke of Edinburgh's Gurkha Rifles (7 GR), Air Platoon 1st Battalion King's Somerset Light Infantry, Air Troop 42 Cdo RM, Air Troop 249 Signals Squadron and Air OP Troop 95 Commando Light Regt RA all in theatre during the year. The Integrated Flight Scheme, however, was already experiencing problems. The manning, management and maintenance of so many small aviation units were placing intolerable strain on the supply chain, equipment management system, maintenance infrastructure and training pipeline. In 1967 therefore, despite it generally being accepted that units had benefitted from operating their own aircraft, it was decided to begin the process of recentralisation.

In May 1967 a sad conclusion was reached to the loss of Auster XK374 that had gone missing on 22 November 1961. The wreckage of the aircraft had been located just across the border in Thailand and a team from HQ 4 Wing accompanied by a Thai Police patrol went to inspect it in situ. The wreckage was photographed, and some pieces were recovered and taken back to Singapore. There it was discovered that the accident was caused by the failure of a union in the fuel pipe; the union having been incorrectly wire-locked. Captain Hills was buried in Terendak cemetery, northwest of Malacca while Captain O'Grady's body was repatriated to Ireland.[48]

48 Anon., 'HQ Army Aviation FARELF (4 Wing AAC)', *Journal of Army Aviation*, 1968, p.98.

Withdrawal from Malaya

There was still no sign of a dramatic withdrawal from the Far East during 1968 despite it being the main topic of conversation. FARELF did, however, seem to be treated as convenient pool of aircraft for operations elsewhere in the world. 130 Flight RCT had single Beavers detached, operating in Laos and Nepal as support over the monsoon season. Air Troop 1st Battalion King's Somerset Light Infantry deployed 3,500 miles from Singapore to Mauritius to support the battalion that had been sent to the island at the request of the Governor with 24 hours' notice to help quell ethnic and religious unrest. The Air Troop, commanded by Captain Peter Reece of the Royal Welch Fusiliers, remained in Mauritius for 110 days at the end of a very tenuous supply chain that stretched back to Singapore. The Sioux that deployed flew over 800 hours, of which nearly 170 were flown at night. At the end of the emergency the Mauritian Commissioner of Police stated that 'the helicopters made one of the greatest single contributions to restoring law and order on the island.'[49] During further reorganisation the 656 Squadron title nominally moved to HQ Army Aviation 17th Gurkha Division based in Seremban in order to keep it alive. The last of the Austers left the Far East and the AAC Flights were each equipped with six Scouts apiece. 11 Flight was based in Kluang, 2 Flight in Seremban, 14 Flight at Sembawang and 7 Flight at Terendak.

1968 began unfortunately with another Scout accident. XT625 of 11 Flight suffered a turbine blade failure while flying near Gerik on 30 January. During the subsequent forced landing the aircraft rolled over in a stream bed. The pilot and one other occupant escaped, but Corporal C. S. Galloway REME who was acting as observer was trapped in the aircraft and drowned.[50] Despite improvements, the Scout's reliability was still a cause for concern. In February 1969 Captain John Elliot of 11 Flight received a Green Endorsement in his logbook[51] following a 'copy book engine off into young rubber trees' after his Scout suffered engine failure. In May Captain Christopher Hunt RE of 14 Flight skilfully handled a forced landing after his Scout had a turbine failure shortly after take off when the aircraft was at just 300 feet in a climb. Captain Hunt also received a Green Endorsement.[52]

1969 was the year when the long-awaited drawdown from FARELF began in earnest, combined with the recentralisation programme. Air Troop 249 Signal Squadron disbanded in March followed by Air Platoon 3rd Battalion, The Light Infantry in April and Air OP Troop 14 Lt Regt RA in November. As 17th Gurkha Division's brigade flights began to disband, beginning with 7 Flight AAC in December (resurrected shortly after in Berlin), there seemed little reason to maintain an aviation branch in Divisional HQ. That spelt the end for 656 Squadron in Malaysia and on 17 November 1969 a farewell review was held for the unit in Paroi. The GOC 17th Division, Major General Derek Horsford praised 656 Squadron's skill and dedication throughout the Malaysian Emergency, the Confrontation in Borneo and beyond. 'The high reputation they have justly earned, and the technical skills they have achieved will I know be perpetuated not only by their immediate successors...but also in any part of the world where Army Aviation is called upon to participate in future operations.'[53] It was not the end for 656 Squadron however, as it was immediately resurrected to take control of Army aviation in Hong Kong.

49 C.H. Chambers, 'Aviation Mauritius', *Journal of Army Aviation*, 1969, p.38.

50 Anon., 'Obituaries', *Journal of Army Aviation*, 1969, p.25; Anon, 'Army Air Corps Aircraft Losses Since 1 September 1957', HQ AAC, 13 June 2019.

51 The Commendatory, or 'Green', Endorsement is the highest available air safety award. Often awarded for recovering an aircraft after a serious malfunction a Green Endorsement is entered in the pilot's log book and personnel record.

52 Anon., 'HQ Army Aviation FARELF', *Journal of Army Aviation*, 1970, p.100.

53 AFM ALB 56: Pamphlet, 'Farewell Review by Major General D. G. T. Horsford CBE DSO at Paroi, Monday 17th November 1969', undated.

Scouts and Sioux of 656 Squadron line up for the farewell review at Paroi on 17 November 1969. (Army Flying Museum)

Despite the drawdown there was still useful work to be done by those flights remaining. On 10 December Kluang experienced catastrophic flooding with water reaching depths of three to five metres. 11 Flight AAC was called on to assist with the rescue effort and in the mid-afternoon a Scout piloted by Major Spencer Holtom RA with Captain Christian Jackson RA as observer began to scour the rooftops for stricken survivors. Over the course of 90 minutes the Scout plucked 78 people to safety, lifting 11 survivors at once on one occasion. Captain Jackson had to keep a sharp look out for TV aerials and the wires that crisscrossed the area. As the Scout approached a roof top loose corrugated iron could be lifted or the entire atap (a small hut thatched with palm fronds) could disintegrate in the downdraught. Every time he leaned out to help a survivor on board Captain Jackson received an unpleasant static shock. The weather was also appalling with high winds, heavy rain and numbing cold. Flying with the doors off, after 90 minutes both pilot and observer were numb with cold and forced to return to base. They were replaced by Captain Garry Daintry Irish Guards flying a second Scout with Lance Bombardier Ritson RA as observer. During their sortie they lifted another 50 people to safety. Over the course of three hours a total of 128 persons were saved, the majority of who were children and infants that would not have survived any attempt to swim to safety.[54]

Such heroics, however, could not halt the loss of AAC flights. At the beginning of 1970 14 Flight was disbanded, followed in March by 2 Flight. 11 Flight remained but moved to Sembawang and had its establishment of Scouts halved. The flying training element of HQ 4 Wing was closed down. 130 Flight RCT departed the Far East later in the year and with it went the Beaver, the aircraft flying back from Singapore to Middle Wallop. The six aircraft took 22 days to make

54 AFM ALB 56: Anon., 'Exercise Flood Tide – Kluang Flood Rescue', undated.

656 Squadron AAC ground crew and REME personnel prepare a 'casualty' during an exercise at Ipoh in 1969. (Army Flying Museum)

the journey stopping at Phuket, Rangoon, Calcutta, Nagpur, Karachi, Sharjah, Esfahan, Tabriz, Ankara, Izmir, Naples, Nice, Bordeaux, Hurn and finally Middle Wallop.[55]

On 11 January 1971 HQ FARELF was disbanded and with it HQ 4 Wing AAC, and the remainder of 11 Flight AAC with its Scouts. Later that year Malaysia, Singapore, the UK, Australia and New Zealand signed the Five Powers Defence Agreements which guaranteed ongoing military cooperation in the region between the signatories. The tangible results of the agreement included the formation of 28th ANZUK Brigade made up of an infantry battalion from each of the three constituent nations. The brigade was based in Singapore and supported by a multinational artillery battery, engineer squadron, signal squadron and aviation squadron. 28 ANZUK Aviation Squadron comprised Australian and UK aircraft and crews. The UK element was formed from the RE Air Troop from the Jungle Warfare School at Tiram, Kota Tinggi, Johore and was redesignated 11 Flight AAC equipped with three Sioux. The Australian element, 182 Recce Flight, also had three Sioux and all were based at Sembawang, renamed Kangaw Barracks. Each flight had one post for a New Zealand pilot to maintain the tri-national character of the brigade. The Squadron was initially commanded by Major Brendan Kavanagh of the Royal Signals (R Sigs).[56]

The Australian Sioux were a direct buy from Bell and a later model than the British Westland version and there were significant differences between the two types. The canopy on the Australian

55 Anon., 'Ferry Flight – The Return of the Beavers of 130 Flight from Singapore to Middle Wallop', *Army Air Corps Newsletter,* No.12, January 1971, pp.5–11.
56 Anon., '28 ANZUK Aviation Squadron', *Journal of Army Aviation*, 1973, pp.53–55.

model was less domed and tinted, although the British canopy was found to be more aerodynamic when climbing vertically from jungle clearings. The battery on the Australian model was located aft in the fuselage, whereas the Westland aircraft positioned the battery forward of the instrument console which meant the centre of gravity differed between the two. The British models employed the external Stokes litters for carrying casualties whereas the Australian aircraft were fitted with more basic flat alloy platforms. Because of the aft location of the battery, loading the Australian platforms with the usual mix of ammunition boxes and jerry cans could result in running out of forward cyclic movement at dangerous moments. Different turbochargers required different engine rundown procedures. The radios on the aircraft differed though they covered the same frequency ranges for UHF, ATC and tactical nets. After much staff work all pilots whether British, Australian or New Zealand were cleared to fly both types. The different radio types however, obliged each pilot to be issued with both British and US flying helmets.[57]

Tasking was largely in support of brigade exercises at the jungle training school in Johore including reconnaissance, resupply and casevac. FAC training was also conducted with RAF Phantoms and RNZAF Skyhawks. Air OP was conducted on the artillery ranges at Asahan, Malacca and naval gun fire support off Pulau Aur in the South China Sea. The SAS was also supported when it visited the Peninsula. One pilot was tasked to drop off a passenger on the Australian submarine HMAS *Onslow* as she steamed up the South China Sea 50miles from the nearest land.[58]

The multinational nature of the Squadron did cause issues. Australian Army Aviation recruited and operated officer only aircrew. On formation of the 28 ANZUK Squadron the Australians objected to Staff Sergeant John Essery RE transferring from the RE Air Troop. Captain Douglas Fox RA was therefore assigned to take his place and was posted to the Squadron as 11 Flight Commander. On arrival however, he found that the Squadron was being operated as a single unit and the flight distinctions had been removed. Captain Fox therefore became Major Kavanagh's 2IC. Once settled in Captain Fox was handed an unusual task:

> I was to fly two RAN [Royal Australian Navy] officers out to HMAS *Melbourne*, Australia's aircraft carrier, when some 20nm [nautical miles] out, to pilot the ship into Singapore harbour. Having successfully located the carrier and obtained radio authority to land on the aft flight deck, I then flew the normal landing pattern only to discover on base leg that the Melbourne was steaming at 20 knots into a 20 knot wind and was now a mere dot. It seemed an eternity on full power and 70 knots IAS [Indicated Air Speed] to catch up and land my passengers, much to my embarrassment. I learned about maritime flying from that experience.[59]

28 ANZUK Squadron proved to be a relatively short-lived enterprise. The Australian Government decided to withdraw from the commitment and so, on 1 February 1974 11 Flight AAC was left on its own in Sembawang with four Sioux and 26 personnel.[60] It was clear that the AAC's association with Malaya and Malaysia, which had spanned more than 25 years, was drawing towards an inevitable close. It was therefore, all the more unfortunate when 11 Flight lost its OC in a flying accident. On 23 October 1974 Major Michael Sharpe AAC, who had taken command of the Flight in April that year, was flying Sioux XT163 during a night casevac exercise. Having just dropped off a passenger at Jemalung, 60 miles north of Singapore, the aircraft struck trees as it departed the HLS. XT163 crashed and caught fire. It took a rescue team two days to reach the wreckage and

57 D. Fox, account to the authors, 29 January 2023.
58 Anon., 'How to Land on a Submarine Without Really Trying', *Journal of Army Aviation,* 1973, pp.108–111.
59 D. Fox, account to the authors, 29 January 2023.
60 Anon., '28 ANZUK Squadron/11 Flight Army Air Corps', *Journal of Army Aviation*, 1974, p.109.

Major Sharpe was found to have died during the crash. Major Sharpe had been a RA AOP officer and saw action in Korea before transferring to the AAC in 1957. He originally converted to the Saunders-Roe Skeeter and became a QHI flying in Canada, Germany, Aden and Cyprus. He was considered one of the most widely known officers in the AAC.[61]

The final year of operational flying in Malaysia saw a resurgence of communist terrorist activity on the peninsula. With the British withdrawal almost complete however, the task of countering the new threat was left to the Malaysian authorities to deal with. At 0900 hours on 21 August 1975 Sioux XT516 took off from Sembawang for the final sortie of 11 Flight and the AAC based in Malaysia and Singapore. It was a routine task providing support to a NZ infantry company operating on the eastern coast of Johore. As the pilot returned to Singapore he considered the significance of that final sortie:

> He pondered deeply as he flew over endless rubber and palm oil plantations, passing the Malaysian Army Pulada School which was known down the years to tens of thousands of Commonwealth soldiers as the Jungle Warfare School, Ulu Tiram. This would be the last time a British Army aircraft based in the theatre would set out on a sortie into the Malaysian peninsula. It would end thirty years of Army flying in the theatre and one of the proudest chapters in the history of the Army Air Corps would be closed. 11 Flight has had the honour to be the last AAC unit to serve in this part of SE Asia. With apologies to the old and bold, a few words of history of the Flight may not be out of place in this newsletter. There are after all many Army aviators nowadays who have never flown East of Suez and who are not now likely to.[62]

61 Anon., 'Obituaries', *Journal of Army Aviation,* 1975, p.60; Anon., 'Army Air Corps Aircraft Losses Since 1 September 1957', HQ AAC, 13 June 2019.

62 Anon., '11 Flight AAC', *Army Air Corps Newsletter,* No.56, December 1975, pp.5–6.

Chapter 4

Borneo

Prior to Malayan independence in 1957 the British Government had considered the idea of creating a greater Malaysia state by merging Malaya, Singapore and two more colonies on the island of Borneo: North Borneo (also known as Sabah) and Sarawak. Within the latter was the tiny Sultanate of Brunei, a British Protectorate with great potential oil wealth. The merger, initially dubbed the 'Grand Design' offered Britain a means of decolonisation in Southeast Asia while retaining access to the strategic Singapore military base by more-cost effective means. After independence the Malayan Prime Minister, Tunku Abdul Rahman, took up the idea as he saw it as a means to forestall Indonesian expansion and gain benefits from the economic importance of Singapore. The British Government under Conservative Prime Minister, Harold Macmillan, were wary of Tunku's motives but was under intense pressure to reduce defence spending following Minister for Defence Duncan Sandys' 1957 Defence White Paper. After protracted negotiations, in 1961 Macmillan came to an agreement with Tunku to form Malaysia. It appeared the best compromise at the time to preserve British strategic and political interests in South-East Asia and it afforded a degree of protection to the vulnerable Borneo colonies, including Brunei. It would also ensure that Singapore remained free of Chinese communist influence, while the complex of bases there continued to be available for British use without the huge expense of direct ownership.

What had been agreed in principle however, proved more difficult to achieve in practice. The proposed Malaysia contained a multiplicity of nationalities and ethnicities not all of whom wished to fall under Malay control. Opinion in the Borneo colonies was particularly unenthusiastic where the population was a complex mix of Malays, Chinese and tribal peoples such as Kadayan, Punan and Dayak. There was also discontent at the plan elsewhere in Southeast Asia and the Philippines resurrected a historical territorial claim to North Borneo. More concerning was the position of Indonesia under its committed anti-imperialist president Sukarno. The southern three-quarters of the island of Borneo was part of Indonesia, known as Kalimantan. The Indonesians however, saw the whole of the island as part a Greater Indonesia and were likely to resist any agreement they saw as a post-colonial ruse to maintain power and influence in the region. Finally, there was the status of Brunei which held a rather anomalous position in the Malaysia scheme. The *Parti Rakyat Brunei* (Brunei People's Party, PRB) had been formed as a branch of the Malayan People's Party in 1956. In early 1962 The PRB leader Sheikh Azahari bin Sheikh Mahmud, better known as A.M. Azahari was elected to Brunei's Legislative Council. The PRB won all 16 of the elected seats in the Council however, with the majority of the seats made by appointment, the Sultan, Omar Ali Saifuddien III, retained autocratic control of the tiny state. Negotiations on the Sultanate's position within Malaysia had broken down in late 1961 because the Sultan was reluctant to share Brunei's oil wealth and since that point trouble had been widely anticipated.[1]

1 For more information on the background to the Borneo Confrontation see D. Easter, *Britain and the Confrontation with Indonesia 1960–66,* (London: Tauris Academic Studies, 2004) and Jackson, *Withdrawal*

See Map 3

The Brunei Revolt

For hundreds of years the Celebes and Sulu Seas to the east of Borneo had been plagued by pirates. The chaotic security situation and the availability of arms and ammunition in the region post Second World War had seen an upsurge in piracy which had reached a peak in the late 1950s and early 1960s. North Borneo recorded 232 pirate attacks between 1959 and 1962, primarily targeting the copra trade but also attacking fishing and passenger vessels and conducting coastal raids on villages.[2]

In 1957 Captain Donald 'Chris' Crouch RTR of 656 Squadron AAC was sent to North Borneo with two Auster AOP9 aircraft, transported by RNZAF Bristol Freighters. His task was to ascertain what support the Squadron could provide to ongoing anti-piracy operations around Borneo. Having landed at Sandakan on the northeast coast of North Borneo, Crouch proceeded to despatch one radio operator and one aircraft fitter to dispersed coastal airstrips at Kudat on the northern tip of North Borneo, Lahad Datu south of Sandakan, Tawau further south again and Jessleton (now Kota Kinabalu) on the west coast. Good communications were maintained between the sites using dismounted No.62 radio sets with skywave antennae with messages even being sent to and received back from 656 Squadron's HQ in Kuala Lumpar, 1,100 miles to the west. Having returned to the Squadron to report his findings, a few months later Crouch took a second detachment to North Borneo to begin working on anti-piracy procedures and tactics. He worked with two companies of 1st Battalion 7th Duke of Edinburgh's Own Gurkha Rifles (1/7 GR) and members of the local PFF. The Austers carried out reconnaissance for joint 1/7 GR and PFF patrols which travelled along the coast in small boats known as *kumpits*. Having been recalled to Kuala Lumpar to take command of 16 Flight of 656 Squadron, Captain Crouch was almost immediately despatched to Borneo once again. Having landed at a makeshift airstrip near Kota Kinabalu his team reassembled their Austers and began work with a Gurkha engineer squadron to site, plan and build a new airstrip close to the 13,000 feet high Mount Kinabalu.[3] 16 Flight's Austers were withdrawn from North Borneo, but in October 1962 14 Flight of 656 Squadron AAC deployed a Beaver 1,100 miles from Malaya to Brunei to reinvigorate the anti-piracy effort. Although it was found the Beaver was not particularly suited to chasing the powerful pirate craft, the experience of maintaining the aircraft over an extended line of communication would prove valuable for what was to follow.[4]

Unhappy with the move towards a greater Malaysia, in the early hours of 8 December 1962 a shadowy organisation known as the North Kalimantan National Army (*Tentara Nasional Kalimantan Utara* or TNKU) began a revolt in Brunei. Around 2,000 to 3,000 TNKU rebels seized police stations, large centres of population and the Shell oil fields in the Sultanate. The airport was, thankfully, held by the North Borneo Police allowing British Army reinforcements to be flown in from Malaya and Singapore in order to try to stabilise the situation. The stated aims of the TNKU were to establish an independent federation of Brunei, North Borneo and Sarawak with the Sultan of Brunei as head of state and A. M. Azahari as Prime Minister.[5] As news of what became known as the Brunei Rebellion filtered out the 14 Flight Beaver immediately moved its attention from anti-piracy operations to counterinsurgency.

---

from Empire, pp.186–210.

2 S. Eklöf, *Pirates in Paradise: A Modern History of Southeast Asia's Maritime Marauders*, (Copenhagen: NIAS Press, 2006), p.38.

3 C. Crouch, 'Borneo et al', *The Chinthe*, Iss.1, 2011, pp.28–30.

4 Anon, '656 Light Aircraft Squadron, Army Air Corps 1961', *Army Air Corps Association Journal*, 1962, p.29 and Warner, *From Auster to Apache*, p.119.

5 Easter, *Britain and the Confrontation with Indonesia*, pp.26–27.

Meanwhile, at 656 Squadron's new home in Kluang, 60 miles northwest of Singapore, there was frantic activity to reinforce the flight in Brunei. The first step was to move the rest of 14 Flight, under the command of Captain Bill Duthoit of the Staffordshire Regiment, out to Brunei as quickly as possible. 14 Flight's AQMS, Len Edgecock REME, described the deployment:

> The plan for the move and the detachment was fairly simple. We were to go as light as possible to support two Austers and two Beavers, one of which was already out there. We restricted ourselves to four boxes for spares and equipment. Personal kit as light as possible, small arms and ammunition to be carried, water bottles filled. The Austers were to fly out in a [Blackburn] Beverley and the Beaver to fly. We were to be resupplied from Kluang.[6]

The Austers had their mainplanes and tailplanes removed to enable loading. Two days after the revolt had broken out the aircraft arrived at RAF Lauban on a small island just off the coast of Brunei for reassembly. The second Beaver flew direct to Brunei Airport. On arrival Captain Duthoit jumped out to report to the brigadier in command and, as he halted to salute, he was 'uncharitably engaged' by a rebel machine gunner positioned on a nearby rooftop. 14 Flight's aircraft were soon in action dropping off and picking up patrols, finding rebel groups, leaflet dropping and dropping flares at night. In the first three weeks in Brunei the four aircraft and four pilots flew a total of 355 hours. The aircrew had to be self-sufficient. On one day Sergeant Nick Nicholls flew his Beaver for nearly 11 hours, carrying out all his own ground handling and refuelling. During one sortie he landed on a remote airstrip and taxied into the middle of a firefight. Brunei Airport became a hive of activity with the Austers and Beavers jostling for space with RAF, RNZAF and civilian aircraft. Once away from the city the terrain was challenging:

> The topography was formidable and by comparison even the worst Malayan territory was easy. There were no roads up country; rivers were the sole means of communication, except for aircraft which could operate to existing strips. Mountains towered up and the weather was treacherous. There were no really accurate maps. However, despite this, the Flight pilots soon became familiar with their new environment and their support was much in demand and well appreciated.[7]

During the second week of December 14 Flight was reinforced by the arrival of 7 Flight. The Flight's Austers were fitted with long-range fuel tanks to allow them to self-deploy but it was a hazardous endeavour with little margin for error. The single-engine aircraft had to cross 450 miles of the South China Sea from Singapore to Kuching on the coast of Sarawak before then continuing up to Brunei. After the successful deployment 7 Flight was immediately pitched into the flying operations to assist 14 Flight:

> Both Flights were very active and flew a vast number of recce hours by night and day (the 4 inch recce flare is used a great deal out here and an introduction to its use might be considered for inclusion in the Wallop training syllabus). Not every hour was spent in recce and the evacuation of European civilians was a task given to the Beavers on occasions. Capt. Carey was seen to land after one such sortie and 4 nuns in white habits disembarked – he would not agree to our proposal that we should put stained glass windows in the Beaver![8]

---

6 Warner, *From Auster to Apache*, p.120.
7 Warner, *From Auster to Apache*, p.122.
8 Anon, '656 Light Aircraft Squadron', *Army Air Corps Association Journal*, 1964, p.38.

By the close of February 1963 however, it was clear that the revolt had run its course and 7 Flight was replaced in theatre by 11 Flight AAC while 14 Flight continued to fly sorties to support the final mopping up of the rebels. Lieutenant Simon Salter RASC had only recently completed his pilot's course when he found himself flying the Auster in Brunei during the tail end of the revolt.

> Flare dropping became a major requirement as the rebels were on the run and believed to be holed-up on a mangrove island in Brunei Bay. In order to reduce their chances of escape nightly missions were flown to drop 3.5 inch parachute flares over the islands. As the technique of flare dropping had not been covered in training, my Flight Commander, Bill Duthoit, sent me off with the AQMS, Len Edgecock, on my first night mission and he would brief me as we carried out the sortie. For the benefit of those who have not flown Austers on flare drops, the loading was as follows: 3 flares were fitted under the fuselage on a light series carrier (the fourth being omitted as it sometimes caught fire in the slipstream) 4 flares were placed between the co-pilot/dispatcher's legs and a further 4 were strapped to the rear facing seat…
>
> We taxied out in a heavily laden door-less Auster which seemed to absorb the entire length of Brunei International runway before becoming airborne. Slowly we climbed to 3,000ft over the target area where the 3 flares under the aircraft were dispatched at random intervals. 'Q' then explained that from now on he would dispatch the remaining 8 flares one by one on my orders. He prepared the first flare by pulling out a short length of static line and tying it to the door jettison handle. To my surprise, in the dull cockpit lighting, I then noticed him pull out the second flare's static line and also attach it to the jettison handle. At this stage he placed the first flare on the doorsill and said he was ready. I told him to dispatch it and, as is normal when throwing an item out, he raised the flare and threw it downwards. To my horror, I noticed the static line whip its way round the static line of the second flare, pulling the two parachutes down the side of the aircraft and firing the flare between his legs! At this stage we had a million candle power cockpit lighting. 'Q' complaining of burnt feet and me saying 'Get the bloody thing out'! Why we didn't catch fire I don't know. As he dispatched the burning flare there was a clunk and the elevator was jammed. In the tail light the parachute could be seen streaming astern. The remaining flares were dispatched free fall as I contacted ATC and explained my problem. When one is on the ground it is easy to give advice. I was told to free the flare chute either by forcing the stick forward or back. My guardian angel thought otherwise and said 'You're still flying – don't do anything'. I didn't and by throttle and flaps made a safe if not dramatic series of landings back at base. After the ground crew had cut the chute from between the tailplane and the elevator horn I was made to taxi the aircraft back to dispersal. I don't think 'Q' wears nylon socks anymore!![9]

The Indonesian Confrontation

At the close of the Rebellion Captain Duthoit, WO2 'Red' Meaton AAC and WO2 Edgecock of 14 Flight were all Mentioned in Despatches for their work in Brunei. Although the security situation had been successfully resolved there were worrying indications that political instability was set to continue. The revolt appeared to have had the support of the majority of the people of Brunei, which could put the British Government in an awkward position if it continued to support the Sultan and his government. It was also clear that Sukarno's Indonesia had aided the rebels, most likely in an attempt to frustrate the formation of Malaysia in the northern Borneo

9 S. Salter, 'Beer Bottle Bombing in Borneo', *The Chinthe,* Spring 2007, pp.24–25.

states. To complicate matters further, a minor insurgency was also being conducted in Sarawak by a group known as the Sarawak People's Guerrillas, later the North Kalimantan Communist Party. The Party's membership was largely ethnic Chinese and the group had close links to the People's Republic of China. The group was usually referred to by the British as Clandestine Communist Organisation. The stand-off between the British Government and Indonesia over the future status of Sarawak and North Borneo became known euphemistically as the Confrontation. In April 1963 it began to take on a military dimension as Indonesia began mounting guerrilla raids into Borneo. On the night of 11 April some 60 to 75 fighters mounted a raid on the police station in Tebedu, Sarawak, killing one policeman and stealing weapons before disappearing back across the border into Indonesia.[10]

These types of incursions would be difficult to counter. The border between Sarawak and North Borneo and Indonesian Kalimantan stretched for some 1,250 miles from the north-west corner of the island to the eastern coast on the Celebes Sea. That border ran through primary and secondary jungle and scrubland, much of it over mountainous regions. On 16 September 1963 the British Government confirmed Malaysian independence (Merdeka) and the formation of the Federation of Malaysia including Sarawak and North Borneo. As Indonesian cross-border activity increased 7 Flight returned to Sarawak to reinforce 11 Flight. Captain Tim Deane RA of 7 Flight gave a vivid description of the ground and the problems that it presented:

> The terrain is characterised by dense virgin mountain jungle in the interior reaching to 9,000 feet and more, falling away to a wide strip of coastal swamp. Most of the country is uncharted and no topographical information is given on any map outside the centres of population .... The only lines of communication are the great rivers which dominate the country. Roads are almost non-existent, except within towns. There are no railways. Rivers are the lifeblood of the country, and the vast majority of tribes live in community longhouses along the banks of these great chocolate coloured turgid streams. In the thousands of square miles of mountainous jungle separating the rivers, no sign of human life is ever seen from the air.[11]

In mid–1963 the British Government resolved to press ahead with the Malaysia project even though, in the short term, countering the confrontation with Indonesia would probably increase defence spending rather than reducing it. Gurkha officer Major General Walter Walker, an experienced jungle operator, was appointed Director of Operations in Borneo. Five Gurkha and British battalions were committed to countering Indonesian aggression. As the confrontation continued, that increased to 13 battalions with Royal Engineers and RA support and the equivalent of a battalion of SAS. With such poor infrastructure in terms of roads the only way of protecting against Indonesian incursion from the south was to establish a chain of FOBs along the border from which infantry patrols could deter any infiltration. Moving troops to and from the FOBs, resupplying them and evacuating casualties could only be done from the air. 7 Flight's aircraft were based out of Kuching where it shared the airstrip with a plethora of RAF and RN aircraft including Hawker Hunters and Gloster Javelins. The personnel were accommodated with 42 Cdo RM in Semengo Camp, where the rattan shelters they lived in were described as 'primitive but reasonably comfortable'. The flights that operated in Borneo during the confrontation imported their tactics and techniques from the considerable experience they had gained during the Malayan Emergency. Nonetheless, that did not mean that everything ran smoothly as recalled by Captain Bill Morgan of the R Sigs:

10 Easter, *Britain and the Confrontation with Indonesia,* pp.45–46.
11 T. M Deane, 'Austers over Borneo', *Army Air Corps Association Journal,* 1964, pp.13–14.

> Flying duties were many and varied. Liaison trips to Sibu and Simanggang were regulars, plus mail drops to ground units in the jungle. Target registration was carried out on border crossing points in the 1st Division, as well as recces for Brigade and Commando staff. Forward air controlling was tried using marker flares dropped from the Auster bomb rack. Jim Dawson had a hang up when the flare initiating cord failed to snap. He spent a perilous 30 minutes or so trying to hook the flare up on a wire tied between two upright poles. Ultimately the flare was hooked off. Jim seemed to lose his enthusiasm for this method of marking reference points for FGA [fighter ground attack] aircraft after this episode...
>
> I was also detached for some time at Simanggang [on the Batang Lupar river in Sarawak], in support of 1/10 Gurkhas, carrying out recces and aerial photography in the border area. I recall taking some 'near vertical' photos of an Indonesian Army Camp at Badau just across the border and discovering the odd hole in the tail plane of the Auster upon my return to base. The CO and I decided that it was better to call these drainage holes rather than admit our proximity to the border![12]

With the announcement of the independence of the Federation of Malaysia, Indonesia began to escalate the confrontation using regular Indonesian Army soldiers to penetrate further into Sarawak and Sabah (formerly North Borneo). On 28 September 1963 a handful of men from the 1st Battalion, 2nd King Edward the VII's Own Gurkha Rifles (1/2 GR) along with approximately 20 Malaysian Border Scouts under the command of Corporal Tejbahadur Gurung were stationed at the forward post in Long Jawai, in Sarawak's Third Division, nearly 30 miles from the border. An Indonesian reconnaissance force infiltrated the village, and were then reinforced by a raiding force of around 200 Indonesian soldiers plus a substantial group of porters. A brief but intense battle took place between the Gurkhas and Indonesians during which one Gurkha rifleman was killed. Corporal Tejbahadur called for reinforcements and then withdrew his small team. The Indonesians rounded up members of the local Border Scouts, tortured and killed them. One managed to escape to Long Linau, where he raised the alarm resulting in Gurkha reinforcements being raced to Long Jawai by helicopter. The reinforcements began to track down and ambush the Indonesians during their long withdrawal to the border, killing 26 of them. Corporal Tejbahadur's team withdrew on foot to their headquarters in Belaga. Two weeks later Captain Deane arrived in Belaga after a 250 mile flight from Kuching and heard first-hand accounts of the Battle of Long Jawai:

> ...on 14th October,1963, I was bundled out of a Wessex helicopter of 845 Royal Naval Air Squadron, on to what must be the nethermost part of the Commonwealth, Belaga Strip – 160 miles upriver in the Borneo interior. Four hundred and fifty yards long and barely adequate for Austers in the heat and humidity, overshadowed by a 2,500 foot mountain ridge on either side. The section commander was Mike Munro...
>
> For the man keen on getting away from it all, Belaga was the ideal spot. Two days journey by boat over treacherous rapids from the nearest sizeable town and surrounded by the dark inaccessible forests of the interior. Supplied by daily Argosy or Hastings drop, the AAC detachment and the Navy Wessex squadron were completely dependent on air supply. The Belaga section were quartered 20 minutes ride in a powered longboat upriver, at a Somerset Maugham type rest house which served as the local administrative building for the village...
>
> Flying Austers on sorties lasting four and sometimes five hours over some of the most rugged and hostile country in the world took some getting used to after the green fields of

12 B. Morgan, 'Memories of Sarawak 1963–1964', *The Chinthe*, Summer 2008, pp.27–28.

Lance Corporal Walton of 14 Flight AAC holding a F95 camera used for gathering tactical intelligence over the jungle. (Army Flying Museum)

Hampshire. Selecting fields and 800 foot points soon became good cartoon material, with that endless green carpet of mountainous jungle intersected with chocolate coloured torrents. Every change in engine note brought an uncomfortable shifting in the seat, and an anxious glance at the gauges, but there was a definite feeling of the wild frontier when we occasionally discovered completely new river systems and mountain ranges, where the map was a virgin white.

My arrival coincided with the mopping-up operations of the Long Jarwi battle. Briefly, a sizeable party of Indonesian irregulars had made their way over the frontier, and using river systems, had penetrated to Long Jarwi [*sic*] on the Balui river, a boat and trading station deep in the interior. They had quickly overcome the meagre defence by sheer weight of numbers, and as quickly started to withdraw back up the rivers to the frontier, several days' journey away. Thanks to Auster recce and the quick use of Naval helicopters, Gurkha cut-off parties were positioned on the exit rivers, and the majority of the enemy were killed in ambushes as they paddled back up their escape routes. This sort of battle really underlined the value of the Auster and the helicopter in this type of warfare, and was perhaps a copybook example of the use of both. With the eventual capture or escape of most of the enemy involved in this incident, affairs quietened down in our sector. Frequent long distance recces, lasting four hours or so, based on intelligence information, now became our general routine.

The majority of these recces were up to the headwaters of the Baleh river, a river so rugged and utterly remote, cleaving through 5,000 foot mountainous jungle, the source perhaps 100 miles from the nearest living human being, that every visit sent an involuntary shiver up the spines of those who flew up it. The weather on these long distance recces was generally the major headache. There was always the threat of a rapid build-up of widespread thunderstorm activity over the passes on the route back, and it was commonly agreed that the nearest diversion airfield was the river below. With the extreme turbulence of tropical thunderstorms, and no radio aids, instrument flying was out of the question and the wearing of Mae Wests [life jackets] at all times a wise precaution. Occasionally we would catch a glimpse on these long distance forays of the weird science fiction wild life of Borneo. Giant hornbills which flew low over the jungle canopy looking like the great pterodactyls of pre-history. When low flying up the rivers, monitor lizards could be seen basking like crocodiles on the shingle banks giving the impression that one was viewing a world untouched for millions of years.[13]

With the Indonesians increasing the rate and depth of their incursions, reconnaissance from the air was called on more and more frequently. 11 and 16 Flights moved out to Borneo while 14 Flight returned to Kluang for rest and retraining. 656 Squadron also received reinforcements in the shape of a section from 651 Squadron based in Middle Wallop. The increased sortie rate required hard work and frequent improvisation from the REME to keep the Beavers and Austers in the air. On one occasion Captain Daniel Jacot de Boinod of the Coldstream Guards made a routine landing at Long Somado, a jungle strip at a small jungle village in Auster AOP9 XK414 in order to contact the local Gurkha commander. On preparing to take off again he found that the aircraft's engine was unserviceable and would require replacement. He radioed his situation back to Kuching and a RAF Twin Pioneer with an engine on board was flown out to Long Somado accompanied by Sergeant John Bennett REME:

I was to take minimal tools, as the troops would render 'all necessary assistance'. The journey and arrival went without hitch, with the local force of Gurkhas unloading the engine on its

13 Deane, 'Austers over Borneo', pp.14–18.

> cradle, and field tool kits. I started to prepare the ECU [engine control unit] for removal, getting as far as I could before dark, as with the operational state at that time there were no lights permitted. The village perimeter was bristling with Gurkhas expecting trouble at any time, and I settled into a wary sleep under a temporary shelter.
>
> Early the next morning I completed disconnections and called for help to remove the lump. With the airframe not having the tail raised on a trestle as was usual, it meant the engine was not only high but, the mountings were inclined upwards! I was given 6 Gurkhas, 3 tent poles and 3 ropes to do the operation! The ropes went round the engine and the poles so that 3 men on each side could lift and move the engine forward, out of the airframe. That's fine, except Gurkhas are by nature of smaller stature than Europeans, and by the time the lift was ready the poles were almost head height to the Gurkhas. When it came to the 'lift' it was indeed a struggle for these very willing men to lift high enough to clear the mountings. Nevertheless it was achieved, and the replacement fitted in reverse order. After reconnection of all parts a ground run was carried out which was satisfactory. The tool kits were loaded into the rear, I strapped the Auster to my back and with J de B in control we trundled down the short strip to climb and complete the air test during the one hour flight back to base.[14]

An incident at the end of 1963 brought home the dangers present for all aircrew and their passengers flying over Borneo. On 18 December Sergeant David Thackeray of the Royal Army Ordnance Corps was flying Auster AOP9 XK408 on a routine sortie dropping mail to a FOB near the border. With him as passenger he had a visiting padre, Wing Commander Melville Ross RAF. Having carried out a successful drop at Kandai the pilot was conducting a recce of the next dropping zone at the village of Biawak when XK408 was engaged by 0.5-inch calibre small arms fire from across the border. Diving over the village the aircraft was struck by the fire. Thackeray was hit by two bullets in his left arm and Ross was also seriously wounded. Sergeant Thackeray took prompt action and considered his options:

> I cleared the village at tree top height and flew north for 2 or 3 miles, then climbed to give a Mayday call, which was acknowledged by Kuching. By this time I had made up my mind to make a forced landing at Kandai (Red 18) because it was very close and I knew the troops there would have medical kit. The passenger was obviously very badly hurt and I was losing a lot of blood and I reckoned that I would not get to either Sematan or Kuching before fainting. I got an acknowledgement from Kuching that I was going to make a forced landing at Red 18. I contacted Red 18 on the B47 [VHF tactical radio] net and told them of my intentions.
>
> I arrived overhead Red 18 and started an engine off forced landing leaving the fuel boost pump and master switches on in case I had to go round again. This in fact was necessary and I over shot. I went to half flap and climbed away and entered a steep turn to port. Whilst in the turn my left arm holding the stick snapped. I regained control of the aircraft with my right hand and decided that I would not have enough control over the aircraft to do a normal engine off landing, so planned an approach with power, intending to switch off all the switches just before landing. I made two unsuccessful approaches because I could not get lined up to my satisfaction, but the third approach appeared to be all right.
>
> I approached at half flap, 55 knots and 1500 RPM aiming for the cleared space between trees on the threshold of the LZ. I trimmed back, held the stick between my knees and applied full flap. I continued the approach until 10 – 15 feet up still at 55 knots, again took the stick between my knees and turned off the fuel, ignition and closed the throttle but did

14 J. Bennett, 'Long Somado ECU Change', *The Chinthe,* Iss.1, 2011, p.25.

> not have time to turn off the master switch. I took control with my right hand and saw small tree stumps sticking up in my path. I decided to try and hit tail first in order to minimise the risk of turning over. I pulled the stick hard back and the tail struck the ground. The main wheels then struck and the aircraft came to a standstill almost immediately, with its back apparently broken as I was looking up at the sky.
>
> The 40 Commando RM detachment at Red 18 came immediately to our assistance. I told them that the passenger was badly hit and told them to get him out gently straight away. Once they got him out some of them came back for me. One of them undid my straps and straightened my left arm and helped me out. I warned them to keep clear of the aircraft because of the fire risk. They then took me to the long house and rendered me first aid and did what they could for the passenger and informed their HQ on the wireless. Then the MO [medical officer] arrived in a helicopter and I was taken to Kuching. I remained fully conscious throughout but the passenger passed out soon after being hit. I should like to compliment the Commando Det at R18 on their prompt and efficient action, especially the sgt [sergeant] there.[15]

Sergeant Thackeray had decided that the only option to save his passenger, and indeed himself, was to attempt a landing at Kandai, a location that did not have an airstrip, only a jungle cleared HLS. Successfully bringing the Auster down on that confined landing would have been a supreme test of skill for any pilot, let alone one partially disabled by a broken arm and blood loss. It was confirmed that Wing Commander Ross had sustained no further injury through the forced landing but unfortunately died of his wounds sustained in the engagement before he could be evacuated from Kandai. Sergeant Thackeray's actions were subsequently written up in a citation that summarised his proficiency and resolve. 'Throughout this action Sgt Thackeray demonstrated courage and determination of the highest order. His calm assessment, skill and airmanship, and the undaunted manner in which he continued to fly his aircraft, including three landing approaches with his arm broken, were in the finest traditions of valour in the air.'[16] Sergeant Thackeray was recommended for the DFM but that was downgraded to a Mentioned in Despatches, which he received on 17 November 1964.

As 1964 began 656 Squadron, back in Kluang, had a number of difficult issues to contend with. First was the introduction of the Westland Scout helicopter and the conversion of aircrew, ground crew and technicians to be able to operate the aircraft. Second was the move towards the Integrated Flight Scheme whereby flights would be devolved to teeth arm and supporting units to form air troops or air platoons. All of this would need to happen while maintaining effective support to operations in Borneo. To better manage the effort 656 Squadron established a regular pattern of sub-unit roulement, with two flights in Borneo and two in Malaysia rotating every three months. The operational tempo remained high as 1964 began. John Everett-Heath was posted to the detachment at Belaga on New Year's Day:

> Within a few hours of John's arrival, we were airborne with a supply drop of compo rations for a lost Gurkha patrol in deep jungle, one hour's flight to the South East. Smoke – a sure means of detection in the jungle, soon had their position pin-pointed; and within minutes they were eating their first food for four days, and reading a message drop grid reference of their location. This type of sortie undoubtedly gave us more satisfaction than any other.

15 AFM 174: 'Statement by 22962874 Sgt Thackeray DJV concerning the incident involving the loss of Auster 9 XK408 on 18 Dec 63', undated.

16 TNA, WO 373/139, 'Recommendations for Honours and Awards', 10 February 1964.

The wreckage of Auster XK408 near Biawak having been crash landed by Sergeant David Thackeray on 18 December 1963. (Army Flying Museum)

> Affairs now began to warm up, and on 6th January we learned that an Indonesian fishing vessel had landed a party 30 strong on the coast 25 miles to the West. The next few days were intensely busy as infantry, helicopters and Austers hounded the enemy deep into the swamp – perhaps the most unpleasant fighting terrain in the world. Gurkha troops spent hours wading thigh deep in the impenetrable mire, and evoked tremendous admiration from all supporting arms. Within a week the entire party had either been killed or captured, thanks to really smooth inter-service and inter-arm cooperation.[17]

While the Auster continued to operate in Borneo one of its chief tasks reflected the original purpose of the aircraft during the Second World War, control of artillery fire as an AOP. It was a critical task along the Sarawak and Sabah borders where a Forward Observation Officer's view on the ground was restricted by the mountainous terrain and thick jungle. Captain Morgan of 7 Flight recalled the purpose of the AOP sorties:

> I carried out some target registration tasks in the border area between Kuching and Lundu. This was to preregister targets that we thought were likely to be used by the enemy as forming up or crossing points. The aim was to be able to quickly call down artillery fire should the troops on the ground need support in any instances of incursion. I think there were a couple of 105[mm] pack howitzers and one 5.5 inch howitzer.[18]

17 Deane, 'Austers over Borneo', p.19.
18 Warner, *From Auster to Apache*, p.127.

Supporting fire was not confined to land-based artillery as recollected by Lieutenant Simon Salter.

> During the flying course at Middle Wallop we had a day at RM Poole to learn about maritime ops with a brief on Naval gunnery. [While in Borneo] I made a visit to the guard ship HMS *Lincoln* and we carried out the first of a series of NGFS shoots. A lot of time was spent registering targets on the border crossing points using the 105[mm] pack battery based on Sebatik. It was during one of these flights I foolishly mentioned I was doing Naval shoots. Two days later a Captain RA arrived from Hong Kong – a real NGFSO [Naval Gun Fire Support Officer]. I pointed out to him that he would have difficulty carrying out his role without an aerial platform and if I was to fly it I was to do the first shoot – he left two days later...
>
> The greatest of all NGFS shoots was when the Australians arrived in HMAS *Vendetta*, a D[aring] Class destroyer carrying six 4.5 [inch] guns. After an initial 12 ranging shots I asked for four guns main armament six rounds gun fire to which the answer was 'negative' six guns six broadsides (three salvos were in the air at the same time). That night I visited the ship and was offered my gin and tonic in a tin mug, I always knew Australians were crude, only to be told that they did not have a crock left on board. Dinner was served on tin plates![19]

In February 1964 the crew of Auster AOP9 XK379 had a lucky escape during a sortie from Tawau on the east coast of Sabah. Second Lieutenant Richard Pugh AAC had been tasked to check on suspected gun running activity near the island of Sebatik. He elected to take along newly arrived Corporal Brian Hill REME to gain experience of the local area and to act as an observer. Shortly after take off and flying at around 50 feet above the sea the crew spotted a suspect boat and Pugh put the Auster into a wingover turn to starboard. The aircraft stalled and as the pilot attempted to recover it the wheels hit the water and XK379 flipped on to its back. Hill takes up the story:

> My recollections after that are waking up under water upside down and trying to exit, with no result. Believe or not the [pre-sortie briefing] started to run through my mind: "Don't try and get out with the plane still attached to you! Hand down, twist buckle, release and away!" This was indeed successful and I emerged from underwater quite quickly. Fortunately the door had been removed before take-off or I would not be writing this now! I surfaced alongside the wreckage along with the pilot and we swam to the 'suspect' canoe which picked us up, and that was my last physical movement for about two weeks! I was casevac to Labuan, and then on to Singapore by RAF [Handley Page] Hastings – which did not improve my thoughts on flying when it developed problems and shut down both port engines before diverting to Kuching![20]

In both Malaysia and Borneo there were high hopes pinned on the arrival of the Scout in the Far East. The Scout's entry into service was delayed due to issues with the Rolls Royce Nimbus engine and when it did arrive many problems persisted. The engine life was notoriously low, with only four to six hours before a change was required not being unusual during the aircraft's early service. The cockpit ergonomics were not ideal in some areas with the cabin heater switch being mounted next to the fuel cock and both operating in the same sense. The Scout's handling during autorotation was also challenging. Captain Bill Morgan recalled the arrival of the Scout in Borneo during 1964:

19 Salter, 'Beer Bottle Bombing in Borneo', p.26.
20 B. Hill, 'The First Flight', *The Chinthe*, Spring 2008, pp.15–16.

> The Scout was a disaster from day one, the 'fracto' Nimbus 102 engine was unreliable and we had two Rolls Royce Engineers attached on site to keep things going. They were carrying out major engine rebuilds in the field. Whilst in UK on training we were not allowed to fly over woods, yet here in Sarawak it was 99.5% woods in the shape of solid jungle. We never got anywhere near the target 150 hours engine life out the Nimbus and, to add insult to injury, one arrived flown in from UK, marked 'temperate climates' only.
>
> The later Nimbus 105 post Mod 664 engine was hardly any better. We also had a problem called lateral shake, whereby the helicopter's tail boom twitched. Damper units were fitted, but this made little or no practical difference. We spent many hours doing flight idle descents from about 10,000 feet to try and adjust the dampers. Much later it was discovered that there was a design fault in the free turbine which caused the vibration.
>
> Scouts were deployed at forward locations with units near the border, crewed by a pilot and a mechanic. The crew stayed forward until the aircraft needed a 30 hour service when they were flown back to Kuching. There were not enough tools to cover these deployments, so we had to go to Kuching Town and buy extra spanners etc out of the PRI funds....
>
> Often when all the Scouts were grounded I flew in the second seat in the Hastings. The RAF need two pilots when they flew within range of the Indonesian AA guns on the border. They only came out with one pilot per aircraft, so they offered to basically train any AAC ex fixed wing pilot to bring the Hastings back to Kuching a land it in an emergency. For a couple of months I had more Hastings hours than helicopter ones.[21]

Pilots also flew as second pilots in RAF Whirlwinds when the Scout was grounded. On 15 July 1964 the shortcomings of the Scout precipitated a tragedy. XR596 was on a training sortie near Kluang when it suffered engine failure and crashed killing the crew Captain Jacot de Boinod and 656 Squadron's QHI WO2 'Hutch' Hutchings AAC, a man known for his meticulous standards. It was recorded at the time that the entire AAC in Malaya turned out to assist in the search for XR596, but it was three days before the aircraft and its unfortunate crew were located.[22]

In mid–1964 the confrontation in Borneo and the operations to counter it were escalated. Major General Walker launched Operation Claret in an attempt to wrest the initiative from the Indonesian cross-border incursions. Claret actions involved special forces reconnaissance parties and infantry fighting patrols pushing more than 6 miles into Kalimantan to find and attack or ambush Indonesian forces. The patrols were supported by mortar and artillery fire, but aircraft were not authorised to cross the border. In August 1964 the Indonesians began amphibious infiltration operations across the Straits of Malacca to land troops on the southern coast of Johore in Malaya, either side of Singapore. On the nights of 1 and 2 September 1964 three Indonesian Lockheed Hercules carried paratroopers towards Malaya. One of the aircraft crashed into the Straits while trying to avoid interception by a RAF Javelin. The remaining two however dropped 98 paratroopers around the town of Labis, 100 miles north of Singapore. The operation to counter the airborne incursion was led by 1st/10th Princess Mary's Own Gurkha Rifles with 1st Battalion Royal New Zealand Infantry Regiment. The Gurkhas successfully engaged the Indonesian force killing 32 and capturing 62. 10 and 16 Flight supported the successful operation. In all, the Indonesians attempted 38 incursions into Western Malaysia, none of which were remotely successful thanks to the efforts of British, Commonwealth and Malaysian forces.[23] During this

21 B. Morgan, 'Memories of Sarawak 1963–1964', *The Chinthe*, Summer 2008, p.29.

22 Anon, 'Obituaries', *Army Air Corps Association Journal*, 1965, p.11.

23 N. Wells, 'Operation Flower – West Malaysia 29th October 1964 – (The Battle of Sungei Kesang), *Foundation News*, September/October 2007, pp.3–6; Anon, '656 Squadron AAC Headquarters and Workshops', *Army Air Corps Association Journal*, 1965, p.46.

period of menace to Western Malaysia 14 Flight AAC was slowly getting to grips with the Scout while based in Sarawak supporting the West Brigade from Kuching.

> West Brigade with five battalions on the border is responsible for aiding the civil powers in keeping order not only against the bands of Indonesian infiltrators on the border itself, but also against a large number of the C.C.O. or Clandestine Communist Organisations, Chinese immigrants who now live and work in the territory. From the map it can be seen that roads are few and far between; the local population inland consist of Land or Sea Dyaks – the Ibans, former head hunters who unfortunately appear to be returning to their old sport in parts of Sarawak, taking the lead from the successes of our own troops on the borders. On our side this time!
>
> The Brigade covers the First, Second and Third Divisions of Sarawak but the Flight has spent most of the time operating in the former. This Division has few mountain ranges, most of the area is or has been cultivated so there is a great deal of secondary jungle and scrub. The primary jungle – the main forest – is found in the hills, a dark green carpet of giant trees up to 200 ft. high!
>
> The only recognisable features, as you fly over this Division, are the hills and often these are great slabs of limestone with vertical faces, full of caves and stalagmites (and vice versa). A large number of clearings and helicopter pads have been made in the border area; all these are numbered and it took us days to plot them all on our maps, not to mention the inevitable amendments.
>
> The Scout is performing the tasks for which it was designed, namely: liaison flying, visits and general communication duties; reconnaissance for commanders and other specialists including surveyors; air OR for the guns; photo reconnaissance for pilot briefing and identifying installations across the border and lastly casualty evacuation which includes serious cases from the local population.
>
> For the professionally minded, the Flight aircraft are tasked through the joint Army/RAF. tasking authority in Brigade H.Q. in the evening prior to the flight. Every task is passed in detail to the units concerned and given a mission number. That task is then referred to by this number until it is completed, this includes the aircraft call Sign for reporting purposes. Therefore Mission 272 could be either a Scout on recce, Single Pioneer on a casevac or a Javelin escorting a Hastings on air airdrop![24]

14 Flight's edited diary reveals numerous sorties over the jungle to carry out casevac requests, AOP and the movement of commanders. The diary entry for 23 August gives a flavour of the incidents for which the Flight was called on for a response:

> There were two border incidents last night. At Dukit Knuckle in the 1st Green Jackets area, the base was attacked for about three hours by about 100 enemy. The Green Jackets fired a large number of rounds in reply and apparently inflicted four or five casualties, helped by a 105mm gun of 176 Battery RA firing over open sights at 200 yards! Ian Dawson flew the Colonel in at first light but the follow-up was unsuccessful. Leslie Bond, on the other hand, chose the Tepoi incident. He left at first light, collected the Colonel of the Ulsters en route and arrived in time to see three Indonesian bodies being brought in slung under bamboo poles. They had been caught running away from the village in the light of flares and killed by a section of Ulsters in a short exchange of fire.[25]

24 Anon, 'Headhunting with 14 Flight in Sarawak', *Army Air Corps Association Journal,* 1965, pp.51–53.
25 Anon, 'Headhunting with 14 Flight in Sarawak', pp.54–55.

105mm Pack Howitzer of 70 Battery RA in its gun pit at border Fort Serikin south-west of Kuching with a 14 Flight AAC Scout on the HLS in 1964. (Army Flying Museum)

The Flight also took part in joint sorties to insert 'stop parties'. A Scout would fly to a prearranged area close to the border and carry out a close reconnaissance to identify suitable HLSs or 'rope-in' clearings for infantry to abseil into. The Scout would then loiter and identify the site for RAF Bristol Belvederes or Whirlwinds which would insert a patrol before returning to Kuching. It was felt that the Scout was particularly vulnerable to fire during those sorties and that a cabin mounted GPMG would be useful for defensive purposes. All 14 Flight pilots experienced small arms fire from the ground during their tour of Borneo and RAF supply drop aircraft were hit on several occasions. The Scouts also worked out a method of adjusting the fire from Alvis Saladin armoured cars. The burst of the shell could not always be observed in the jungle, so a technique was worked out where the aircraft hovered 'in the line of fire'. The trace of the shell was quite distinctive and the trajectory flat, so the point of impact could be estimated if not seen and adjustments made over the radio.[26]

14 Flight's spell in Borneo overlapped with that of 10 Flight AAC who were based forward at Long Pasia (usually referred to as Long Pa Sia at the time) in the southwest corner of Sabah. The FOB was remote and the living and operating conditions for aircrew, ground crew and technicians was austere as described in great detail by one member of 10 Flight:

> Pa Sia itself is a small long house village or kampong on the Sungei (river) Padas tucked into the south-westerly corner of Sabah, about four miles east of the border with Sarawak and eight

26 Anon, '14 Flight Army Air Corps', *Army Air Corps Association Journal*, 1965, pp.63–64.

miles north of the border with Indonesian Borneo, although the Indonesians insisted that we were only three miles north of their border – a fact we patriotically chose to ignore! The original strip was constructed by the Japanese during the war, and was some 300 yards in length, but by the beginning of this year this was hardly recognisable, as the jungle had taken over some 19 years previously and had made quite a good job of restoring the place to its original state.

This, however, did not deter 68 Independent Gurkha Field Squadron who moved in in January to start to clear a Twin Pioneer-length strip, and build a camp to accommodate the Flight. Their task, although appearing practically impossible, was made a little easier by the bulldozer that was dropped to them by the R.A.F., who also ferried in in pieces a Ferguson tractor and a light motorised roller. All supplies were airdropped to the Engineers during their five month stay, and by the beginning of June, the project was completed and we moved in to start operating our aircraft in support of Central Brigade in the Fourth and Fifth Divisions of Sarawak and the Interior residency of Sabah.

The strip at Pa Sia is situated some 400 yards north of the kampong in the middle of dense primary jungle, and is only accessible by air, unless one is prepared to accept a three week walk from the nearest civilisation! It is surrounded by mountains on all Sides which rise up to 6,000 feet, and the runway itself has an elevation of 3,275 feet, which makes it a very pleasant climate to live in. Fresh water is permanently available from the Padas which rushes past the strip, and although this water is a bright orange in colour, it becomes quite palatable when diluted with a little of John Haig's taste and colour-removing elixir! Fresh meat can be obtained by issuing one of the villagers with a couple of 12-bore cartridges, though this practice could have had its perils as the locals are apparently fervid pro-Indonesians whose sole interest in the British Forces and Malaysia was one of finance. However, even though the villagers had proved unreliable in the past and were viewed with some suspicion by us while we were there, they gave us little cause for complaint. The occupation of Pa Sia by the Army Air Corps went off with hardly a hitch, though some people found it hard to believe that they were seriously expected to live there for three months…

The camp itself is typical of the jungle forts which dot the border in Sabah and Sarawak, and is built entirely of weed, bamboo and sandbags, with tarpaulins for the roofs of all the bashas. The initial construction work was to build an Ops Room, and this, together with improvements to all the other bashas, a water supply for the Officers'/Sergeants' Mess Basha, communications trenches, bunker rebuilding, new latrines, a basket ball court and sundry other building projects, occupied most of the people for most of the time. In fact living at Pa Sia is very like painting the Forth Bridge!…

All supplies arrived at the end of a parachute, and air drop days – normally a Friday – were the highlight of the week. On the whole, the accuracy of the R.A.F. was of the highest standard, and not only had they to hit the strip with the one-ton containers, they also had to miss the camp and the parked aircraft. Apart from the odd one-tons worth of Avtur [aviation fuel] that landed irrecoverably in the ulu, the only real tragedy was when our N.A.A.F.I. supplies and fresh rations free-fell into the jungle one morning, and all that was salvaged were a few pencils and a roll of film. Beer and spirits of course, arrived in this way and thankfully were always recovered. Spirits were dropped in one gallon oil tins to avoid any possibility of breaking the bottles, and a can of Tiger beer would sustain a fairly hard blow before splitting. This was most fortunate as we felt we really couldn't declare it operationally vital to send a Scout air dispatch to Brunei to supplement our 10 cans per man per week!

It was after a month or more of occupation that we managed to get some form of defence for the aircraft in the shape of barbed wire coils around our "dispersal," but the main defence of the fort was fairly adequate when we arrived. These defences consisted of nine bunkers set into the hill side at about 20 yard intervals around the camp, all inter-connected by crawl trenches. Outside these bunkers was a wire defence system with sharpened slivers of bamboo – pangees

Airtrooper Thomson and Gunner Wilson from 14 Flight AAC on routine patrol near Serian in 1965. (Army Flying Museum)

– set into the ground to impale any would be attackers. On the east side of the camp was thick jungle and the defences here were augmented with sundry trip flares, with both trip wires and electrical means of detonation, plus several 36 [fragmentation] grenades which were arranged on trees about 4 feet from the ground and could be detonated from inside the camp. Improvements to the defences were, of course continually in progress, and after a few weeks in site the Flight considered itself pretty impregnable. Central Brigade's Gurkha Battalion maintained a company base in the kampong of Pa Sia across the river, and the main responsibility for guarding the Flight rested with this company. A Gurkha section of 12 men came across every night, and their presence inspired considerable confidence in those asleep in their bashas. Their efficiency was such that more than once has a pilot nearly been shot by a member of his own guard!

As we were entirely isolated from the outside world – the nearest road being some 75 miles away – all messages of any priority came to us over the C11 [high frequency tactical] wireless set and two magnificent Gurkha signallers loaned to us by 247 Gurkha Signals Squadron. These two signallers had such a control of the net that we never suffered any real difficulty with communications. All urgent tactical and technical messages came over this link, and the operators even mastered such tongue twisters for them as "tail rotor gearbox output oil seal" and "accelerator bleed screw adjustment!"[27]

27 Anon, '10 Flight in Borneo', *Army Air Corps Association Journal*, 1965, pp.57–60.

10 Flight had five Scouts based at Long Pa Sia. Each evening the Flight's tasking for the following day would be received over the radio, often by Morse code. In the early mornings the airstrip, sat at 3,000 feet, was invariably covered by a thick blanket of cloud that lifted and dispersed at about 0815 hours when the day's flying could begin. Storms often broke in the late afternoon and early evening, forming very rapidly and bringing a prompt end to the flying activity. Normally, three out of the five Scouts would be tasked daily engaged on troop-lift sorties, moving commanders around their area of responsibility or resupplying forward bases. All of these meant approaches to and taking off from confined jungle landing sites surrounded by trees up to 200 feet high. This posed challenges for the pilots as the aircraft were habitually at maximum all-up mass and power limited due to the altitude and heat. Nevertheless, the Scout could reach landing sites which the Belvederes and Whirlwinds could not, which made Army aviation invaluable along the border. Whenever border incursions by Indonesian troops or irregulars were reported the immediate action was to lift a patrol and attempt to land it between the raiders and the border to prevent their escape. These sorties often occurred towards the end of the day and presented numerous challenges to the aircrew flying in response:

> One incursion in particular comes to mind when the enemy were reported to be approaching Long Pa Sia, and the call came in for two aircraft to lift some Gurkhas in between them and the border. This operation started at 1730 one evening and two aircraft were soon airborne to collect the first load of soldiers. The clearing that was selected for the follow up patrols to go into was situated a few miles south of Pa Sia and was known to be insecure, though it was not known whether the enemy had occupied it on the way through or not! The clearing was situated at the bottom of a very steep valley whose floor was at 4,500 feet A.M.S.L., with a one-way, low, snaky approach up the valley culminating in a deep hole in the trees. On arrival and initial landing (disgorging Gurkhas at the rate of about one a second), the pilot found to his profound relief that there were no enemy in ambush for him, and two rather relieved pilots returned for the next load. On the second trip back to the clearing, someone switched off the sun and it very suddenly got dark. The second (and last) lift into the clearing that night was carried out on memory and the landing lamp, and there were eventually 16 soldiers on the ground, armed to the teeth, but with the prospect of a night out in the ulu with only a few rounds of 7.62mm ball for a bed. The pilots' problems were by no means over, however, as a return journey to Pa Sia and a landing had to be carried out. The strip was finally sighted thanks to a hastily laid proportional flare path in the black jungle below, and after the first attempt at a landing had caused a pretty rapid instrument take-off in a cloud of sand, the aircraft slipped down the side of the strip and landed.[28]

Even in daylight flying was demanding. The poor mapping and the featureless terrain of endless indistinguishable ridges blanketed in jungle canopy made navigation difficult and the position of the border unclear. Any accidental incursion over the border into Indonesia could prove extremely hazardous.

> This was unluckily confirmed one day by S. Sgt. Purcell who was flying three Gurkha platoon commanders on a recce for a patrol they were to carry out and his task took him as far as a village marked on his map about half a mile on the friendly side of the border. Actually this village was incorrectly printed on the map and its ownership still disputed, a fact which was emphasised by the sudden appearance of several bullet holes in the aircraft. Staff Purcell

28 Anon, '10 Flight in Borneo', p.62.

> executed a quick 4G turn and retired in good order. Splintering glass had caused a superficial cut on the pilot's forehead, and three shaken passengers were flown to the nearest landing site by a pilot with rivers of blood running down his face. A passing R.A.F. pilot who witnessed this return journey now insists that the world speed record for a 5,000lb helicopter is now comfortably in the hands of the Army Air Corps, and he showed some surprise that "your aircraft seem to have a propeller on the front, rather than a rotor on the top like ours"! Fortunately this was the only sortie to achieve direct contact with the enemy, although all the pilots had their moments one way or another.[29]

By the time they handed over to 11 Flight, 10 Flight had flown the Scout on well over 1,000 sorties during their tour in Borneo totalling 455.5 hours. During that time, they carried 1,664 passengers, 30 casualties and moved around 9,000lbs of freight. While managing the relentless tempo of operations in Borneo the AAC in the Far East was also having to cope with the process of moving to the Integrated Flight Scheme. On 1 October 1964 16 Flight AAC was disbanded and reformed as the Air Troop of 4 RTR in Seremban. On 8 April 1964 30 Flight RASC was formed from existing Beaver sections in 656 Squadron AAC and commanded by Major John Riggall RASC. On 15 July 1965 30 Flight was placed under the command of 3 Army Air Supply Group and was renamed 130 Flight RCT. It continued to operate as the Far East theatre Beaver flight with detachments in Brunei and Sabah. During the year further air troops and platoons began operating in Borneo including 1st Battalion, the Gordon Highlanders, 1st Battalion, Scots Guards and several Gurkha battalions and RA regiments. The air troops and platoons flew the Westland Sioux, although Air Troop 4 RTR continued to fly the Auster while it waited for the arrival of its helicopters. When the Sioux did arrive the regimental tradition of naming tanks was extended to the aircraft and Denise and Delilah deployed to Brunei. The units with integral aviation soon found their contribution invaluable, as recorded by 1st Battalion, the Gordon Highlanders during 1965:

> Operating as we were in deep primary jungle along an 80 mile front cut by swamp and rivers, air travel was virtually our only means of movement other than on foot. For this purpose we had call on Whirlwinds, Wessex and Alouette helicopters for personnel and freight lifts, but these were tasked centrally and hard to come by. Our Sioux made us independent of outside help. This independence was woven throughout the fabric of the Battalion and it is hard to imagine life without them.
>
> Our Air Platoon did not form with us until we actually arrived in our operational sector, and this could have been disastrous. However, we were lucky. Not only did we have two excellent and likeable pilots in Captain Alan Bower PWO [Prince of Wales Own Regiment] and Lieutenant Simon Lytle Royal Irish Fusiliers, who being both experienced infantry men and having previously done service in the Far East, fitted into our Battalion life without notice, but because they in turn were supported by a very good ground team headed by S/Sgt Sheather, R.E.M.E. Sheather was not only a tower of strength in keeping the pilots under a very tight rein, but also was quite invaluable in the Sergeants' Mess! The serviceability rate of aircraft was high, and only when one small 2/6d spring gave out and had eventually to be flown in from Italy, did we have an aircraft off the road. The Battalion felt like an amputee until the second Sioux was flying again.
>
> I understand that our predecessors, who were not so fortunate, had to rely on thumbing a lift whenever they wished to visit any of their outstations. This is hard even to imagine. I was airborne almost every day, visiting Companies, Detachments and even on occasions, Patrols.

29 Anon, '10 Flight in Borneo', p.58.

> Our Medical Officer was continuously on the move, holding sick parades, advising isolated posts on hygiene and the like. We had at least six casevacs and many more medevacs, and there was no doubt that the soldiers derived great comfort from the fact that, should they ever require a quick evacuation due to wounds or sickness, they could be back in hospital bed within the hour, irrespective of where they were. The airborne stretchers were comfortable and safe.
>
> Mail – probably the greatest single morale factor – was entirely dependent on our Sioux. Every Detachment, however isolated, got one mail run in and out, daily. The Signal Officer was able to visit out stations at regular intervals and could, in an emergency, send out spares or experts to deal with substations which went off the air. The versatility of the little Sioux was high-lighted when we had to suspend our gallant Signal Officer at the end of a 100 ft. rope, so that he could replace the antenna on top of an extremely high tree! This procedure – later with a proper harness as the original makeshift affair very nearly carved our Signal Officer in half – became routine…
>
> When I left the Battalion we had only had the Air Platoon for four months, but it already had become, because of its versatility, and because of our great luck in having such an excellent team, an integral part of the Battalion. It is not overstating the case to say that the Sioux may well prove to be the greatest single contribution that has been made to any unit, particularly to the Infantry, since the [Second World] War. In civilised country it makes life quick, efficient and comfortable; in difficult country it is indispensable. It is hard to imagine life without one's Air Platoon, I hope I shall never have to![30]

In April 1965 the QDG arrived in Borneo. The regiment deployed a squadron which was based at Engkilili on the Kuching to Sibu road, with detachments 15 miles south-west at Batu Lintang and 40 miles west by road at Sungei Tenggang on the Kuching road. The QDG were equipped with Alvis Saladin and Saracen and Daimler Ferret armoured cars and were tasked with protecting the road network adjacent to the border and providing support to other units operating in the area. The QDG Air Troop also deployed, initially equipped with a single 'long suffering but much loved' Auster, commanded and flown by Captain Francis Chamberlain QDG. In July 1965 the Auster was replaced by two new Sioux under the command of Captain Peter Blaker of the Royal Green Jackets until he was relieved by Captain Mike Richards QDG. The Air Troop consisted of seven personnel with Captain Richards and Sergeant Ken Kendall as pilots, supported by Sergeant Clarke and a REME team of Corporals Fee and O'Neill, with QDG Troopers Bennet and Lugg as observers and handlers. Captain Richards recalled the conditions the Air Troop operated from:

> Our life was governed by Sioux XT158 that tried and XT157 that didn't. They lived on psp [pierced steel planking] helipads outside the crewroom in an atap, a palm leaf basha on stilts. This was situated next door to the rather more elegant RAF establishment for the twin Whirlwind detachment that shared the strip and provided spasmodic local ATC. They kept much to themselves but flew well when we could actually get them tasked. 50 yards away lay our AVGAS [aviation fuel] fuel dump on wooden rails under a flimsy palm roof. Snakes loved the cool under the drums so refuelling could become a brush with the nasty side of nature.[31]

30 Anon, 'The Air Platoon in Borneo', *Journal of Army Aviation*, 1966, pp.35–36.
31 M. Richards, account to the authors, 16 December 2021.

Apart from the British and Commonwealth units the Air Troop supported, they also worked with the locally raised 1 Ranger unit. Captain Richards had great respect for the Rangers and frequently worked with them from the jungle forts they manned.

> The Rangers were mostly local Iban tribesman with a big platoon of Dayak jungle trackers. They were small, tough and generally covered with the most spectacular tattoos. Some had filed teeth and the imagination did not have to stretch far to accept their headhunting heritage. On the whole they were not the sort of people to meet on a dark night! Their English skills were limited but this did not affect their great sense of humour and friendliness…
>
> A new fort was established in the north of the area at Nanga Mepi by the Malay infantry company attached to 1 Ranger with an enthusiastic Company Commander. I landed there just after they had revised the perimeter. As I was shutting down on the helipad I noticed to my horror that the Claymore anti-personnel mines which they had laid to protect the area were back to front with the business sides pointing straight at the aircraft. I took off rather gingerly and informed the post that it had a little sorting out to do.
>
> Tasks were wide and varied calling for all the hours we could get out of the aircraft. Because we were a scarce resource, liaison support for QDG suffered and they had to cover far too much by road in addition to their patrol and escort tasks. Our time was spent mainly on jungle reconnaissance and liaison for 1 Ranger and other local forces including the SAS who were busy in the sector. We also did a good deal of helicopter training for the Iban soldiers. AOP tasking and training was frequent with infantry mortars, Malaysian 105mm batteries and at times, the 76mm main armament of QDG's Saladin armoured cars.
>
> A good relationship was built with the RAF FAC team at Kuching who provided us with valuable training and open tasking access to any RAF [Hawker] Hunter missions in our area, effectively giving FGA on call. As a consequence, FAC work increased for us and the Fleet Air Arm joined in with their Sea Vixens. On one mission we had two Gloster Javelin delta wing fighters; due to their design they had to fly their attack approach inverted to see our smoke and the targets, rolling upright to fire. An impressive sight….
>
> The other big Ranger forts were in Luboc Antu and Batu Lintang. Being located in border enclaves with hostile anti air weapons close by, approach and exit was done at low level on concealed routes as was much of the close border flying. In the last two months of the tour hostile AA incidents increased significantly. This was mostly from 12.7mm machine guns, some mobile. One Indonesian position at Nanga Badau was particularly hostile with reports of 30mm cannon being installed there although we never suffered from them. Our response to this hostility was a rapidly delivered combination of artillery fire and fighter ground attack, where our links with Kuching paid off. By the time we left things had quietened down.
>
> Hearts and Minds missions were flown, usually taking the Rangers doctor to remote longhouse locations whereas pilots we helped with his ministrations. These longhouses, full of families, with livestock beneath the building's stilts, provided fascinating glimpses into a truly ancient culture. While we were there the only access was by foot or sometimes, water. I am told that their first sight of a Land Rover caused more excitement than any helicopter! …. The Rangers were generous hosts and …. capped it all by taking us to a Dayak cockfight.[32]

Aside from the air troops and platoons, during 1965 3 Flight, 7 Flight, 10 Flight, 11 Flight and 14 Flight AAC all found themselves on operational tours of Borneo. Despite the changes in organisation and aircraft the nature of the flying over Sarawak and Sabah remained constant. Technical

32 M. Richards, account to the authors, 16 December 2021.

A 3 Flight AAC Scout providing temporary shade for a group of Punan people in Sarawak in the mid–1960s. (Army Flying Museum)

problems and the weather often intervened leaving aircraft and their crews in danger of being kept on the ground at an isolated jungle outpost. It often required a great deal of skill, ingenuity and even some rule bending in order to get the aircraft back into the air. On Thursday 4 February 1965 Captain L. C. Bond RA from 7 Flight AAC was on a reconnaissance sortie in a Scout with the COs of the Royal Ulster Rifles and the Argyll and Sutherland Highlanders. At around 1700 hours Captain Bond landed the two colonels at a remote company base close to the border so they could get a brief from the resident company commander. Twenty minutes later when it was time to depart the Scout refused to start and Captain Bond diagnosed the fault as a failure of the high energy ignition unit which ignited the Nimbus engine. He required advice and assistance:

> A quick call on the Battalion net from the company command post bunker brought Bat-Wing (Battalion Air Tasking Officer) to the set. He had heard of Scouts refusing to start before and offered such useful advice over the net as "Hit it with the crowbar!"
>
> Help was soon on the way in the form of another Scout with David Pitkin driving, bringing Sgt Hildyard and Cpl Dorrian, spare battery and other bits and pieces. This aircraft had to land on the lower LZ near the village because the pad in the base was 'occupied'. The air mechs rushed up the hill to my position and arrived exhausted while David took the COs back to base.
>
> At this stage it was about 1815 hrs and last light was at 1915 hrs. We were at least 40 minutes from base and a mile from the border. A quick check by Cpl Dorrian proved that it was the H.E. [High Energy] Unit at fault and that we had no spares at base. This is where we started to break the rules. Cpl Dorrian suggested that we recalled the other aircraft and used his H.E. Unit to start mine once he was restarted! I contacted David on the radio and told him the story while Cpl Dorrian and Sgt Hildyard removed my H.E. Unit. David came back

> to the lower pad at about 1830 hrs and closed down allowing Cpl Dorrian to loosen his H.E. Unit sufficiently to remove it once he had restarted. This he did successfully and the H.E. Unit was soon on its way up the hill to me while the other aircraft flew direct to base.
>
> Daylight was now getting short and the weather deteriorating with low cloud and showers. The unit was quickly installed on my engine and we started first time, watched by most of the company with their fingers crossed. I landed at 1900 hrs in failing light and the ground crew spent the next hour wire locking and checking their hurried work.[33]

The following day the weather was poor with low cloud and continuous rain. 7 Flight received a call to put their sole serviceable Scout on standby as the Brigade had run out of RAF Whirlwinds. At 1630 hours Captain Bond was tasked to fly out to a patrol of the Scots Guards who were on a border ridge at about 1,500 feet with a young Dayak boy they had found who had suffered a broken femur. The Scout had to fly the 45 miles to the ridge and back again through driving rain and, although the boy was successfully flown to the hospital in Kuching, the rain had an invisible effect on the aircraft. The following day Captain Bond was flying the same aircraft again all day on reconnaissance sorties. At 1700 hours he was tasked to take a REME team of gun fitters to work on a 105mm Light Gun at the same company base where he had experienced problems two nights before. Having dropped off the REME team once again the Scout refused to start:

> Here we were in the same place, with the aircraft within 6 inches of its position on Thursday night, with exactly the same fault. The only difference was that it was an hour later and the nearest spare was 400 miles away.
>
> Another call on the radio and a Whirlwind of 225 Sqn [RAF] came to my rescue piloted by Sqn Ldr Peter Bulford and bringing Sgt Townend our Flight Artificer and Cpl Dorrian again. It returned with my three passengers, taking off just as it got dark. Sgt Townend said he would let me know when the aircraft would be ready – there was no mention of 'if'! I went down into the radio bunker, all blacked out by this time, to clear my flight back after dark. The Air Defence Battery guns on the airfield are free [to fire] after 1800 hrs. Meanwhile the two magicians dismantled the H.E. Unit showing no lights so as not to upset our hosts. They took the H.E. Unit down into one of the bunkers, took it to pieces and dried it out over candles for nearly two hours. Anyway, they refitted it at 2030 hrs and tested it – we were ready to go.
>
> By this time I had obtained clearance to fly back, together with the Kuching weather, track and safe height, all produced for me by David Pitkin and by phone and then the Battalion net. I started up successfully at 2045 hours leaving all lights off until the last moment. I used the landing lamp for about half a minute to get me out of the clearing and climbed out to safe height. I was in contact with Kuching straight away and saw lights over the hills from about 3,000 feet. I was very lucky to have a clear night and a new moon. Still in contact with the company I thanked them very much for their help and bid them good night before landing at Kuching at 2110 hrs after a simple night flight.
>
> The real credit behind these two main incidents goes to the REME Repair Teams that came down to rescue me and especially Sgt Townend and Cpl Dorrian who had to work with no lights and with the knowledge that there were no spares available. One thing was certain – they were not going to spend the night at a forward company base![34]

33 AFM 210: Captain L. C. Bond RA, 'From the Sarawak Scouts', undated.
34 AFM 210: Captain L. C. Bond RA, 'From the Sarawak Scouts', undated.

To an extent Captain Bond was lucky to suffer his technical failures while on the ground. If the failure occurred in flight, then the pilot had an entirely different, extremely dangerous set of issues to deal with. Even if he got back on the ground successfully his problems were far from over. Should the worst happen, and an aircraft had to make a forced landing in the jungle then the aircrew had to be ready to survive and extract themselves on their own initiative. Sergeant Tony Markham R Sigs was flying Scout XP905 on a recce from Padawan on 4 July 1965. He was carrying the orders group (O Group) from A Company 60th Battalion King's Royal Rifle Corps to the border to assist them in preparing for an intended attack on an enemy base:

> The weather was perfect and the brief and pick up at Padawan short and sweet. After an hour into the recce, the GPO [Gun Position Officer] asked to examine the site for his guns on top of a mountain, which had been prepared by sappers blowing off the top level which was close to border with Kalimantan for an area the size of two tennis courts. It was also to be the HLS for the insertion troops of A Coy. On being told to head for base, I turned onto heading and initiated a descent, only to hear the engine wind down completely. Entering autorotation for range I wanted to get clear of the 200 foot trees of primary jungle, which would certainly have killed us.
>
> After warning the O Group to strap tight, confirmed the engine had stopped, transmitted a Mayday call, and carried out Immediate Actions, I realised a ridge was looming towards us. I pulled back on the cyclic and used some of the kinetic energy of the rotor to skip over the ridge, and when clear, resumed range auto to recover the rpm. Shortly after another ridge required the same treatment. Fortunately the initial height had been about 3000 feet but as we had descended quite a bit I knew I could not clear the next ridge. Looking along the ridge I spotted a lighter green area and adjusted the approach and speed to reach it, flaring the forward speed off, and as the rotor blades started hitting trees, kept a very high nose up attitude and used the remaining kinetic energy of the rotor to create cushioning lift.
>
> We flopped the last few feet into secondary jungle, which completely covered the aircraft and left all four rotor blades on one side. I remember it was eerily silent. After checking that all were OK, we disembarked and began clearing the secondary undergrowth. It's amazing how training kicks in during these situations, so I organised two of the officers to look for water to generate carbide gas in order to inflate a signal balloon, the others to cut a clearing, piling the debris for a signal fire, and set off a rescue beacon. One of the officers had a camera, which I used to photograph the aircraft, especially the areas of damage.
>
> Whilst waiting for the water seekers return, I made some tea from my emergency kit to relieve their symptoms of shock. We were all grateful for it. The next problem was to dissuade the Coy OC from marching out, which would have taken days. The contents of the survival packs were checked and flares readied, just as we heard to sound of a Whirlwind approaching. A couple of flares later they were winching up the first man, the GPO, who had hurt his back on landing. The Whirlwind returned some minutes later, but could still only take the OC. On its next return it could take two, so I was left alone.
>
> I used the waiting time to remove radio frequency cards, stow all kit in my pack, search the area and aircraft for anything of value, and found the complete op-order behind the OC's seat cushion. Stuffing this inside my shirt I was subsequently winched up and taken to Padawan.[35]

Sergeant Markham also took XP905's doors with him back to Kuching and the grid reference of the stricken Scout which was subsequently recovered by RAF Belvedere. Shortly after Sergeant

35 T. Markham, 'Service with 656 Squadron AAC in FARELF', *The Chinthe*, Iss. 2, 2012, pp.28–29.

Sergeant Tony Markham (left) around the time of his forced landing in XP905 in July 1965, discussing flight test results with Sergeant Dorrian REME. (Army Flying Museum)

Markham was air testing a Scout when the engine flamed out, forcing him to make another engine off landing, this time on an airstrip at Terendak. Despite improvements, the Scout, and in particular its Nimbus engine, was still causing problems. During their tour beginning in May 1965 10 Flight AAC elected to withdraw from Long Pa Sia and operate from Brunei Airport in a bid to improve serviceability. It was judged that the move would improve the flow of spares and allow 24-hour servicing and maintenance which was not possible at the forward base. The move did not significantly impact on the Scouts useful tasking time for the troops on the ground. It took time but the improved servicing facilities at Brunei eventually began to bear fruit.

Scouts from 3 and 14 Flights AAC flying low over the jungle heading for Kuching in the mid–1960s. (Army Flying Museum)

> Nimbus is an appropriate name for the power plant of our aircraft as up until now it has been little more than a cloud cast over the successful operation of the Scout. However, during this tour in Borneo, a landmark in the history of the Nimbus in 10 Flight was reached. For the first time we had an engine which became life-expired at 200 hours – and still serviceable. Previously all our engines gave up the ghost before this magic figure was reached, sometimes on their own and sometimes with a little assistance from grit, rubber gaiters and jungle hats. It would be nice to be able to record honestly, that on this achievement of 200 hours, a short ceremony was held which was attended by selected members of the Brigade Staff. The engine was admired by all and a small commemorative plaque was attached to the airframe of XR602 from which it had been removed.[36]

The Sioux meanwhile, was proving remarkably robust in terms of serviceability but the weaknesses in the Integrated Flight Scheme soon began to have an effect. Pushing aircraft out to battalions and regiments inevitably spread REME support very thin and stretched the spares supply chain to breaking point. 1 SG recorded that they were initially spoilt by a serviceability rate of 85 percent. They were soon disillusioned however, as components began to require replacement and the supply system failed to keep up. The Air Squadron 4 RTR exacerbated the problem by splitting into two air troops, one based in Seremban, Western Malaysia and the other at Brunei Airport supporting the Central Brigade. This arrangement caused predictable issues in trying to split the already light REME and tools allocation across nearly 1,000 miles. Despite these issues the units involved found their air troops and platoons invaluable with the aircrew quickly establishing themselves as experts on their own area of responsibility and familiar with the indigenous people that lived

36 Anon, '10 Flight A.A.C.', *Journal of Army Aviation*, 1966, pp.73.

there. 1 SG recorded that their Sioux were vital in allowing their commanding officer to exercise proper control of his battalion. 4 RTR reported that the Sioux had made a most favourable impression and that Denise and Delilah had 'earned their keep twice over'.[37]

The Beaver was gaining praise for its work in Borneo with 30 Flight RASC. At the end of the Confrontation its performance and record received official praise:

> The inherent reliability of the Beaver and its constantly high rate of serviceability compared with army rotary winged aircraft has made it the workhorse of Army Aviation and proved conclusively that a rugged and reliable fixed wing aircraft is an essential component in the organisation of Army Aviation for the foreseeable future.[38]

Such was the reliance on fixed wing aircraft that in July 1965 a detachment of two Austers returned to Borneo, provided by 20 Flight AAC in Hong Kong, which provided great service through to the end of the campaign. Even with its extremely short take off and landing capability, airstrips suitable for the Beaver were few and far between in the Borneo interior and Beaver pilots sometimes had to stretch the limits to complete their tasks as described by Sergeant Barrie Davies REME:

> The Beaver has 3 fuel tanks underneath the pilot/passengers compartment, 2 of 28 gallons each and 1 of 20 gallons. 76 gallons of high octane petrol in total. Giving a range of 3½ hours at a consumption of about 20 gallons an hour or 3 minutes per gallon.
>
> On a busy day in Borneo I had a 'round robin' sortie, calling in at a number of short landing strips plus a bit of target spotting for the artillery. Not all landing strips had Avgas 100/130 octane (petrol), most only had Avtur (turbine fuel) for the Scout Helicopter's turbine engine. So balancing fuel load with payloads, length of landing strips and availability of the correct fuel en route was a very important part the day's planning for Beaver pilots.
>
> Normally we worked on a 20 to 30 minute fuel safety margin to cover the chance that the end destination was closed due to weather and the need to use an alternative landing place some distance away. There were often times when the safety factor had to be used, mostly due to bad weather but sometimes because of the actions taking place on the ground. That's why we had a safety factor!
>
> So, after that busy day, I arrived back in Brunei having burned 2 tanks down to the red 'empty' light and the other hovering on empty as indicated by the fuel gauge. Relaxing with a cup of tea in the crew room I was surprised by a mechanic coming in, standing there with his hands on his hips saying "And how, Sgt Davies, did I manage to pump just over 75 gallons of Avgas into your aircraft during re-fuelling? That left you with less than 3 minutes safety margin"!
>
> It took a while to live that one down apart from the 'four eyes chat' the boss had with me in his office!
>
> In Borneo one did what one had to do![39]

When the Beaver, in particular, did encounter a technical problem, it could cause a real headache in Borneo's environment. In April 1965 Beaver XP817 suffered a heavy landing at the remote airstrip at Bario and was declared unserviceable. Bario was around 6 miles from the Indonesian

37 Anon, 'The Guards' Sioux' and Anon (Ironside), 'Denise and Delilah in Borneo', *Journal of Army Aviation*, 1966, pp.36–40.

38 AFM 210, 'The Joint Report on the Borneo Campaign', 27 January 1967, p.38.

39 B. Davies, 'Borneo 1963–1966; A Day to Remember', *The Chinthe*, Iss.2, 2011, p.16.

border in central Borneo and around 60 miles from Brunei as the crow flies. It was surrounded by jungle-clad mountains, rising in places to over 7,000 feet and had no means of reaching it on the ground except by foot. The largest aircraft that could land at Bario was the RAF's Twin Pioneer but, even if the Beaver was broken down, the main components, fuselage, mainplane and engine were too large to be lifted out by that particular aircraft. The only practical solution, therefore, was to under-sling the main components and carry them out under a helicopter. The RAF's Belvedere twin-rotor helicopter was the immediate candidate for the task. The Beaver's stripped fuselage, at 1,100lbs was well within the Belvedere's capability, the crew however were wary of how the bulky load would behave in the downwash. Not long before a Belvedere had been carrying the under-slung fuselage of a Wessex that had swung wildly in flight and smashed the Belvedere's stabiliser, forcing the pilot to jettison the load. The RAF crew gave the chance of success with the Beaver as less than 20 percent and so the RAF vetoed the task.

Major Dick Woodbridge AAC of 11 Flight AAC was convinced the task could be completed using a Scout and began to make plans for the Beaver's extrication. He coopted Sergeant Chester RE, also of 11 Flight and Sergeant Jones REME from 30 Flight into his team and presented his plan to recover XP817 which was approved by the hierarchy:

> Flight planning and positioning of fuel along the route was completed and the Scout was meticulously adjusted and air tested to check altitude performance. The aircraft had been lightened by removal of doors, rear seat and freight floor. It carried a pilot's 30lb survival pack (standard squadron equipment) plus the aircraft's P22 pack and marker balloon. The power margin at 5,200lbs AUW in a 45 foot hover, the power required at 40 knots and the fuel consumption were all carefully checked. The Scout then left for Bario on 21st March.
>
> At Bario a number of cautious trials were carried out with the fuselage to discover the best method of suspension. The solution was to put a mild steel bar through the rear lift point and to suspend the fuselage horizontally fore and aft from both this point and from the Beaver lift bracket at the front. A combination of 4 x 1,200lb nylon ropes, reinforced with webbing straps from supply parachute harness packs was used, attached to a long strop. This strop was made up from a 6,000lb, 8 ft nylon strop with swivel connected to a 5,000lb chain to make a total length of 30 feet. The webbing straps served to take some of the 'spring' out of the nylon ropes, thus reducing bouncing in turbulence and keeping the fuselage level, while the 30 feet strop kept the load hanging clear of the main rotor downwash in forward flight.
>
> On the morning of 23rd March, the escort Whirlwind took off to check the weather over the Bario Ridge and at 1000 hrs it was decided to go. With Woodbridge flying the Scout and Chester observing from the co-pilot's seat of the Whirlwind they climbed out to the north, aiming initially for Bakalalan or, if all went well, on to Long Semado 44 miles away. Turbulence caused the load to bounce and porpoise and control was difficult but the Scout worked its way up to 5,600 ft to clear the Bario Ridge, dodging some cloud on the way and being forced very close to the border. Tail rotor drift caused the aircraft to crab sideways to starboard and turns to port were difficult to control except at a gentle 5 degree bank.
>
> Frequent fuel checks and calculations, confirmed by the escort, indicated Bakalalan could be overflown, so the Scout flew on towards Semado. The load was swinging badly at times and the continuous effort was arduous work for the pilot. The approach into Semado valley had to made at a very gentle rate of descent as any increase made control more difficult. On nearing Semado the escort went ahead and its crew marshalled in the Scout with its load still firmly in position after a flight of 45 minutes.[40]

40 AFM 210: 'Annex A to AVN 1190', 7 May 1965.

The first and most difficult leg of the journey was complete. The Scout was refuelled and Woodbridge briefed Sergeant Chester on the handling characteristics of the Scout with its unorthodox load. Chester then piloted the Scout for the next 50 mile leg from Semado down the Trusan valley to Kimabong. Chester encountered rain and turbulence in the valley but 50 minutes later he arrived safely in Kimabong. After another refuel Woodbridge once again took the controls for the leg out of the Trusan valley and on to Limbang. While negotiating a 2,000 foot ridge along the route he encountered a storm and, although he skirted the worst of it, the Scout was subject to severe turbulence with the rate of climb and descent indicator fluctuating wildly between 1,000 feet per minute descent and 1,000 feet per minute ascent and the wind speed varying between 30 and 65 knots as 'the load tried to fly the aircraft'. Nevertheless, Limbang was reached successfully after 35 minutes. With more fuel taken on, Chester joined Woodbridge in the cockpit for the final, relatively straightforward route down the coast to Brunei. They finally put the XP817's fuselage down safely after a journey totalling nearly 125 miles that had taken a total of two hours and 35 minutes flying time.

During the following week Sergeant Chester returned to Bario to recover the rest of XP817. The engine was a comparatively normal underslung load weighing a little over 750lbs. The mainplanes, stripped of their control surfaces and flaps were strapped directly to the belly of the Scout which flew 'quite well' at 60 knots. Once all the parts were safely in Brunei they were packed into trucks and then ferried across to RAF Lauban. From there they were flown out to Singapore for repair and reassembly.

The Scout had performed beyond expectations during the remarkable recovery of XP817, but its frailties were still a cause for concern and perhaps a contributing factor to a major incident later that year. During the night of 20 September 1965 Sergeant David 'Doc' Waghorn of the RAMC was flying Scout XR599 carrying Rifleman Sukbahadur Gurung of 1/2 GR who, in turn, was guarding an Indonesian prisoner captured during recent fighting. Having departed from Lundu the 7 Flight pilot, aircraft and passengers failed to reach their destination of Kuching. On 25 September Whirlwind XP327 of 225 Squadron RAF, which had been tasked to search for XR599, crashed 50 miles from Kuching killing all five crew and passengers on board. Despite an extensive search and the loss of XP327, XR599 and its occupants were never found. It is thought that the aircraft possibly crashed into the sea and small parts of wreckage were netted by local fishermen.[41] A few days after the double tragedy a major reorganisation of Army aviation in the Far East took place. On 1 October 1965 HQ 656 Squadron reorganised to form HQ 4 Wing AAC at HQ FARELF in Singapore. After 17 years continuous service in Malaya 656 Squadron was designated the Aviation HQ Borneo and moved to Lauban to take up its new responsibilities. HQ 4 Wing took on all training, support and flight safety responsibilities for the AAC and integrated flights across the Far East from Singapore to Sarawak and Sek Kong in Hong Kong. 656 Squadron was then able to concentrate its attention towards the support of the ongoing operations to counter the Confrontation in Borneo. There was no let-up in the operational tempo or the associated dangers. The Scouts and Sioux were in constant demand to support troops on the ground as recalled by Sergeant Markham:

> An Indonesian incursion took place on Sarawak side of the border, and during one at 21 Mile Bazaar [near Beratok], I was tasked to take the CO of the Police Field Force Unit, plus bodyguard, and the Brigade Commander to observe a contact and engagement, with several enemy pinned down. We could clearly see the patrol, but not the enemy, and I was ordered to land

41 Warner, *From Auster to Apache*, p.140 and Anon, *Army Air Corp Aircraft Losses since 1 Sep 57*, HQ AAC, 13 June 2019.

A 14 Flight AAC Scout on the HLS at the relay station at Gunong Serapi west of Kuching. (Army Flying Museum)

> behind the patrol, which I duly did. Remaining with the aircraft ready to go, the officers went forward for a briefing. On rejoining the aircraft, I found I was struggling for power to take off and would need a running take-off technique to get airborne. I used the full length of the area available, turning immediately once clear of trees, when the area ahead of me lit up with a wall of tracer. I turned rapidly first one way, and then the other, until it ceased and without being hit, called out a reserve Coy to act as a blocking action, and then acted as pathfinder and recce until they were in position. The action was successful, but the patrol lost one man when storming the enemy position. I was thanked for drawing the enemy fire, and saving more of the patrol, but if only they knew why.[42]

On 1 October 1965 in Jakarta, the capital of Indonesia, events took place that would affect the direction of the confrontation in Borneo. In the early hours of the morning the self-proclaimed 'Thirtieth of September Movement' (or *Gestapu*) assassinated six senior Indonesian Generals as the opening move of a coup. At 0700 hours a radio broadcast announced that Gestapu, an internal army organisation, had taken control of strategic locations in Jakarta with the assistance of other military units. The coup, however, had been poorly planned and prepared, and by 1900 hours Suharto, the commander of the Indonesian Army's Strategic Reserve, had regained control of all-important installations. President Sukarno gave Suharto the authority to take command and control of Jakarta. The circumstances that followed the attempted coup allowed Suharto

42 T. Markham, 'Service with 656 Squadron AAC in FARELF', pp.30–31.

to gradually consolidate his own power and marginalise Sukarno. At the same time an anti-communist purge further reduced Sukarno's power base. Suharto's increasing power eventually allowed him to establish a new government, and in March 1967 he formed a new cabinet that excluded Sukarno. During that intervening period of instability there was a reduction in the scale of Indonesia's campaign into Sabah and Sarawak although it was not necessarily immediately evident on the ground.

Violent tactical encounters remained frequent along the border. On 21 November 1965 Sergeant Markham was carrying forward the CO of 1/10 GR to observe a counter-incursion operation. Landing short of a ridge line, Sergeant Markham accompanied the CO on foot to the crest to observe his troops. From there he witnessed the action that resulted in the award of the Victoria Cross to Lance Corporal Rambahadur Limbu. The Gurkha lance corporal had gone forward with two men to recce an enemy position, but when they were only 10 yards from a machine gun position, the sentry opened fire on them. Limbu rushed forward and killed him with a grenade which resulted in the remaining enemy combatants opening fire on the small party, wounding his two colleagues. Under heavy fire, Limbu made three journeys into the open, two to drag his comrades to safety and one to retrieve their Bren gun, with which he charged down the enemy position killing several in the process.[43]

Early in 1966 the Life Guards Air Squadron took over from 4 RTR, retaining the split between Seremban and Brunei. Pushing their Sioux out to the FOBs, the Life Guards Air Squadron found that aircraft could operate independently for about five days before having to return to base. This allowed one or two aircraft to be deployed with only one REME technician to remote areas, where the local IO advised on the areas to be patrolled. There was also an emphasis on 'hearts and minds' activities in the border areas to try to ensure that the villages did not fall under the influence of Indonesian propaganda and intimidation.

> Almost every kampong in Sabah and Sarawak had been visited at some time by our helicopters and an incalculable number of hours have been flown, and stores of various natures been carried. Considerable assistance has been given to the local authorities by way of transport for doctors and nurses, as well as our own limited medical help. One man was found in a remote village where he had lain paralysed for seven months after a fall off a tree. After an hour's persuasive talk with his family and a further hour pushing and pulling he was lowered from the longhouse and into the Sioux where he had to lie across the seats with his head in the pilot's lap. All this was done without either speaking the other's language.[44]

The rotation of AAC flights between Borneo and Malaysia continued. 2 Flight AAC reformed in Netheravon on 3 January 1966 and six months later found itself taking over from 3 Flight in Sibu. Having noted that the newly completed accommodation was excellent in that location the Flight was moved to Lauban where it had to make do with unserviceable Aldershot shelters and scrounged caravan trailers. They also had to build their own pad from tarred paper and pierced steel planking to cover the expanse of mud that passed for a HLS on their arrival. Although 2 Flight were new to Borneo 7 Flight AAC had been supporting the campaign through frequent roulement from Malaysia since the beginning of the Confrontation. A plethora of helicopters were constantly operating over the jungle from the AAC, RAF, RN, Malaysian and Commonwealth forces. There were also fixed wing transport, reconnaissance and ground attack aircraft sharing the same airspace and it was more through luck than design that there were no mid-air collisions.

43 Markham, 'Service with 656 Squadron AAC in FARELF', p.32.
44 Anon, 'Air Squadron Life Guards', *Journal of Army Aviation*, 1967, pp.54–55.

7 Flight were happy to report however, that by 1966 a system of airspace control had been established to ensure that all aircraft could continue to operate safely in each brigade area:

> With the continuance of the good joint service relationship between Army and R.A.F., which was a feature of operations in Borneo, we now have at Brigade Headquarters the air cell tacked on to the Ops Room. This air cell contains G3 Air, Brigade Flight Command and R.A.F. helicopter tasking agency. All helicopters and fixed wing transport aircraft are in radio contact with this cell on their different sets. On the central map board all the aircraft positions are marked. A glance at the board inform all air users what the situation is in the area in which they are currently interested. The integrated air platoons/troops check in on the flight net only on take-off giving their intentions and are warned at the same time of any air activity or impending air strikes in their area of operation. They also clear with the air cell on closing down at the termination of sorties. The operation at this central air agency gives:-
>
> a) Great flexibility in responding to urgent tasks, e.g. casevac, picking up sitreps etc by nearest aircraft.
> b) Positive control without imposing any height limits.
> c) Ability to lay on strikes without stopping helicopter operations.
> d) Air safety and quick initiation of search and rescue operations over jungle terrain.
> e) Artillery and helicopter activity can be coordinated as the artillery commander is also located in the ops cell.
> f) Saving of valuable flying hours by more flexible tasking.
>
> In this manner we find that we can control aircraft which may he operating from secure bases up to 50 miles away because of the insecurity of the forward areas and give all pilots an up-to-date in the air briefing as they come forward to work. Working continuously with the same teams and aircrews we find that what appeared at first to be major differences of operations are not so, and that inter-co-operation between flights and squadrons is feasible and desirable.[45]

As the AAC and Army aviation continued to improve its doctrine, tactics and procedures in Borneo, 300 miles away and across the South China Sea US forces were also attempting to control a hugely complex air and aviation operation in Vietnam. Occasionally atmospheric conditions meant that US and British aircrew found themselves talking over each other while using the same frequency. 7 Flight recalled one such situation which resulted in the following conversation: American pilot, "Will you get to . . . off this frequency; we're fighting a war over here." Sharp reply, "So are we bud, but we're winning!"[46]

Although the popular image of the helicopter in combat is synonymous with the American war in Vietnam, at times the aviation effort in Borneo was equally significant although applied in a distinctly different matter. On 14 February 1966 a large mixed party of Indonesian Army and Chinese insurgents crossed the border into Sarawak in the area of Tebedu. On the following day the group were sighted by a border scout who discovered their tracks and camp in the jungle. On 16 February there was contact between the party and the security forces in which two enemy were killed. Two days later there was a second contact and another enemy fighter was killed, and the group appeared to be splitting into smaller parties. A major operation was launched in an attempt

45 Anon, '7 Flight Army Air Corps', *Journal of Army Aviation*, 1967, p.91.
46 Anon, '7 Flight Army Air Corps', p.89.

to prevent the insurgents returning to Indonesia and to kill and capture as many of the enemy as possible. On 19 February Operation Mixed Bag was launched.

Six infantry companies and a PFF company were deployed under the command of Lieutenant Colonel 'Runce' Rooney, the CO of 2nd Battalion 7th Duke of Edinburgh's Own Gurkha Rifles (2/7 GR). It was unfortunate that Mixed Bag coincided with a serious fault involving the torque tube on the Sioux. As a result, only one aircraft in 40 Light Regiment RA AOP Troop was available during the first three days of the operation. Captain Tim Deane RA, Captain David May RA and Captain David Canterbury RA kept the aircraft in the air from dawn to dusk with all servicing taking place at night. The Sioux was placed under the direct command of Lieutenant Colonel Rooney as were further Sioux as they became serviceable during the operation. Ninety percent of the Sioux's sorties involved Rooney being on board. During the three weeks of Mixed Bag he spent 82 hours in the air, 52 hours of which were during the 10 days of most intensive action. From his airborne vantage point and using the aircraft's ARC 44 VHF tactical radio Rooney was able to effectively deploy and redeploy his troops. The Sioux also conducted extensive recce sorties and although the dense jungle often prevented any sighting being made, the presence of the aircraft just above the canopy had the effect of slowing the enemy's progress and allowing the security forces to catch up. This effect was confirmed when it was found noted in a captured enemy diary. Rooney also asked the Sioux to fly slowly over suspected enemy positions in order to try to draw fire. The aircraft carried out that task several times 'with a certain reserve, flying as tactically as possible.'

The Sioux also allowed Rooney to land next to his troops and receive situation reports and to deliver instructions face to face. On several occasions a suitable landing site could not be found and the CO disembarked and re-embarked from the hover. The manoeuvre had been anticipated and rehearsed in a safe area. Rooney kept his headset on when disembarked, connected by an extension lead so he could remain in contact with the pilot. 'The ability of the CO to do this greatly increased the intimate control of the battle.' The single Sioux available at the beginning of the operation flew 40 hours in six days and made over 200 landings. Twenty-five of those landings were made on landing sites that had to be specifically created for that occasion, something a Gurkha troop could accomplish in 20 minutes.

The Scouts from 7 Flight AAC were also kept busy during Mixed Bag with up to four tasked at any given time. They were employed extensively to move troops and stores, conduct recces, and as a command and control platform. The Brigade Commander was a frequent passenger in the latter role although the Scout was hampered by the lack of an ARC 44 radio which made direct communication with troops on the ground more difficult. A typical command and control sortie during Operation Mixed Bag on 7 March 1966 was recorded by Sergeant Markham. He was flying a Scout configured with armoured seats in the front, all doors removed and with Sergeant Markham wearing an armoured vest. Beside him sat the CO 2/7 GR armed with a Sterling submachine gun and several smoke grenades for target marking. In the rear right seat sat the OC PFF Company, Superintendent 'Jock' Guillan also armed with a submachine gun. In the rear left seat was positioned a PFF marksman with a self-loading rifle. A group of insurgents had been reported near the Kuching to Serian road but contact with the enemy had been lost. Superintendent Guillan's company had been tasked to locate them, but their position was also unknown. Sergeant Markham was tasked with conducting a recce to find both the PFF company and the insurgent group:

> From the air I saw a PFF section pinned down in a paddy field by small arms fire from a low hill feature to their front. We orbited out of small arms range watching for some time trying to contact the patrol on the radio but had no reply. In fact this patrol was separated from its radio and other sets in the PFF were in short supply. The OC PFF told me which reinforcements were known to be entering the area. I chose an LZ in a safe and concealed valley where

> we could RV [rendezvous] with the ground commander. It was in a low scrub between rubber trees and the PFF secured it shortly after we landed. The LZ was about 300 to 400 yards from the contact and I did a concealed approach shielded by a low hill. I remained on the ground in flight idle [with engine and rotor blades still running] ready for an immediate lift off for about five minutes while the PFF commander briefed his patrol commanders and I could hear firing going on nearby.
>
> When my passengers returned from the briefing I got airborne again using trees to cover my exit. I pulled up at a safe distance to observe and the tracer from two automatic weapons from two separate points 200 yards away from the initial contact formed a curtain of fire in front of the aircraft. We now knew that there was more than one enemy group. I was at about 150 feet AGL when the firing started. I discovered later that the weapons were 7.62mm Kalashnikov automatic rifles firing one in three tracer. I took violent evasive action and this prevented the CO 2/7 GR from dropping a smoke marker. The curtain of fire was aimed in front of the aircraft not at it. I (and my passengers) decided it was not safe to return to the area. I was short of fuel and the time was then about 1700 hours.
>
> The CO told me that the section was still pinned down and no covering fire was available. I did not have GPMGs fitted otherwise I might have been able to help but I knew there were two Ferrets on the road so I flicked to the Armoured Car net and talked them in. Later, they were forced to withdraw without firing due to oncoming darkness.[47]

Sergeant Markham returned to Semengo Camp. On the way the OC PFF and CO 2/7 GR worked out a plan and radioed it ahead to the Gurkha company at Semengo. The plan was to immediately lift the Gurkha standby platoon into a back-stop position behind the insurgents. Once in position the PFF section would be informed and would assault the low hill feature forcing the insurgents into the Gurkha platoon's position. Following a fast refuel at Semego Sergeant Markham was ready to lead three RAF Whirlwinds back to the scene of the contact to insert the Gurkha platoon. The Whirlwinds however, were delayed as they tried to fit Bren LMGs to the aircraft because the sortie was considered a 'full precautions lift'. To save time the rest of 7 Flight AAC was alerted, and two Scouts were immediately available to begin lifting the Gurkhas forward with Sergeant Markham leading them in his third aircraft. On route the Scouts were ordered back to Semengo in order to allow the now armed Whirlwinds to go first. After a quick consultation with Lieutenant Colonel Rooney it was decided that as the Scouts each had a Gurkha Bren gunner sat in the rear door they should ignore the order to avoid unnecessary delay. Sergeant Markham led the Scouts into the HLS and then did the same with the Whirlwinds. All aircraft returned to Semengo to bring in further lifts until the platoon was complete on the ground. The two Scouts flying two sorties brought in 28 men whereas the three Whirlwinds flying two sorties only managed 24. With the platoon in place and Sergeant Markham in overwatch the PFF assault went in killing two Indonesians and driving the rest off the hill. The group broke up and unfortunately, most managed to get away in the dark.

The action on 7 March 1966 was an example of what could be achieved through the fast decision making and coordination that the airborne platform afforded ground commanders. It also demonstrated the skill of the pilot who aside from flying and navigating in a hostile environment was also operating two radios. The UHF was in constant use talking to RAF Operations, the Whirlwinds and between the Scouts while the B47 was used to speak to ground units with

47 AFM 210: P. E. Collins, '"The Front was Every Where", some lessons on Army Aviation in Borneo', undated but reproduced in edited form under the same title in the *RUSI Journal*, May 1967.

Major Dick Woodbridge stands beside Scout XP888 on a log platform HLS at the Gunong Sepadang patrol base west of Kuching. Note the rotors are running with no one in the cockpit. (Army Flying Museum)

constant 'flicking' and tuning between three frequencies required to speak either to the Gurkhas, PFF or armoured cars.

The 20 Flight AAC Austers also contributed to Operation Mixed Bag providing continuous observation over the operational area during daylight hours. The capture of an enemy diary indicated that the presence of the Auster often hindered the movement of enemy parties. The Auster also gave valuable service acting as an airborne radio relay station, allowing units on the ground to remain in contact with each other when the terrain prevented direct communications. Superintendent Guillan was a frequent passenger in the Austers from which he could direct his PFF patrols on the ground. Due to its reliability and loiter time the Auster also made an effective airborne FAC platform. On 18 February during the early stages of Mixed Bag a group of insurgents were isolated and surrounded in the area of Suburan and Tapah. Despite being outnumbered they continued to 'give a good account of themselves' and it was decided to employ FGA against them. Due to the close proximity of friendly troops the strike leader was flown over the area beforehand in an Auster from which the 20 Flight pilot, who was to be the FAC for the attack, was able to brief his RAF passenger on the situation on the ground. Disappointingly, the strike was called off, but it was a good illustration of the ongoing value of a reliable light aircraft flown by pilots with intimate knowledge of the ground and progress of an operation.[48]

48 AFM 210: Collins, "The Front was Every Where".

Reflecting on Operation Mixed Bag the Commander Army Aviation Borneo, Lieutenant Colonel P. E. Collins RA made an oblique reference to the US aviation operations currently taking place in Vietnam. Sergeant Markham had lamented the fact that without GPMGs on his Scout he was unable to provide fire support, but Lieutenant Colonel Collins put forward a different point of view:

> During MIXED BAG the local population were put to a minimum inconvenience by the movement of our own ground forces and they were never in danger from our own fire. This is an essential prerequisite of operations against an insurgent enemy and the possibility of 'browning' a large area in the hope of destroying some enemy that does not exist. Under these conditions a helicopter armed with a suitable offensive weapon would be a great asset but it should not be at the expense of tactical lift and must therefore be a button-on or pick-up-and-carry type of weapon. It must also be fully appreciated that the arming of a helicopter with this (or any other) type of weapon does not reduce its vulnerability to SA [small arms] fire and does not give it the ability to land with impunity in areas which are not secure.[49]

By mid–1966 the situation in Borneo appeared to be reaching a resolution. On 27 May an Indonesian political delegation made a goodwill visit to the Malaysian capital, Kuala Lumpur. Three days later the Malaysian Deputy Prime Minister, Abdul Razak Hussein, met with the Indonesian minister of foreign affairs Adam Malik. By 1 June they were able to secretly exchange notes which essentially constituted an end to the Confrontation. Indonesia undertook to cease incursions over the border between Kalimantan and Sarawak and Sabah and in return Malaysia promised to allow Bornean voters the chance to decide if they wished to remain part of Malaysia. It was not, however, the end of hostilities. The agreement had to be ratified by both governments. The struggle for power between Sukarno and Suharto in Indonesia meant it was unlikely that military action would come to an end in the short term.[50] Although Operation Claret was suspended on 2 June there was no such move on the Indonesian side, as confirmed by 10 and 11 Flight AAC's experience:

> After the somewhat 'Ruritanian' peace agreement between Malaysia and Indonesia life began to hot up in the Central Brigade area. The signing of the peace pact seemed to have been the cue for large incursions to take place and the cries of "I must have an aircraft" were heard echoing in earnest throughout the Brigade .... The Scout was a very popular aircraft with RE survey teams and many hours were flown helping to map the country. A lot at this involved maintaining a lengthy free-air hover at heights in excess of 5,000 feet while the surveyors took readings from their "black boxes" in the back. The remarkable performance of the Scout in this aspect of flight enabled the job to be completed with relative ease. A great many hours must have been flown by Scouts in Borneo on survey work.[51]
>
> The month opened with a bang, with an incursion of Indonesians in the Bario area. Once again, the speed of the Scout demonstrated itself when two Scouts were sent from Brunei, one as the Brigade Commander's personal OP, the other for possible casualty evacuation. Time and again, the Scout has proved itself far superior to the Whirlwind 10 in any particular role in this theatre, carrying at least the same payload, but some 30 knots faster.

49 AFM 210: Collins, "The Front was Every Where".
50 Easter, *Britain and the Confrontation,* pp.183–184.
51 Anon, '10 Flight Army Air Corps', *Journal of Army Aviation,* 1967, p.92.

> SSgt Chester [RE] made front page news in the local paper, with Capt Cemm [R Sigs], for the rescue of a "World-wide Helicopters" pilot, Mr. Brian Danger. He certainly lived up to his name when, after escaping from his Whirlwind III, which crashed and burned out in primary jungle giving him severe burns to head, arms and legs. He was then in the second "World-wide" Whirlwind which rescued him and also crashed on take-off, with no further casualties. Happily we can record that Mr. Danger survived the critical first stages of his recovery, and is now well on his way to complete recovery. Incidentally, SSgt Chester will never be seen now flying without gloves, or with his sleeves rolled up – a visual lesson that won't be forgotten![52]

The incursion at Bario also received support from 14 Flight AAC as part of Operation Double Cross. The Flight assisted with the redeployment of 2/7 GR to the tactical area of responsibility of 4 Royal Australian Regiment along with a Mobile Air Operations Team to control the dense air activity overhead. The rapid redeployment of teams from 2/7 GR proved extremely successful, sometimes moving the same team three or four times a day to create stops to prevent the Indonesians re-crossing the border in what was described as 'a very fluid situation'. Staff Sergeant Kenneth Hall of the Royal Welch Fusiliers was tasked with recovering the bodies of insurgents killed during the action and was on the ground in the process of doing so when a further contact was made with the enemy by D Company 2/7 GR just 70 yards from his aircraft. The Flight recorded that Staff Sergeant Hall immediately took off and 'a very rapid upward twizzle away from the enemy was made in record time!'[53]

As 1966 progressed the intensity of the Confrontation did eventually begin to wane and, with cooperation between Sukarno and Suharto, a full peace treaty was signed between Malaysia and Indonesia on 11 August 1966 and was ratified three days later.[54] A rapid withdrawal by British and Commonwealth forces followed and 14 Flight was soon the last AAC unit operating in Borneo during the Confrontation. With the whole of Sarawak and Sabah as their area of responsibility and very few troops on the ground it was a nervous period for 14 Flight aircrew as they covered and assisted the security forces withdrawal. The Flight's final task was in support of 84 Survey Regiment RE, flying a challenging 90 hours during a two week detachment. At the beginning of October 14 Flight's advance party made the move back to Singapore. By the end of the month the entire Flight had left Kuching on RFA *Sir Lancelot* marking the end of over three years of AAC and Army aviation operations in Borneo.[55]

Many lessons were gained by Army aviation during the three and a half years of the Borneo Confrontation. In particular the Integrated Flight Scheme was considered a great success as noted in an official report:

> Those COs who got an air platoon rapidly came to regard them as indispensable and found that by using the Sioux they were able to exercise better personal command and control of their units. The success of the introduction of the unit light aircraft [ULA] has been largely due to making them an integral part of the battalion where the aircraft are available with an immediacy and flexibility which no tasking system could achieve. At present there is no doctrine for the tactical employment of the ULA and in fact it is only by having and using the aircraft that new concepts are developed and the possibility of new tactics are seen. There are obvious advantages in having pilots who are members of the unit and who are not only

52 Anon, '11 Flight Army Air Corps', *Journal of Army Aviation*, 1967, p.93.
53 AFM 183C: '14 Flight AAC Operations Record Book', 1–30 June 1966.
54 Easter, *Britain and the Confrontation*, pp.191–192.
55 Anon, '14 Flight Army Air Corps', *Journal of Army Aviation*, 1968, pp.101–103.

A Scout from 3 Flight AAC in a jungle clearing near Belaga having landed the Commander 5th Infantry Brigade, Brigadier Rollo Pain in 1966. (Army Flying Museum)

> constantly aware of the tactical situation on the unit front but are also in current practice as operators on the battalion net and can frequently be used as a relay between distant stations.
>
> In an emergency the helicopter as a Command Post, as a means of rapid transport for personnel or stores, or as an aerial observation post is always an immediate requirement.
>
> Although the Sioux did much good work in Borneo, it is questionable whether, owing to its size, it is the right aircraft for a unit air platoon. It is too small to carry either a properly staffed Battalion Command Post, or a patrol or small reserve which a CO might wish to position rapidly in his battalion area.[56]

A common theme in post operational reports throughout the first 60 years of the AAC's operational experience frequently highlighted the advantages of employing Army pilots in battlefield aviation. At the end of the Borneo campaign, before the AAC was even 10 years old, it was already felt that this opinion was self-evident.

> This topic is almost a hackneyed one in service journals but it would be an obvious omission not to mention it here. From the author's experience in Borneo the job can clearly be done equally well by airmen, soldiers or sailors, but unless the airman and the sailor can be trained to think like a soldier they will be at an acute disadvantage compared with the soldier pilot when it comes to providing intimate tactical support for the ground forces. And, if the pilots and machine are not clearly under the operational command of the ground force commander he also will be inhibited from using them to their fullest advantage.[57]

56 AFM 210: 'The Joint Report on the Borneo Campaign', 27 January 1967, p.37.
57 AFM 210: Collins, "The Front was Every Where".

Perhaps however, the final word on Army aviation operations in Borneo should go to Brigadier William Cheyne, the Commander, West Brigade. 'I command my Brigade with a Scout helicopter. I can not do it any other way. There is no other way.'[58]

Support to Brunei

Although it was the end of operations in Borneo the AAC and Army aviation would continue to operate in the tiny Sultanate of Brunei for many years to come. Brunei had been a British Protectorate since 1888, oil was discovered in 1929 and its extraction and export brought great wealth to the tiny nation. After Japanese occupation during the Second World War the British returned to Brunei and in 1959 declared it a self-governing state, while its foreign affairs, security, and defence remained the responsibility of the United Kingdom. With the end of Confrontation, the Air Platoon of 1/2 GR found itself the only aviation unit for 600 miles, based at Seria on the Brunei coast, close to the Shell oil fields. During 1967 a Romney hut was brought up from Kuching and erected by a Gurkha engineer squadron. It acted as a hangar for the platoon's three Sioux and also contained two air-conditioned offices and a crew room. The only disadvantage of the location was the proximity to the sea, just 50 yards away, as the salt-laden air did not benefit the aircraft. The Air Platoon's time was largely absorbed supporting battalion exercises in the jungle, a pattern of activity that would be repeated for decades to follow.[59]

In 1967 the decision was taken to return to a centralised organisation for providing aviation to the Army. That meant control of Army aircraft returning to the AAC and the steady demise of the Integrated Flight Scheme which had proved to be inefficient in terms of repair, supply, training and safety. In 1970 aviation support in Brunei was taken over by C Flight 656 Squadron AAC, flying the Sioux. The Squadron was based in Hong Kong by that time and personnel were detached to man C Flight on a six-month rotational basis. While in Brunei the AAC personnel were looked after by the resident Gurkha battalion and operated from what became known as Scout Base in Seria.[60] In September 1978 C Flight's Sioux were replaced by the Scout. Moving aircraft spares to Brunei via Hong Kong was difficult. with only two civilian charter flights each week covering the route. Larger spares such as engines, main rotor blades and main rotor gearboxes were particularly problematic, although the judicious build-up of a stock of high-value spares commonly required in Brunei helped alleviate the issue.

The Flight was established with three Scouts and 23 personnel, including three Gurkha soldiers and two civilians. They continued to work from the small hangar on the beach at Seria in support of the resident Gurkha battalion and the Brunei Training Team. C Flight's Scouts had a number of modifications to assist their work in the jungle environment. Two of the aircraft were fitted with air hoists, although the 75-foot cable restricted their use in primary jungle where the canopy could be 200 feet above the ground. Nevertheless, the Gurkha battalion and exercising troops were trained to cut a winch-hole in secondary jungle so that casualties could be extracted, or personnel or stores delivered. All Scouts were capable of being fitted for troop abseiling and long-range patrols and the jungle rescue team was trained to insert by that method. VHF FM homing greatly assisted crews in getting to isolated patrols in the jungle. The aircraft were stripped of anything not required, including doors and seats to give a useful payload of between 600 and 800lbs. Despite the modifications it was recognised that the Scout was far from ideal for the role, but replacement

58 AFM 210: Collins, "The Front was Every Where".
59 Anon, 'Air Platoon 2nd Gurkha Rifles', *Journal of Army Aviation*, 1968, p.105.
60 D. Swan, 'Hong Kong 1975 to 1977', *The Chinthe*, Spring 2010, p.32.

with the Gazelle was considered a retrograde step and Lynx was too complex to be supported effectively. It would be a long time however, before a better solution was found.[61]

Despite not being an operational theatre the AAC still provided valuable support to infantry on the ground, particularly in the role of casevac. The Scout, despite its shortcomings, still provided a lifeline for troops in the jungle, but it often required great skill and persistence from the aircrew to ensure a successful outcome to life-saving sorties. On 23 October 1979 Staff Sergeant Raymond Portlock AAC was called out on a casevac sortie. A freak storm had hit infantry on exercise in the Labi area, causing a large branch to shear from a tree, severely injuring a soldier as it fell to the ground. Staff Sergeant Portlock got airborne at 1800 hours with a medical officer on board. In appalling weather, he was forced to fly at low level above the canopy to stay below cloud. He then had to enter cloud and use his instruments to climb over the Labi Ridge. Through expert dead reckoning using his compass and stopwatch he let down on the far side of the ridge and broke cloud, locating the exact position of the stricken patrol in the dark and stormy weather. On reaching the casualty, Portlock found there was no possible HLS and, with no winch fitted to his Scout, he had no choice but to make the difficult flight back to Seria.

> On his return and using his own initiative he was eventually able to organise a winch aircraft of the Royal Brunei Malay Regiment. Having refuelled and re-roled his Scout, he then made an RV [rendezvous] with the Royal Brunei Malay Regiment aircraft and successfully guided the aircraft to the winch site in similar conditions and with no horizon. He directed the winch helicopter and maintained all communications throughout the 35 minutes it took to evacuate the casualty, eventually guiding the winch helicopter back to the hospital at Seria. He then subsequently became involved in the identification of the patient and, having reported the details of the case and completed his preparation for the following day's tasks, was able to get home at 0245 hours, physically and mentally exhausted.
>
> Throughout, he displayed extreme devotion to duty, courage of the highest order and outstanding flying skill. He also showed a sense of duty and initiative well above that which one would expect of an NCO of his rank. His action in successfully carrying out this task in the most difficult and hazardous conditions imaginable which required him to fly and navigate in a mountainous area deep in the jungle for over two hours in darkness in extremely bad weather is well worthy of recognition.[62]

For his endeavours Staff Sergeant Portlock was recommended for the AFC but was awarded a QCVSA on 30 June 1980. With a winch fitted the Scout could provide more immediate assistance but it was no less demanding or dangerous for the aircrew involved. On 27 April 1983 Sergeant Michael Haverson AAC was providing aviation support to 6 GR on Exercise Pahilo Palta when, at the end of the day's flying, he received a request to provide urgent casevac support to an isolated patrol in the jungle. Sergeant Haverson quickly gathered together an air crewman, a winch operator and an abseil qualified medical assistant. The team took off at 1915 hours in darkness and very poor weather.

> By a combination of dead reckoning, VHF homing and skilful flying he managed to locate the casualty's position which was by then marked by four small fires. By this time there

61 Anon, 'Survival Scout (Adaption of an Aircraft)', *Army Air Corps Newsletter*, No.119, September – October 1984, pp.8–10.

62 TNA, WO 373/183/11: 'Army Form W3121, Recommendations for Honours or Awards', Staff Sergeant R.P, Portlock AAC, 23 November 1979.

3 Flight AAC Scout XP850 coming in to refuel at Nanga Gaat south of Belaga. (Army Flying Museum)

was considerable thunderstorm activity inland and the pilot's task was made more difficult by strong gusts of wind and sudden changes in the level of illumination due to lightning. After a thorough reconnaissance Sergeant Haverson slowly descended over the jungle until his landing light picked up the jungle canopy. The aircraft was then skilfully hover taxied across the tops of the trees to a point above the casualty and the duty medical assistant abseiled down.

The medical assistant informed Sergeant Haverson by radio that the casualty should be lifted out by winch and flown to the Medical Reception Station at Seria. Sergeant Haverson then had to find a suitable winch point close to the casualty. The Scout winch has only 70 feet of cable and the trees in the area were between 60 and 90 feet high. By extremely skilful and delicate handling in the dark and despite the danger to the aircraft Sergeant Haverson descended among the tree tops and located a suitable point approximately 150 yards from the casualty's position. Sergeant Haverson then held the aircraft in a steady hover whilst the injured man and the medical assistant were winched up into the aircraft. As soon as both were on board the aircraft Sergeant Haverson flew to Seria Medical Reception Station, where the casualty was admitted and later recovered. Throughout the sortie Sergeant Haverson showed extremely high standards of captaincy, airmanship, flying skill and courage in very difficult night conditions over extremely hostile terrain. As a result of his professional approach, bravery, and skilful execution of the sortie Sergeant Haverson carried out a successful life-saving mission which was in the best tradition of Army Flying.[63]

63 TNA, WO 373/183/26: 'Army Form W3121, Recommendations for Honours or Awards', Staff Sergeant M. Haverson AAC, 1 July 1983.

For his actions Sergeant Haverson was recommended for the AFC but was awarded the QCVSA on 8 November 1983. Nevertheless, despite being singled out for recognition, Portlock and Haverson were just two of many similar incidents in the jungles of Brunei. During the first six months of 1984 alone 32 critical casevac sorties were conducted by C Flight.[64]

In 1994 Army aviation support in Brunei underwent a number of changes. With the handover of Hong Kong to the People's Republic of China the AAC, along with all British forces departed the colony and, as a result, C Flight in Brunei became fully independent. During the same year 7 Flight AAC based at Gatow in Berlin was disbanded and the title was transferred to the flight in Brunei. Also in 1994, the venerable Scout was finally withdrawn from service. A replacement aircraft was required for both the Scout in Brunei and the Gazelle in Belize. Major Chris Hyslop AAC, in his capacity as SO2 G1/G4 Personnel & Logistics at Aviation Branch HQ LAND was responsible for staffing the procurement of the replacement aircraft. A contract was drawn up with FB Heli-services to provide the Bell 212 for 7 Flight in Brunei and 25 Flight in Belize.[65] The Bell 212, a derivative of the Bell 205 UH-1 'Huey', with exceptional hot and high performance and an underslung load capacity of 5,500lb, was ideally suited to the rigours and requirements of both locations.

The Bell 212 soon proved itself a reliable workhorse in the jungle. Despite 7 Flight being a non-deployable training unit, as the only British aviation anywhere in South-East Asia, it was occasionally called on to provide operational support further afield. On 26 December 2004, an earthquake shook the Indonesian island of Sumatra, destroying buildings and killing a large number of people. Half an hour later, as a result of the earthquake, a tsunami hit the north-western coast of Sumatra wiping out the Aceh Provincial Indonesian capital of Banda Aceh and destroying everything in its path along over 200 miles of coastline. With the agreement of the Sultan, the British Foreign and Commonwealth Office offered up assets in Brunei to assist with the relief of the disaster. 7 Flight personnel were recalled from leave and the OC, Major Peter Suddards AAC and Mr Phil Brown, the deputy chief engineer of FB Heli-services, departed on a recce to Banda Aceh. In Brunei the rest of the Flight were working through the issues of moving a non-deployable unit, its aircraft, personnel and equipment over 1,200 miles to the northern tip of Sumatra. Five days later they were ready to move. There was some trepidation over whether the Bell 212s would fit in the RAF's new Boeing C-17 Globemaster which had been tasked to move 7 Flight. In the event two Bell 212s, a Bedford 4 Ton truck and water trailer, a Land Rover and trailer, a Toyota 4x4 vehicle and 17 personnel were loaded and flown to an airfield close to Banda Aceh from which the Flight would operate. Living conditions were austere, in tents on a swampy piece of ground close to the main runway where heavy-lift aircraft were flying in and out almost constantly round the clock. Temperatures were in the low forties degrees centigrade, and humidity reached 85 percent during the day, without much respite at night.

The helicopters were ready to fly as soon as they were unloaded from the C-17 but operations were not undertaken for 24 hours to allow personnel to rest from the intensive preparation period. When flying did begin the devastation to the landscape, the infrastructure and the people of Banda Aceh was immediately apparent. The tsunami had completely devastated all semblance of civilisation along 600 miles of coastline and up to 5 miles inland. All critical infrastructure had been destroyed and the survivors were in critical need of food, water, medical treatment and shelter. Over 100 fixed wing movements per day in and out of Banda Aceh airfield were being controlled by the Indonesian Air Force. At the same time an Australian ATC unit had established itself

64 Anon, 'Survival Scout (Adaption of an Aircraft)', *Army Air Corps Newsletter*, No.119, September – October 1984, pp.8–10

65 C. Hyslop, account to the authors, 14 February 2022.

and was controlling 400 to 500 helicopter movements a day from 15 different nations, including French, Swiss and Japanese. The American contribution on its own provided 90 aircraft at its peak. The challenges and rewards of relief operations were recorded by Captain John Bushell AAC:

> The task in hand meant that it was essential to fly with the winch operator and an engineer. Their role became mainly loading and offloading the aid and the protection of the aircraft and people around it. Most of the landing sites had ravenous people who had never been around helicopters and so didn't know any better. The skills of the AAC Rear Crewman came to play as they had to deal with diverse situations from delivering aid, to providing medical aid to casualties, some of whom had the worst imaginable severe abdominal wounds and leprosy, to calming irate TNI [the Indonesian Army] personnel. The remoteness of the area with no back-up facility meant that flying with an engineer was immensely useful, not only as an extra pair of hands but in the running repairs and immediate assistance which was vital to maintain the tempo of operations.
>
> The weeks were very busy once the daily routine was in place: two aircraft, four crews alternating morning and afternoon sorties. Initially the crews were flying between four and five hours before changing over, which meant we were flying nine to ten hours off each aircraft daily and lifting 12,000–15,000lbs of aid per day.
>
> The regular flying sorties were in areas along the west coast. One town was called Lamno; it once consisted of 24,000 people but now had 6,000 dead, 12,000 displaced of which 2,000–3,000 were orphaned children. Every time the aircraft landed it was mobbed by children who initially needed food but also had nowhere else to go. It wasn't long before the crews were handing out sweets, some of which were donated by the NAAFI and others donated by Garrison families who desperately wanted to send something. The wide grins of these children when they realised what they had been given is something that will stick in all our memories.
>
> This little bit of obvious happiness became a regular pastime, much to the delight of the children and the heartache of the local police. LCpl Charlie Hooley got himself photographed by the *Daily Telegraph* giving out sweets and SSgt Jules Dawes became religious in his preparation of bags of sweets for the kids. He made dummy bags full of the spare rations so the big kids would foolishly fight for that bag, quickly followed by a swift pincer movement of the bag full of proper sweets handed to the little kids; it gave them a head start a least. There were kids rugby tackling each other everywhere; it was complete chaos and then you would see a small girl or boy having hidden the sweets and made it to freedom with a huge grin…
>
> On one particular occasion the aircraft were fully laden, the weather started to turn and the lead crew decided that both aircraft should initially land and wait it out. The site chosen was completely clear but on landing the first aircraft was mobbed by hundreds of people within seconds. The people were so hungry for food they fought each other viciously to grab anything they could. They did anything to get at the aircraft, often not aware of the danger they were in. LCpl Davey Allardyce just managed to rugby tackle two children who were about to run into the tail rotor and kill themselves.
>
> The aircraft couldn't take off because people were swarming all over it. The pilot, Sgt Kief Khanlarian RE, had to get out of the aircraft to replace Lance Corporal Allardyce (who had now stationed himself at the tail rotor) to drag people out of the aircraft who had clambered in to rip open the boxes. Mr Pete Bird (aircraft engineer) was being mobbed on the other side of the aircraft and was trying to maintain some control of the distribution of aid but getting a good beating for his troubles. Having seen this, the second crew into this site sensibly stayed in the air and threw everything out from about 15 feet, but the ensuing fight for food was not pleasant to watch….

> Then there were the sobering flights – the times when you pick up a casualty who is literally falling apart and bleeding all over the aircraft. Or the old lady who looks completely exhausted mentally and physically, she cannot stand without help and is going to hospital with what's left of her family of twelve, just her son and grandson. Or the lady who was returned to Chalang (with the ITN crew on board) terrified of flying but desperate to return even though the whole town is destroyed and nine out of ten of her neighbours are dead. She just wants to rebuild her life and move on, but when we leave her she stands alone on a field with not one of the town's buildings standing.[66]

7 Flight remained in Banda Aceh for five weeks. During the four weeks flying over that period the two Bell 212s flew 209 mission sorties including seven serious casualty evacuations, moving 59 displaced persons and over 110 tons of aid to areas along the coast before returning to Brunei.

At the beginning of 2007 7 Flight was still at Scout Base with three Bell 212s and an establishment of two officers and 16 warrant officers, NCOs and soldiers. On 31 July 2007 the OC, Major Toby Jarvis AAC received a request from the Defence Attaché in Brunei to carry out an unusual casevac. The coxswain of HMS *Sceptre*, a Swiftsure class nuclear submarine, had suffered anaphylactic shock over 75 miles off the coast of Brunei. The distance to the submarine meant that the Bell 212 would only have 12 minutes on station to execute the rescue. The rendezvous was set for 0100 hours the following morning (1 August). The predicted sea state meant the deck of the submarine would be awash and therefore, the casualty would have to be lifted from the conning tower. That required a highline transfer, which the Bell 212 was not equipped with. A member of the RAF Safety Equipment Section managed to improvise one from an abseil rope following a method copied from RN regulations and sent over to 7 Flight by fax. While that modification was underway the rendezvous was postponed to the late morning. At 0630 hours two Bell 212s were ready to depart, each with a full crew of aircraft commander, pilot, winch operator, winch medic and a doctor, and both carrying MS-10 life rafts. Captain John Morton-Race AAC takes up the story:

> Weather in Brunei is notoriously difficult to predict, low cloud cover is common over the jungle early in the morning, but hardly ever over the coast. As we approached our coasting out point (near Lumut) we identified low cloud out to sea as far as the horizon, we were forced to return to base. The spectacle of a tropical storms aside, one benefit of being approximately four degrees north of the equator is that low cloud disperses quickly. We departed Scout Base, for the second time, just before 0800.
>
> Coasting out on a heading to rendezvous felt unusual. I had done this many times before between Northern Ireland and mainland UK, but never in Brunei. Arguably less risky than flying over the 200 foot jungle canopy, but it felt worse, not because of the risk of the sharks either, just different!
>
> The boat was spotted by Staff Sergeant Den Woollams AAC cruising along at around 10 knots. The personnel on the conning tower were visible and attempts were made to contact the sub on HF but to no avail. The OC's aircraft with WO2 Green and Corporal Tokalau were the crew chosen to conduct the task, whilst my aircraft was in reserve with Corporal Rob Hughes AAC. The sea state was between moderate and rough in my estimation and as indicated by the swell as it passed over the sub's main deck.
>
> Highline dispatched and boat hook used to gather it to the conning tower, the transfer of the winch medic began. Keeping the aircraft steadily moving forward on station and in

66 J. Bushell, 'After the Wave', *Army Air Corps Journal*, 2005, pp.17–27.

> unison with the sub was a challenge for the front seat crew. As equally challenging was the task of ensuring the winch cable had sufficient slack between the ac and sub as it moved up and down in the swell, it was vital this had no impact on the ac, but also ensured the absolute safety of the winch medic now suspended between a submarine and a helicopter approximately fifteen miles out to sea in a tricky sea state.
>
> The conning tower only had space for four persons, with the casualty, doctor and person holding the highline this left little space for the winch medic, precise manoeuvring and capturing him was vital. The medic worked fast to secure the casualty in the sling, gave the thumbs up and the aircraft manoeuvred to a plumb-line above the conning tower and, with swift operation of the winch as the boat dipped in the swell, the two were clear of the conning tower and winched on board. The aircraft returned the casualty to the garrison medical facility where he was prepared for transfer by aeromedical evacuation to Singapore. I understand he made a full recovery.[67]

7 Flight continued to provide the Forward Aeromedical Evacuation for all troops operating and exercising in Brunei and beyond. There were existential concerns for the Flight however, as it had to contend with difficult decisions and circumstances concerning the basing of Flight when toxic chemicals and gases were found under Scout Base when it began extensive refurbishment in 2007. There were protracted discussions over the long-term future and location of 7 Flight. Additionally, CAA regulations and European Aviation Safety Agency requirements caused concern over the engineering aspects of the day-to-day running of the Flight. Nevertheless, as the AAC reached its 60th Anniversary in 2017 the small, dedicated team of officers and soldiers in 7 Flight continued to provide critical aviation support to exercising troops and the wider community in Brunei. Perhaps more significantly, the aircrew, ground crew and technicians working with 7 Flight represented the last repository of 60 years hard-won jungle flying experience, gleaned from previous aviation operations in Malaya, Borneo, Belize and British Guiana and continue to operate in the best traditions of all those who went before them.

67 J. Morton-Race, account to the authors, 8 August 2022.

Chapter 5

Hong Kong

Hong Kong Island was formally ceded to the United Kingdom in the 1842 Treaty of Nanking following China's surrender at the end of the First Opium War. The Chinese were again defeated during the Second Opium War a decade later and were forced to give up Kowloon Peninsula and Stonecutter's Island in the Convention of Peking. In 1898 the colony was further expanded when Britain obtained a 99-year lease over the New Territories. Despite the expansion, the Crown Colony of Hong Kong was no more than 400 square miles of land spread between more than 200 islands as well as a portion of the Chinese mainland. Hong Kong experienced rapid economic growth through the second half of the nineteenth century and became an important military outpost, particularly for the Royal Navy protecting trade routes in the Far East. Nevertheless, the Japanese Imperial Army successfully invaded Hong Kong on 8 December 1941, the same day as the attack on Pearl Harbour. The colony was occupied by the Japanese for nearly four years before control returned to the British on 30 August 1945.[1] Four years later, in 1949, the Chinese Communist Party under Mao Zedong took control of mainland China. Following a confrontation in the same year between the Chinese People's Liberation Army (PLA) and the Royal Navy on the Yangtze River the British military presence was increased in Hong Kong in order to protect the colony. The rugged terrain and sub-tropical landscape across the archipelago meant there was a requirement for light aircraft to assist with liaison, reconnaissance, and mobility. 1903 AOP Flight was detached from 656 Squadron RAF in Malaya and deployed to Hong Kong on 1 April 1949. The Flight, with its Auster AOP.6 aircraft, redeployed to Korea in 1951. 1900 AOP Flight deployed to Hong Kong in 1953 and on the formation of the AAC in 1957 was renamed 20 Independent Recce Flight AAC.

See Map 4

20 Independent Recce Flight AAC

20 Flight was based at Sha Tin in the New Territories on the southern edge of Tolo Harbour. Over the Lion Rock ridge to the south was RAF Kai Tak on the edge of Kowloon Bay. 20 Flight initially operated four Austers with six pilots. The runway was a 350-yard concrete strip that was only a few feet higher than the high tide level. Nissen huts were used as crew rooms, offices, stores and workshops and there was a purpose-built concrete control tower. A mile away on the other side of Sha Tin village was the main flight camp built on the edge of the Sha Tin Hoi harbour entrance. Another quarter of a mile beyond the main camp was the Officers' Mess in an old school building.

1 For more information on the history of Hong Kong see S. Tsang, *A Modern History of Hong Kong*, (London: Bloomsbury Academic, 2007).

The Sha Tin Hoi, or cove, was home to dozens of sampans, flat-bottomed wooden boats, usually with a small shelter over part of the deck, full of Chinese families living just yards away.[2]

20 Flight was, in effect, a satellite of 656 Squadron in Malaya. The Squadron customarily provided the pilots for the Hong Kong Flight, usually towards the end of their Malayan tours, although the Flight Commander was generally posted direct from the UK. From the beginning Hong Kong was a popular posting with aircrew, ground crew and REME technicians alike. Kowloon, just a few miles away was a 'beautiful, dynamic and vibrant place', with many attractions, particularly for the single officers and soldiers. Operationally, the Flight's main role was to assist the Army and RHKP, which could make for interesting flying at times as recalled by Lieutenant Philip Pettyfer RA:

> The most interesting work with the police was on the two days of the year celebrated by the communists on the one hand and the nationalists on the other. There were flags all over the city and the possibility of riots. We flew with an inspector, who was in contact with police HQ, to report on signs that a crowd was gathering. It was fascinating to be authorised to fly at rooftop level over a large city and observe the many people going about their business there. It was a rare opportunity indeed, reminiscent of scenes from Mary Poppins.[3]

One of the Flight's primary tasks was to assist in preventing incursions into Hong Kong territory by Chinese illegal immigrants (IIs). Mass incursions were orchestrated by the Chinese Communist Party and the Flight's pilots were frequently tasked to fly along the border to conduct recce and surveillance sorties, as described by Captain Peter Williams AAC:

> On occasions there was trouble with the Chinese across the border. At one point they persuaded their peasants to cross into the New Territories. I was detailed by the C-in-C [Commander-in-Chief] Far East to take photographs of the border incursion.
>
> Immediately afterwards I was ordered by the Governor of Hong Kong NOT to take any photos as it might upset the Chinese! After taking council [*sic*] I obeyed my superior officer. We loaded our photographer with a huge WW1 style camera and flew the border area.
>
> At the river mouth end on our side of the security fence I was astonished to see the ground was blue and not green or brown. This was due to thousands of people dressed in blue denims covering the ground. They were then herded back to the other side of the border by the army over the only bridge. This was designed to put pressure on Hong Kong.[4]

In one such move in May 1962 the Chinese Communist Party temporarily opened the border to Hong Kong allowing a flood of 60,000 immigrants into the colony in the space of a few weeks. The British Army troops and RHKP struggled to find and intercept the illegal immigrants in the mountainous and densely wooded ground of the New Territories close to the border, as experienced by men of 1st Battalion, Queen's Royal Surrey Regiment:

> We soon found out for ourselves how hard was the terrain and how exposed to the searing heat of the sun were the mountainsides.... It did not help that you might be able to see a Chinese across the valley. By the time you had clambered down and then back up through the thorny undergrowth, he had probably gone to ground in some inaccessible thicket and

2 P. Williams, '20 Indep Recce Flt, 1962 to 1963, Hong Kong', *The Chinthe*, Iss 1. 2013, p.23.
3 P. Pettyfer, 'Austers in Honkers', *Army Air Corps Journal*, 2003, p.149.
4 Williams, '20 Indep Recce Flt, 1962 to 1963, Hong Kong', p.24.

20 Flight AAC Auster WZ731 on internal security duties over Hong Kong harbour. (656 Squadron Association)

> you had a tedious search on your hands. It was hot and tiring work and your water had to be carefully conserved.[5]

Working in such conditions meant that the support from the air provided by 20 Flight was invaluable, however flying in Hong Kong was not without its challenges. The air space was very busy with civil and military traffic, but the main issues were presented by the environment and the landscape. The ground was extremely rugged and either heavily vegetated or urban and if there was an in flight emergency there were very few options available to an Auster pilot to get his aircraft safely on the ground, as recalled by Lieutenant Pettyfer:

> I had an engine failure at 800 feet over the city of Kowloon when returning an aircraft to Shatin [*sic*] after a service that included a fuel tank replacement. In those days we had just the one tank in the starboard wing. It was later discovered that, during the service, the disc that covers the lower exit from the tank had not been removed, and so the engine was starved when the fuel level dropped below the level of the upper exit. By raising the nose to increase my gliding distance, I just made matters worse. It was a sweaty few minutes, but I opted to stretch for the airfield rather than go down into the harbour and fortunately there was no other traffic about that afternoon.
>
> Even more seriously, one of our pilots made a low level pass over a Centurion tank. You will appreciate the term 'low level' when you read that his starboard wheel caught the commander's hatch and (so I was told) slammed it shut. The commander, being sensible rather than daring, had ducked in time, but the wheel sheared off and the pilot made an excellent landing

5 M. Felton, *China Station: The British Military in the Middle Kingdom*, 1839–1997, (Barnsley: Pen & Sword, 2013, Kindle edition) chapter 10, para 21.

> on the other one, without further consequences. I don't think that there was even a Board of Inquiry.[6]

The weather could be unpredictable, particularly during the monsoon season that extends from July to September each year. The typhoon season of 1962 was particularly active and during the night of 31 August Typhoon Wanda hit Hong Kong with very high winds and torrential rain. Sha Tin was particularly hard hit as Tolo Harbour endured a huge storm surge. 20 Flight was warned that Typhoon Wanda was due to strike the colony and took precautions. Three of the Flight's four Austers were tied down in the hangar at Sha Tin. With no further room, the fourth was flown to Kai Tak, which had already been closed to commercial traffic, and tied down in one of the RAF's typhoon-proof hangars. Flight personnel were then sent to their homes and accommodation to ride out the worst of the typhoon. The following day, when the winds dropped members of the Flight returned to Sha Tin to survey the damage as Captain Williams described:

> The nearest part of the flight was the airstrip which was a shambles. The only intact building was the control tower. At the height of the storm the sea had risen up to cover the airstrip to a height of around six feet and was driven by a wind of over 150 mph. (The wind measuring equipment had broken at over 160 mph). The sea came into the typhoon proof hangar through the back wall and washed the three Austers through the metal doors and on to the security fence. The Nissen huts had been washed around and some days later I found my flying clothing and log books under the remains of a different hut from the one I had left them in. The top of the control tower became the refuge for the guard and the normally fierce German Shepherd guard dogs as well as some six or so Chinese. The flight personnel were unhurt but there were a number of dead on the airstrip.
>
> It was then on through the remains of Shatin [*sic*] village where only the well built houses remained standing, the rest had been flattened by wind and water, as usual it was the poor who, without the protection of a strong house, suffered most. The fleet of sampans and the attendant families had disappeared. At the main camp the Nissen huts including the guard room had been flattened. One member of the guard floated out on his mattress fast asleep before being rescued by his mates! A large car was sitting on the roof of a three ton lorry and none of the vehicles were where they had been parked. The cellars where most of the flight's equipment had been stored were flooded but a great deal of the equipment had been rescued by the people on site at the height of the storm.
>
> At all the three sites the people there at the time and without outside help had done a wonderful job saving themselves and a lot of equipment. It took us months to get reorganised and re-equipped. The aircraft flown to Kai Tak was the only one to survive. Eventually we got other aircraft from store to discover they were unmodified (no tropical modifications) and with only one fuel tank.[7]

Auster's WZ791, WZ720 and XN411 were all damaged beyond economic repair during the typhoon. On further inspection the amount of work required at Sha Tin meant that 20 Flight would be unable to operate from the site for a very long time and perhaps permanently. The Flight therefore moved to Kai Tak where it fitted in with the RAF units operating from the main airfield. With no casualties 20 Flight had been fortunate. Typhoon Wanda caused 434 deaths in Hong

6 Pettyfer, 'Austers in Honkers', p.150.

7 Williams, '20 Indep Recce Flt, 1962 to 1963, Hong Kong', pp.24–26.

20 Flight AAC's Austers destroyed by Typhoon Wanda at Sha Tin on 31 August 1962. (Army Flying Museum)

Kong and left 72,000 people homeless. The move did not unduly affect the Flight's operational capability as Captain Conan Carey RASC recalled:

> 20 Recce was part of 48 Gurkha Infantry Brigade, deployed in the New Territories on the Chinese border, with its HQ at Sek Kong. Although the Flight was only 15 minutes flying time away, the field force atmosphere was absent at cushy Kai Tak. Brigade exercises and operations could be supported from base, although occasionally we set up under canvas at Sek Kong to refresh field skills.
>
> Despite the restrictions of terrain and air space, Hong Kong was not bad for flying training. We did photo reconnaissance on the border, message and supply drops to patrols on the hills and islands and plenty of low flying. Supplies dropped included straw and hay for the RASC's mules in remote places and much of the best low flying was at water-skier level over the sea. There was Air OP on the sea-ranges at Port Shelter and also live Forward Air Control with 28 Squadron RAF's Hunters. The latter liked to play seek and evade games with us, using a cine-camera to record the action. With just one Hunter in pursuit, the Auster 9 had a good chance of escaping with low level, tight turns, but a pair of Hunters did better. The de-briefs with film were interesting.[8]

On 25 July 1963 members of 20 Flight flew to Sek Kong for a 48 Gurkha Infantry Brigade study day including Captain Carey:

8 C. Carey (CJC), 'Hong Kong Memoir 1963', *The Chinthe*, Spring 2010, pp.21–22.

> I had set off early for Sek Kong in XN409 with John Valenzia (JFV) on board. Ian Stevens followed later, delivering Peter Richardson and Ian H-C, and then returning to base. The Study ended in the late afternoon and I left with JFV for Kai Tak as XN420 arrived to collect the other two. Just before Tai Mo Shan ridge, looking back towards Sek Kong, I could see 420 taking off, a few minutes behind me. Having landed at Kai Tak, we still had not heard 420 checking in, but thought that they might be returning by a longer, low-level route. In dispersal, I checked with air traffic control but they had heard nothing from 420. As this seemed odd, we got airborne again, made a few calls and, no contact made, raised the alarm. Even with radio failure, 420 should have appeared by now. Back at Kai Tak, we re-fuelled, removed the doors and took off to search. Failing to spot anything on the more likely routes, after a while I began some slow, half-flap descents down the gulleys running off the Tai Mo Shan ridge, with JFV hanging out of the open door peering into undergrowth not far below. On one of these, he spotted the wreckage, just off the top of route TWISK (Tsuen Wan into Sek Kong). We reported the position to Kai Tak. There were no signs of the crew but there was no fire either, so we remained hopeful. We saw Gurkha patrols starting to climb the ridge from the Sek Kong side and vehicles moving up TWISK and in the dusk tried to guide them. Eventually, the wreckage was reached by a Brigade rescue team, which found that no one had survived.
>
> The day had been very hot but the flying conditions not too bad. These can change quickly, however, and the enquiry, I believe, concluded that 420 had encountered turbulence or a downdraft near the ridge. The impact signs and damage indicated that the aircraft had stalled and spun in.[9]

The pilot of XN420, Captain Ian Stevens AAC was killed in the accident along with the Flight Commander, Major Peter Richardson AAC, and Captain Ian Horsley-Curry AAC. At that point in the Corps' history, it was the biggest loss of personnel it had suffered in an aircraft accident.

20 Flight frequently operated in a joint environment, working closely with the RAF and the RN. Forward Air Control sorties were flown in support of 28 Squadron RAF's Hawker Hunters and in February 1963 three Austers flew 60 miles out to sea to rendezvous with the Illustrious-class aircraft carrier HMS *Victorious.* All three aircraft made successful deck landings with many of the ship's crew lining the flight deck to watch the unusual spectacle. Back on land the Flight was frequently engaged on photographic surveys to help colony's mapping to keep pace with the rapid land reclamation and urbanisation that were a constant feature in Hong Kong.[10]

With the Integrated Flight Scheme in operation, 20 Flight were not the only Army aviation asset in the colony. 6th Gurkha Rifles (6 GR) Air Platoon operated from the airfield at Sek Kong and in April 1965 49th Light Regiment Royal Artillery's (49 Lt Regt RA) Air Troop, commanded by Captain Tim Deane RA, arrived in Hong Kong with Sioux helicopters. Captain David Canterbury RA recalled that the arrival of the Sioux, '…created quite a stir, from the Governor downwards, through HQ HKLF and 48 Gurkha Inf Bde plus medical services both civil and military, one has to wonder how they got by before without rotary wing help…'[11] By August of that year the Air Troop was well established in Sek Kong and the Sioux had begun to demonstrate its utility. Locally manufactured litters were fixed to the side of the aircraft to assist with resupply and troop movement sorties. Three complete OP parties could be lifted from Sek Kong to the top of the 2,000-foot Castle Peak, 9 miles to the south-west, in 25 minutes.[12] In July 1965 aircraft from 49

9 Carey, 'Hong Kong Memoir 1963', p.23.
10 656 Squadron Association Archive (656 AA) 187: '20 Flight AAC reports', February and April 1965.
11 D. Canterbury, account to the authors, 29 August 2020.
12 656 AA 187: '49 Light Regiment Royal Artillery, Flying returns and newsletter', August 1965.

Lt Regt RA Air Troop were involved in a search and rescue operation during Typhoon Freda for one of the Troop's own soldiers, Bombardier Mahabier:

> This was our first experience of such extreme weather. He had been sailing out of Gordon Hard at Castle Peak, the base of a Gurkha engineer squadron, in one of the regimental enterprise dinghies without checking in with anyone and it was only when he was declared absent later that evening when the search was initiated. Obviously, we were put on standby for a first light departure. Now, for those brought up on 'rigid rotors', the Sioux had very heavy teetering blades which were prone to sail on start-up and chop off the lattice tail boom, so our OC, Captain Tim Deane, made the command decision that the risk would be reduced if we flew them out of the hangar, so we did just that. Unfortunately, the REME did not agree. At the early stage of our new home being built, there were no offices, so the technicians had all their paperwork on 6ft folding tables. Happily, as we flew out, so did all their paperwork! Nevertheless, it became a local SOP [standard operating procedure] to use this option in winds of over 30kts, hopefully with better management of their paper. On our first sorties, all our observers either felt or were sick so Tim decided that future flights in typhoon conditions would be two pilots up. This arrangement worked well but in the major turbulence we found that whoever was actually doing the flying felt fine while the one observing, particularly while using binoculars felt very queasy. It was a little embarrassing, after 6 or 7 sorties, the missing Bombardier Mahabier was found by an RAF Hunter on the far South West corner of Lantao, about halfway between Macau and downtown Hong Kong. I was refuelling back at Sek Kong at the time, the aircraft was duelled up [with two sets of flying controls] which would have taken too long to revert so I set off on my own. The RAF had given a particularly good fix so I found him without difficulty with the mast and a few shreds of the mainsail still attached, while the hull itself was just matchwood. He appeared quite pleased to see me and I got one skid onto a piece of rock and he clambered in without difficulty and was then promptly sick on the way home.[13]

Captain Canterbury also recalled early operational tasking with the Sioux of the Air Troop 49 Lt Regt RA:

> As the mid–60s were the time of the little red book of Chinese doctrine, the attitude towards the British was quite sensitive. We had a series of watch towers or entrenchments right on the border and all teeth arm officers had to do a week's duty at one of these about every 6 months. I am not sure that anyone saw anything of a world changing nature, but it was quite interesting to watch, what was then, a very primitive way of life. In addition to continuous ground recce, we were tasked with flying photo experts with telephoto lenses about 2ft long to take photos well into China from 10,000ft. Some were a bit diffident about flying at all while others got vertigo, but it was understandable. Naturally, we had to fly with the doors off so that the cameras could stick out and we had to keep up a respectable flying speed up as there was no way we could keep a stable hover at that height, so the aircraft was flying nose down with no sensation whatever of forward speed. I think the feeling was best expressed when one of my passengers said, "It was like being in Heaven sitting on a [folding] chair..."[14]

13 D. Canterbury, account to the authors, 29 August 2020.
14 D. Canterbury, account to the authors, 29 August 2020.

Captain John Charteris in command of 1/10 Gurkha Rifles Air Platoon prepares for a sortie with officers of the Hong Kong Police. (656 Squadron Association)

In July 1966 20 Flight AAC was sent from Hong Kong to Serembam in Malaysia to assist the British effort in the confrontation with Indonesia. The Austers were to be taken out of service and the Flight was to be re-equipped with Scouts. The confrontation came to an end however, and before it was established in Serembam, 20 Flight was returned to RAF Kai Tak and re-equipped with three Sioux in October 1966. The following month the Flight carried out its first operational sortie when Lieutenant Simon Lytle, Royal Irish Fusiliers flew the commanding officer 1st Battalion, Welch Regiment in Sioux XT226 on 11 November 1966.

During the second half of 1967 large-scale disturbances, organised by the clandestine local branch of the Chinese Communist Party, disrupted life and business in Hong Kong for a prolonged period. Frequent demonstrations, marches and protests took to the streets in the heavily populated colony. On 8 July a detachment of Chinese militia crossed the border near Sha Tau Kok and attacked a police station with small arms fire killing five members of the RHKP. Over the following four months police and army bomb disposal experts (one of whom was killed) had to deal with over 8,000 improvised explosive devices of which nearly 1,500 were found to be viable. In response the police and armed forces fought back against the communist activists. An RN Wessex from HMS *Hermes* landed police on the roof of an activist stronghold, who then entered the building by force, disrupting the group's plans and activity.[15]

656 Squadron AAC

During internal security operations in 1967 the RHKAAF operated two Aérospatiale Alouette IIIs and the value of the helicopter in assisting the security forces to maintain law and order in an

15 Tsang, *A Modern History of Hong Kong*, pp.183–188.

urban environment became apparent. It also became clear that, to counter similar disturbances in the future a more coordinated approach to military operations would be required. Along with the demise of the Integrated Flight Scheme, this led to a consolidation of Army aviation assets in Hong Kong under a unified squadron command. The Hong Kong Aviation Squadron was established at Sek Kong in February 1969. It brought together 20 Flight AAC, which moved from Kai Tak, 6 GR and the recently arrived 7 GR Air Platoons and the Air Troop of 25 Lt Regt RA (which had replaced 49 Lt Regt RA).[16] In October 1969 it was decided to give the Squadron a permanent title and in December that year, once the final drawdown in Malaysia was complete, it was formally named 656 Squadron. The Squadron was initially equipped with 12 Sioux but in January 1970 Scouts arrived from Malaysia and 656 Squadron was reorganised into a Sioux Flight and a Scout Flight, each with eight aircraft, initially under the overall command of Major Peter Ralph AAC (promoted in post to Lieutenant Colonel) and subsequently under Lieutenant Colonel Alan Woodford AAC. The Squadron also had C Flight detached in Brunei, manned on a six-month rotational basis from Hong Kong, and had command over 11 ANZUK Flight in Singapore.

The Squadron provided support to both 48 Ghurkha Inf Bde based in Sek Kong and 51st Infantry Brigade (51 Inf Bde) in Kowloon, with a flight allocated to each and tasked directly by the respective formation HQs. During periods of high intensity internal security operations the Squadron could deploy two radio rebroadcast (rebro) stations and liaison officers to the Joint Air Tasking Cell and 28 Squadron RAF. Alternatively, the Squadron could be deployed with the Sioux Flight split between the two Brigades for direct tasking and the Scout Flight held centrally for tasking by the Joint Air Tasking Cell through Squadron HQ.[17] As 656 Squadron began to establish itself operationally in its new home, Lieutenant Colonel Ralph described the challenges and opportunities presented by flying around Hong Kong at the time:

> The flying conditions vary with the season but are always interesting. The often strong winds make operations in the mountainous areas reminiscent of North Wales in the later stages of pilot training, except that it is a lot hotter and humid. In Kowloon and Hong Kong the air traffic density is similar to the circuit at Wallop. It is not uncommon to see four or five helicopters of all three services trying to cross the runway of the civil airport with a stream of Boeing 707s landing or taking off. Night flying in the urban areas is an awe-inspiring sight for which tourists would pay a fortune to see. The lights are many and colourful, making London's Leicester Square look like a blackout zone. Landing points are varied, ranging from a jetty surrounded by Chinese territorial waters, to roof tops in the centre of Kowloon, to observation posts a few hundred yards from the Chinese border with Guangdong Province, from whence comes "provocations" and "shows of force" by the million strong People's Liberation Army.[18]

Due to the amount of flying conducted over water some of the Sioux were fitted with float landing gear which provided added interest, with engine off landings into Tolo Harbour followed by hover-taxiing around on the surface. Captain Rob Welsh REME also recalled the operational flying conditions and tasking that faced 656 Squadron from the start:

16 There is a discrepancy in the designation of the RA Air Troop that became part of 656 Squadron; Warner, *From Auster to Apache,* lists it as 25 Lt Regt RA (p.157) whereas AFM 215: 'Army Aviation in Hong Kong' records it as 18 Field Regt RA (p.3).

17 AFM 215: D. Knight, 'Army Aviation in Hong Kong', 18 December 1972.

18 Warner, *From Auster to Apache*, pp.157–158.

> We developed a new and additional role of a 'Quick Reaction Force' where a Sioux would provide overhead command and control for a formation of Scouts, carrying troops, to enable them to quickly place troops on the ground at tactical locations in the event of an Internal Security situation developing. This procedure was practiced and developed both by day and night until we became quite slick at it. It seemed pretty effective as we could move a small force of men quickly from one place to another as situations required. Thankfully it was never used in anger in my time…
>
> The general flying was in some ways quite challenging as we had to contend with hot and high conditions and the various wind affects in and around the many high hills and rocky outcrops. In addition to this the typhoon season produced its own batch of violent storms and strong winds/gales. Some of the HLS locations were more challenging than normal – for instance Victoria Barracks on Hong Kong Island had a landing platform built into the side of a steep hill and which jutted out towards the sea. There was only one approach which was directly towards the hillside and thus there was no chance of an overshoot! If the approach had been downwind then the take off was best accomplished by a transition backwards with a controlled right pedal turn into wind.
>
> We (the Scout pilots) did quite a lot of troop lifting which worked very well. The requirement was usually to move something like a Company of Infantry a short distance from the base of a steep hill to some point further up which, in the heat, would be both difficult and tiring to climb taking many hours with the troops exhausted at the end. It took us only a matter of a few minutes with each load to cover some really difficult terrain. Sometimes the RAF with Whirlwinds would work with us and it was weird to see the much bigger Whirlwind helicopters just lifting 2/3 troops whilst the Scouts were taking 4/5 soldiers.[19]

As 656 Squadron developed its relationship with the two resident brigades the range of operations that could be conducted became more complex.

> Our Internal Security operations range from the normal and well known tasks, such as reinforcement of police stations by heliborne troops and night surveillance etc to Vietnam type helicopter assaults, complete with pathfinders and gunships. We practise these regularly by day and night, including live fire-plans from guns and mortars which have been initially lifted into their fire positions. The night ops are particularly interesting with flare illumination for two pathfinder Scouts engaged in roping down to the landing point, and the securing troops being provided by another pair of Scout/Sioux stationed above them. During exercise Eastern Triangle we flew 450 hours, including fifty at night on ten aircraft in ten days. Our friends in the RAF flew just under 400 hours on eight Wessex in the same period.[20]

The main day-to-day issue for the security forces in Hong Kong remained intercepting illegal immigrants attempting to immigrate to the colony from mainland China. In order to attempt to stem the influx in November 1974, during Lieutenant Colonel Greville Edgecombe's tenure in command of 656 Squadron, the Hong Kong Government adopted the 'Touch Base' policy. The new approach allowed immigrants from mainland China who reached the urban areas of Hong Kong (defined as south of Boundary Street in Kowloon) and met their relatives to register for a Hong Kong Identity Card. Those who were intercepted in the Frontier Closed Area (north

19 R. Welsh, 'Memories of 656 Sqn in Hong Kong – Dec 1969 to Oct 1971', *The Chinthe,* Spring 2010, pp.26–27.

20 Anon, '656 Aviation Squadron in the Far East', *Army Air Corps Newsletter*, No.32, December 1972-January 1973, p.4.

of Boundary Street) would be repatriated back to China immediately. The policy had little effect and the Squadron remained as busy as ever assisting the interception of illegal immigrants in the New Territories.

In April 1975 Lieutenant Colonel David Swan AAC took over command from Lieutenant Colonel Edgecombe and soon had to cope with a number of major changes for 656 Squadron. It had been decided to concentrate the AAC's anti-tank capability in the BAOR and so all Scouts were being moved to Germany. They would be replaced in Hong Kong by the Aérospatiale Gazelle, which had entered operational service the previous year, despite protests that it would be unsuitable to take on the tasks usually carried out by the robust Scout. Nevertheless, a conversion programme was initiated for the aircrew, with technicians and the first three Gazelles arriving in Hong Kong in November 1975 and the same number of Scouts departing on the return flight.

The Scout had at least one more surprise to spring before it entirely departed Hong Kong. On 3 November 1975 Sergeant Tim Smith RTR was scheduled to fly a dawn sortie to collect a Gurkha battalion company commander and fly him over a series of hilltop OPs where his men had spent the night observing likely landing beaches waiting to intercept any IIs that came ashore. The company commander was to communicate by radio with each OP in turn as the Scout flew overhead. On signing for his Scout at Sek Kong Sergeant Smith was informed that it had returned late from the workshops and had no flotation gear fitted, but he was not unduly concerned. He picked up the company commander and checked in with OPs at A Ma Wat, Fan Shui Au, and Kuk Po in the north-east of Hong Kong. With all well Smith turned his Scout to descend and look at the beaches:

> Crossing the final ridge, the lever was lowered to descend straight down the centre of the inlet. Somewhere between my feet, the landward end of the inlet, chock-a-block with trees and a skinny bit of sand. Ahead the open sea and headlands, a good place to start our patrol. We had barely descended a thousand feet and were levelling off when it happened…
>
> The Major asked, "What was that?" A note of mild interest in his voice. "I think we've lost the engine, Sir" A cautious summary of my thoughts.
>
> Steep rocky hillsides to the left of us. Audible warnings beeping, high pitched, shrill, nagging. Magnetic Indicators flashing, crosshatched black and white, skeletal reminders of impending doom. Steep rocky hillsides to the right of us. The instrument panel had dissolved into a shaking blur. Vibration like an out of balance washing machine on fast spin. Three Maydays on three different frequencies. At 350 feet there was no chance of anyone hearing them. Good form though.
>
> The only bit of flat ground in the whole world was stuffed full of coconut trees, and most decidedly behind us. So, well and truly stuck between a rock and a wet place, there was no turning back. And no floats, no flying over water. I'd get a slap on the wrist for this one!
>
> I have wondered if auto-rotating in a Scout was actually flying, more a case of submitting to gravity. And the 700 didn't say, "No Auto-rotation Over Water." The bright blue water of the inlet looked awfully deep. There was nothing for it but to try and stretch the glide, find something shallower. Stretch the glide in a Scout? The Queen of The Skies auto-rotated like a brick.
>
> So we stretched it, and stretched it until there was nothing more for it but a hard, deep flare and select a piece of watery unreal estate near the rocky beach. Hold the flare and dig the tail in, plough a furrow and slow things down. We need near to zero speed before levelling off. No time for more than a jolly good "Brace, Brace, Brace," and settle into the warm water of Mirs Bay. Pull on the rotor brake and wait for the blades to stop, failing miserably to set them at ten o'clock and two o'clock!
>
> The water crept inexorably up the doors. Over my boots, up my legs, slowly to my knees, and on, upwards to the seat of my pants, hips and waist. We had settled on the bottom. Post

656 Squadron AAC Scout XW797 on patrol over the southern end of Route TWISK looking south over Hong Kong harbour. (656 Squadron Association)

> landing drills were near useless as the windscreen wipers self-started and continued to self-destruct without my help.[21]

A month after Sergeant Smith's exciting sortie the three remaining Scouts were replaced by Gazelles. Lieutenant Colonel Swan explained some of the early issues with the Gazelle that had to be managed:

> Squadron flying operations in Hong Kong consisted of border patrols, joint training with the RAF Wessex squadron and the usual liaison flying. It was amazing how many visitors "required" an airborne familiarisation trip around the island and the New Territories. The border patrols along the Chinese border were predominantly dawn sorties searching for and intercepting illegal immigrants seeking their fortune in Hong Kong. The aircraft usually carried 3 armed Gurkha soldiers whose role was to disembark and attempt to catch the culprit(s) when spotted by the aircrew. This was not always successful as the Chinese became very adept at hiding in the undergrowth, crossing the water at night using all sorts of inflatable devices and hiding up during the day. Once they reached the built up areas they would be impossible to trace...
>
> To avoid damage to the Gazelle airframes, which are a little more delicate than the Scout, walkway strips were fitted on the skids and door frames and rubber mats in the boot and on the passenger floor. Locally manufactured freight floors were designed and produced before flying began.[22]

21 T. Smith, 'No Turning Back', *Army Air Corps Journal,* 2006, pp.97–99.
22 D. Swan, 'Hong Kong 1975–1977', *The Chinthe, 656 Squadron Association Journal*, Spring 2010, pp.33–34.

Despite the teething problems the Squadron was declared operational in its new Gazelle configuration in January 1976. The internal security remit of 656 Squadron had parallels with the developing aviation operations to counter insurgent and terrorist activities in Northern Ireland during the Troubles. Equipment that would become familiar to aircrew in Northern Ireland was also trialled in Hong Kong. The Nitesun searchlight and the Skyshout public address system were both trialled on the Gazelle in Hong Kong in the mid–1970s before the former was adopted in the colony and put to effective use. Another aspect of the British Army's operations in Hong Kong that would also become familiar in Northern Ireland was the manning of isolated observation posts along the Chinese border in the New Territories. Here soldiers could come face to face with members of the Chinese PLA as described by a member of 1st Battalion, The Black Watch:

> At some of these observation posts we had, you could see the Chinese soldiers going about and what they used to do…They always had, sort of like, green boiler suits, a green hat with a red star in the centre and the white canvas shoes and carrying the Russian Kalashnikov rifles.[23]

The isolated border posts and the soldiers that patrolled the difficult country between them relied on the aircraft from 656 Squadron for support, flying in supplies, and for evacuation in case of a medical emergency. However, flying into the small, forest clearing landing sites that served the OPs required skill and perseverance, particularly at night, as demonstrated by the actions of Sergeant Ivor Johnstone AAC.

> Sergeant Johnstone was the Duty Pilot of 656 Squadron Army Air Corps in Hong Kong from 0830 hours on 15 October 1976 to 0830 hours on 16 October 1976. He was on 90 minutes standby during this period in support of the border battalion for operational tasks.
>
> At 0125 hours on 16 October 1976 he received a request from the Brigade Duty Officer to carry out the casualty evacuation of a soldier on a ridge in a remote part of the New Territories in Hong Kong. Sergeant Johnstone accepted the task and detailed Sergeant D Clark REME, the Duty Technician to prepare the aircraft and accompany him on the sortie. Sergeant Johnstone took off at 0200 hours in a Gazelle helicopter with Sergeant Clark.
>
> The casualty was situated on top of a ridge approximately 2,000 feet high. He was suffering from exhaustion and exposure and had a series of blackouts. He was attended by Corporal Bishop and three soldiers of his platoon of 1st Battalion, the Light Infantry.
>
> En route to the site Sergeant Johnstone checked the weather and found virtually complete cloud cover at 1700 feet, which was 300 feet below the height at which the casualty was located. He contacted Corporal Bishop on the radio who advised him he had laid out a landing aid with torches and the landing site was clear of cloud but that it was very dark, visibility was down to 50 metres and there was light rain. The landing site was just below the ridge line on a small plateau of small rocks and shale. Around the plateau and on the ridge itself were large boulders.
>
> Because of the cloud Sergeant Johnstone was unable to climb high enough to see the landing aid. He therefore made a successful approach to the ridge approximately 500 metres to the West and below the cloud base using only the landing light of the helicopter. He then hover taxied up the ridge through the cloud using his landing light and assisted by Sergeant Clark in the co-pilot's seat and Corporal Bishop giving directions over the radio and flashing his torch. He was able to bring the helicopter to the landing site where Sergeant Clark loaded the casualty into the helicopter.

23 Felton, *China Station*, Kindle edition, chapter 10, para 34.

> At this stage the weather deteriorated and the landing site became covered in cloud. Sergeant Johnstone took off again and hover taxied back down the slope until he was clear of the cloud when he was able to carry out a normal take off and flew the casualty to the British Military Hospital.
>
> Sergeant Johnstone showed exceptional skill and courage in carrying out this casualty evacuation where any error in judgement or skill could have placed the helicopter and its occupants at serious risk. By his actions the casualty received medical attention far more quickly than would otherwise have been possible and in doing so he acted beyond the call of duty.[24]

For his actions Sergeant Johnstone was awarded a QCVSA. Alongside conducting such essential tasking, the Squadron also continued to develop its methods of operation as Lieutenant Colonel Swan explained.

> Another of our activities was developing air mobile operations deploying Gurkha infantry around the colony in conjunction with the RAF Wessex. We practised whenever the time and opportunities arose and wrote Standing Operational Procedures (SOPs). Basically 656 Sqn did the recces and the RAF did the troop lifts. Our mutual activities attracted the attention of the Joint Warfare Establishment in UK whose team paid us a visit while on tour and were quite complimentary about our humble efforts.[25]

In December 1976 48 Gurkha Inf Bde and 51 Inf Bde were amalgamated to establish the Gurkha Field Force. 656 Squadron continued to work closely with the new formation on counter illegal immigrant tasking, on exercises to rehearse new procedures and techniques, and providing casevac cover. Following a night casevac sortie during an exercise in February 1976, Captain Simon Fogden AAC was returning to the HLS on Pak Sha Tau Island in Tolo Harbour. It was a still night with mist and a glass-like surface to the sea over which Captain Fogden had to make his approach. He became disorientated and struck the water short of the shoreline. The aircraft sank, severely damaging Gazelle XX409 although the crew escaped safely. A subsequent board of enquiry found that the landing aid had been set up on sloping ground, giving the pilot (without the aid of a radar altimeter) the impression that he was higher above the water than was actually the case.[26]

Soon after the incident 656 Squadron's tenure in Hong Kong came to an end. Major Dick Whidborne AAC took over command from Lieutenant Colonel Swan in June 1977. Within weeks his command was reduced to flight strength and renamed 11 Flight, although it retained responsibility for C Flight in Brunei. Nevertheless, the cloud caused by the cut did have a silver lining. The introduction of the Westland Lynx across the AAC allowed the Scout, now superseded in BAOR, to return to Hong Kong, an opportunity grasped by Major Whidborne.

> When I arrived in 1977 the squadron (then 656) had just re-equipped with the Gazelle. It was soon obvious to me that the Scout was far better suited to the role of border patrolling and insertion of small parties of Gurkhas, complete with their bulging Bergen rucksacks, kukris and SLR rifles which would not easily fit into a Gazelle. I persuaded the HQ DAAC [Director Army Air Corps] and the Hong Kong Government (who funded 75% of the garrison) to

24 TNA, WO 373/182/107: 'Recommendations for Honours or Awards', Sergeant Ivor Derek Johnstone AAC, 26 January 1977. *London Gazette,* 3 May 1977.

25 Swan, 'Hong Kong 1975–1977', p.34.

26 S. Fogden, 'Hong Kong December 1974 – July 1977', *The Chinthe*, Spring 2010, pp.36–39.

> return the Scouts. For me this required a conversion to type which, unusually, I was allowed to complete in theatre. The most suitable area for engine off practice was the polo field, but our failure to replace some impressive divots earned us a yellow card from the horse fraternity, and we were confined to the much rougher grass adjacent to the runway, or in some cases the runway itself.[27]

Less than three years after its arrival, the Gazelle was phased out of the colony and 11 Flight was equipped with six Scouts in Hong Kong and three in Brunei, as well as two Sioux in Hong Kong. A year after its reduction to a flight the status of Army aviation in Hong Kong was reprieved and on 1 August 1978, amid a large-scale renumbering of squadrons across the AAC, Major Whidborne's command was retitled 660 Squadron AAC. By the end of the year the final Sioux sortie was flown in Hong Kong and 660 Squadron became a solely Scout equipped unit, with its establishment expanding to 12 aircraft, six of which were eventually fitted with emergency flotation gear. For everyone concerned, in hindsight, the return of the Scout was timely, as explained by Captain A. F. Thomson REME, the commander of 660 Squadron's LAD:

> Either some far-sighted Staff Officer actually got it absolutely right, or fortune looked kindly on what was then 11 Flight AAC when the decision was made to replace the Gazelle by Scout. At that time the illegal immigrant problem was minimal and the majority of the tasking was for liaison flying, for which the Gazelle was ideal. However, the Gazelle would have been quite unsuited to the sort of flying the Squadron are now [late 1979] doing. The LAD did not really welcome the return of the Scout with its much higher workload; it was a question of back to the basics of tracking, vibration problems, tiedowns and the inevitably high husbandry content for such elderly airframes. Fortunately the Flight, which on 1 August 1978 miraculously became a Squadron, had some 9 months to get used to the Scouts before their full worth was realised.[28]

Operation Culex

In December 1978, following the death of Mao Zedong and the end of the Cultural Revolution, the Chinese Communist Party initiated a policy of reform and increasing openness. It allowed Chinese citizens to visit relatives in Hong Kong and *vice versa*. It also established a Special Economic Zones (SEZ), including in Shenzhen in Guangdong Province across the border from Hong Kong, which allowed a greater degree of economic freedom. Radios and televisions also became more prevalent in China which, if in range, could receive transmissions from Hong Kong. A combination of these factors led to a rising awareness in Guangdong of events and conditions in Hong Kong and to the realisation, particularly among the younger population, of the disparity between their own lives and those living across the border in the colony. This increased the flow of illegal immigration into Hong Kong. In 1979 the Hong Kong Government conceded that its 'Touch Base' policy had done little to stem the influx of illegal immigrants and decided it would introduce a new policy the following year. Immigrants coming directly from the mainland on or before 23 October 1980 would be required to register for a Hong Kong Identity Card in a three-day

27 R. Whidborne, 'Hong Kong Revisited 2007', *Army Air Corps Journal*,2007and D. Whidborne, account to the authors, 27 June 2020.

28 A. F. Thomson, 'Some unique features of 660 Squadron Army Air Corps LAD REME in Hong Kong', *Army Air Corps Journal*, 1980, p.56.

period of grace (24–26 October 1980). Illegal immigrants arriving on or after 24 October 1980 were to be repatriated immediately. It would also become compulsory for Hong Kong residents to carry their identity cards in public areas. The announcement of the new policy caused a huge surge in the number of immigrants trying to take advantage of the 'Touch Base' policy while it remained extant and to reach Hong Kong before the 23 October 1980 deadline. The dramatic surge led the British Government to declare Operation Culex in June 1979 which allowed for a reinforcement of British armed forces in support of the Hong Kong Government and security forces, including aircrew, ground crew and technicians from the UK to bolster 660 Squadron.

The Sham Chun River formed the western edge of the boundary between China and the New Territories, the northern bank of the river being the official border. South of the river's estuary was a large area of wetland known as the Mai Po marshes made up of extensive mud flats, mangrove swamp, and fishponds. From the Sham Chun it was as little as two kilometres to the Castle Peak road, where it was possible to catch a bus or a taxi into Sheung Shui where trains could be taken into Kowloon and south of the Boundary Road demarcation line. Crossing the Mai Po marshes became a favourite route for the new influx of illegal immigrants, using the cover of darkness. The local fishermen used a wooden ski to negotiate the marshes and harvest mussels. The ski had a handle like a child's scooter and the desperate immigrants adopted the crude vehicles to assist their efforts to infiltrate Hong Kong. WO2 Squadron Sergeant Major Chris Wilson AAC described how the illegal immigrants used the ski for the *South China Morning Post*, 'If it's thick mud the illegals stand on the skis and push with their feet and they can shoot across the mud and water at tremendous speed. If they cross thin mud or water they lay down and put out one leg and make a swimming motion and they can travel very fast.'[29]

A police post with an HLS (Purple 1) was located in the centre of the marshes and acted as the HQ for whichever infantry company was responsible for the area at the time. Ops were situated throughout the marshes that could direct patrols towards any illegal immigrants they spotted. It was extremely difficult, however, for infantry on foot to cross the marshland to intercept illegal immigrants and so 660 Squadron developed techniques to assist. Scouts would work in pairs with one acting as the observer, using Nitesun if required, to locate immigrants. It would then descend to hover beside the Chinese immigrants from where the aircrewman, standing on the skid would kick them off their skis into the glutinous mud. The second Scout, carrying an aircrewman and often a Gurkha soldier, would then approach the stricken illegal immigrants and hover with its skids just clear of the marsh. The aircrewman's task was then to haul them into the aircraft where they were handcuffed ready to be flown to Purple 1 or another suitable site to be arrested.[30]

Groups of up to 200 illegal immigrants would try to cross the Mai Po marshes on a nightly basis with 660 Squadron and its attached aircrew picking up over 30 immigrants a night. The interception task was given the task number 273 but known colloquially as the 'Mudlark'. The Squadron moved to seven days a week operations. Three aircraft would deploy to the Mai Po marshes each day at dawn when it was most likely to locate and intercept the immigrants. At 1300 hours the commitment reduced to two aircraft and at dusk they withdrew to Sek Kong, with one Scout with Nitesun on call overnight. The aircraft were modified for Mudlark sorties with the rear seats removed and a specially designed canvas cover across the floor, the back of the pilots' seats and rear bulkhead. A cargo net was secured from skid to skid across the freight floor. The Eagle harness was redesigned to allow the aircrewman to fly 'legs out'. The fit prevented mud and seawater damaging the aircraft and assisted the aircrewman to drag the Chinese men, women and children on board. Nevertheless, the Scouts returned to Sek Kong from Mudlark sorties covered in mud and seawater

29 *South China Morning Post*, 18 June 1979.
30 AFM 215: Manuscript, 'Why the Illegal Immigrants are trying to get into Hong Kong', undated.

A 660 Squadron Scout fitted for Mudlark sorties with scramble nets over the skids and protective canvas in the rear cabin during Operation Culex. The aircraft is also fitted with flotation devices above the doors. (J. Walker)

and required an immediate wash down to prevent corrosion. The tasks were tiring for the pilots who had to fly at extreme low level for protracted periods, in close proximity to vulnerable people, and the aircrew would end up plastered in mud.[31] Sergeant Dick Kalinski Para explained some of the difficulties and dangers of flying Mudlark sorties in the Mai Po:

> The mangrove bushes were about 10 feet high with thick mud between them. The IIs would wade through the mud during the night and hide until the coast was clear. Their footprints in the mud would last for a couple of hours, so it was just a case of following the footprints to where they were hiding, then hovering above the bush with the downdraft parting it and exposing them. The tricky part then was to hover the aircraft just above the water so that the aircrewman could pull them in and plasti-cuff them.... Because of the short flight time, we were usually light on fuel so that we could carry more IIs. Normally we would pull in 2 or 3 at a time but it was not unusual to have many more than that. Being small people with only what they stood up in, weight was not a problem as long as we could fit them in. The most I carried was 8 plus the aircrewman although I believe other pilots got into double figures! The obvious danger in hovering so close to the water was allowing the tail rotor to make contact with the surface. To my knowledge it only happened once.... The aircraft had to be recovered later by a [Westland] Sea King from a visiting Royal Navy ship.[32]

31 AFM 215: 'News from Hong Kong', undated.
32 D. Kalinski, account to the authors, 18 September 2020.

The aircrew had a particularly arduous role and all pilots arriving in Hong Kong on Operation Culex were expected to fly a sortie in the aircrewman role to fully appreciate the challenges and hazards of the task. Staff Sergeant Sean Bonner AAC, from 656 Squadron AAC in its new home in Farnborough, was a Culex reinforcement pilot in 1980 and recalled his initiation as an aircrewman and the first time he was expected to jump from the rear of the Scout into the mud to grab an illegal immigrant:

> Low hovering was obviously necessary but to this day, when I was ordered to jump by the Sqn 2IC, who was not known for his low hovering, I swear I could have recited the first verse of the Ancient Mariner from leaving the aircraft to impacting, and sinking into, the slimy mud. I had to scream at the IIs to pull me out so I could arrest them.
>
> All captured IIs were flown to Mai Po Police Station where they were thrown into a bund (lake) to wash off the mud then they were processed by the HK Police, held until sufficient numbers made it viable then handed back to the Chinese authorities at Low Wu. From returnee IIs we learned that these returned Chinese were transported to the Mongolian border as far away as possible from Hong Kong to dissuade repeat attempts at fleeing. It did not work, we captured many returnee IIs.[33]

Lance Corporal John Walker AAC recalled not just the physical difficulties facing an aircrewman operating in the Mai Po marshes, but also the moral dilemma that they had to grapple with when intercepting illegal immigrants. A typical sortie undertaken by Lance Corporal Walker took place on 18 August 1980:

> The helicopter XT643 was flown by Captain Llewelyn Davies or 'L Squared' as he was known, and I was in the back. We picked up a Gurkha officer from the marshes OP at P01. He directed us to area where they had spotted IIs during the night. I was sitting behind the pilot so I could lean out and point with my nightstick [hand-held torch] if he could not make out my garbled throat mike transmissions. "Contact 2 O' Clock, 100m, footprints." As the visual cues disappear into the ever-thickening mangroves we rely on other senses like smell. A person who has not washed for weeks smells. Like a dead body there is nothing like it. Most times you hover over an II and they realise the game is up and grab onto the skids and net and drag themselves into the helicopter. The role of the crewman is split between watching the IIs and the tail rotor as the pilot buries the helicopter into the mangroves. Sometimes the pilot or crewman must literally knock the II with the skid or nightstick before they give up. On this occasion a girl must have thought she was invisible, she just stayed where she was. L Squared was now screaming at me to get out (my helmet had disconnected). I got covered from head to foot in mud and somehow dragged myself and the girl into the back. I knew I would get the hose treatment when we got back to base. We carried on for another half hour dragging IIs out of the mangroves until it was obvious they had made it through to the fishponds beyond. L Squared spotted an II running along the bunds (strips of land between the ponds). We gave chase. How do you catch a running man? Well you land and run after them, but not with this pilot. Two feet off the ground we zoomed along the bund until L Squared screamed jump; so you jump. 75kg of crewman hit the II and he was flattened. Back on board the helicopter we now have ten IIs, 150lbs of fuel and it was time to leave the IIs at the OP and go home.
>
> The IIs we caught would be processed by the police and sent back to China. They would be ostracised by their friends, family, community and if they did not end up in some gulag, they

33 S. Bonner, account to the authors, 28 May 2020.

Aircrewman Lance Corporal John Walker stands by his Scout while Chinese illegal immigrants are washed down in a fishpond in the Mai Po marshes during Operation Culex. (J. Walker)

> would have no choice but to try again and again and again. That is not say we were indifferent to their plight. On one occasion I caught a man in military clothing. He could only be a deserting PLA soldier and I thought if we handed him over there would only be one outcome, so I let him go. Another time I was chasing an II through the mangroves when I became stuck. The mud was up to my thighs and still rising. The II saw my difficulty, about turned and pulled me out. We looked at each other and I pointed and said "Yeun Long" (the nearest town) and "dor je" [duō xiè] (thank you). We went our separate ways.
>
> I spent two years flying in the Colony and have talked to several crewman and it seems we all had Nelsonian days when we chose to look the other way. Was that playing God, well yes, I suppose it was. When I look back, I wonder why some of us treated this situation almost like a game of hide and seek. Maybe it was our way of dealing with the horror. I mean what do you do when you are ordered to pick up 28 decomposing bodies that included children. I am sure today we would all be counselled but back then it was your mates and Tiger beer.[34]

As well as the psychological pressure, Mudlark aircrew also ran considerable physical risks in the course of their duties. After an illegal immigrant picked up by a Scout during Operation Culex was found to be carrying an improvised pistol, aircrew were also issued side arms but without any ammunition. The actions of Lance Corporal John Jarman AAC clearly demonstrate the risks that the Mudlark aircrew ran and their dedication to the task.

> LCpl Jarman was aircrewman in a Scout helicopter carrying out a routine dawn anti illegal immigrant patrol in the Mai Po marshes on the Hong Kong/China border on 27 July 1980.

34 J. Walker, account to the authors, 26 August 2020.

> Ten illegal immigrants (IIs) were spotted attempting to swim across a nearby bay. As they were clearly exhausted and in difficulties the pilot decided to attempt an immediate rescue. LCpl Jarman, by hanging below the helicopter and directing the pilot, successfully pulled the IIs aboard at some danger to his own safety as they all frantically clutched at him in their efforts to survive. Having flown them to safe custody, the Scout resumed the Mai Po patrol, locating and capturing 11 more IIs.
>
> LCpl Jarman then spotted, in dense mangrove 3 IIs who repeatedly attempted to evade capture by diving under the muddy water and reappearing elsewhere. Again skilfully directing the pilot, LCpl Jarman apprehended 2 of the IIs, but saw that the third, a woman, was floating face downwards under the mangrove in a particularly inaccessible position. As the pilot brought the Scout as close as possible, LCpl Jarman crawled forward along the helicopter skid until he could reach the woman's hair and lift her head clear of the water. By manoeuvring himself round the skid he supported the woman under her arms while the Scout lifted them to firm ground. There he applied immediate resuscitation and confirmed her pulse, continuing this during the short flight to the medical centre.
>
> The woman was subsequently certified as dead, but this does not detract from LCpl Jarman's courageous efforts to save her. During the course of the 3 hour patrol he had been instrumental in apprehending no less than 23 other IIs, many of whom could well have met the same fate as the woman but for his exceptional initiative, enterprise and brave conduct.[35]

For his actions on 27 July 1980 Lance Corporal Jarman was awarded the QCVSA.

Chinese immigrants were not the only group flooding towards Hong Kong. The migration and humanitarian crises caused by the aftermath of the Vietnam War was at its height in 1978 and 1979 and Vietnamese refugees, who became known as 'boat people', frequently arrived on Hong Kong's shores into the 1980s. The OC 660 Squadron, Major Whidborne, recalled the difficulties in locating the illegal immigrants and the tragic circumstances that many of them found themselves in:

> One morning I was on patrol with my Air Gunner observer and a Gurkha major in the back of the Scout. We searched a small island and finding nothing continued towards another search area. The Gurkha major tapped me on the shoulder and gave me a "thumbs up" sign. Rather than an affirmation that all was OK, I was later to learn that this indicated the presence of "enemy" and so we returned to the little island to find five "IIs" who had emerged from under the bush in which they had been hiding. The Gurkha major had spotted them, but I had not! They were rounded up and handed over to the RHKP at their Fanling holding centre. In most cases they were returned to mainland China where they met several differing fates. If they were first offenders they were shipped well inland or returned to their villages. Persistent offenders were sent further away or imprisoned. Fleeing "IIs" were often hounded by dogs and/or shot by [Chinese] border guards. One very sad incident involved an "II" who had gained the HK shore but then hanged himself from a tree having lost the rest of his family through drowning in the water crossing.
>
> During this time the colony was the chosen haven for fleeing Vietnamese "boat people". They arrived in a variety of vessels and attempted to smuggle themselves ashore. The squadron was often tasked to photograph such incidents when the boat people had been discovered. At Tai O, I photographed a sampan crowded with refugees, who were only able to stand

35 TNA, WO 373/183/13: 'Recommendations for Honours or Awards', Lance Corporal John Patrick Jarman AAC, 29 September 1980. *London Gazette*, 12 May 1981.

> up because of the density of the crowd, whilst the sampan was close to sinking. They had managed to float across the Yellow River estuary. One significant incident involved the *Huey Fong*, a rust bucket freighter crowded with boat people in the most unimaginable squalor. The ship, from Vietnam, was halted by the Royal Navy on the international boundary and lay there at anchor for several days. The RAF Wessex helicopters ferried food and water and other supplies whilst a solution to the refugees' plight was sought. The squadron Scouts carried out surveillance and photography sorties of this pitiful scene. Eventually the refugees came ashore and were accommodated in huts on part of the airfield at Sek Kong, which was of course the squadron base.[36]

While remaining professionally detached during the task at hand it was still difficult to not feel some sympathy for the immigrants and refugees, no matter where they were coming from. It was also, at times, difficult not to admire their dedication and ingenuity as Staff Sergeant Bonner discovered following one sortie:

> I had captured three young students and took them to the Mai Po Police Station. They were thrown into the bund to wash off the mud then escorted to the police station. On this particular day *Soldier Magazine* was visiting and was taking photographs of the event. One of the students, in excellent English, asked if he could have a copy. I explained he was off to Mongolia. He replied that he knew because this was his third attempt. I told him that if he was successful the next time then ask for SSgt Bonner, 660 Sqn, at the Sek Kong Barracks.
>
> I knew I would never see him again and forgot about him. Three weeks later I was enjoying the Friday happy-hour in the Squadron Bar when I was called to the telephone. The Gurkha Guard commander stated he had a Chinaman who knew me and wanted to speak with me. I drove up to the guardroom, went in and was confronted with a well-dressed, young Chinese adult. It was the student I had captured three weeks before. I was nonplussed; how had he travelled from the Mongolian border, evaded all the HK Defences, reached the Immigration office, received his ID Card and managed to turn up in a well-cut suit. He would not say. My word is my bond so I told him to return here at 17:00 next Friday. Friday came and I received a telephone call from the Guardroom, my Chinaman was waiting. I drove to the Guardroom, collected the escapee and brought him to our happy-hour where he was presented with a framed, Squadron-signed photograph of his capture. I hope he still has it.[37]

Another method used by the illegal immigrants to cross to Hong Kong on the eastern side of the New Territories was to pay smugglers to bring them in on high-speed boats, sometimes known as 'snakeboats' operated by 'snake-heads', during the cover of darkness. The advantage of this approach was that the immigrants avoided the mud of the Mai Po and could be landed further south, closer to the Boundary Road demarcation line. The crossing was hazardous however, and there was no guarantee as to the seaworthiness of the boat or the skill and experience of the smugglers. A joint approach between the Army, RN and RHKP was needed to successfully intercept the smugglers and illegal immigrants which was demonstrated to the press at night in Rocky Harbour, south of the Sai Kung Peninsula in the New Territories. First the converted Ton-class minesweeper HMS *Yarnton*, operated by the RHKP, picked up the simulated smugglers' boat using its radar and then guided a Scout from 660 Squadron over the target using radio in the darkness. Once in position the Scout illuminated and tracked the fast-moving boat

36 Warner, *From Auster to Apache,* p.167.
37 S. Bonner, account to the authors, 28 May 2020.

using Nitesun. Flares could also be dropped by the aircrewman as an alternative method of illumination. The intruder was then intercepted and stopped using RHKP Launch 53 and two Searider Rigid Inflatable Boats from the RM 3 Raiding Squadron.[38] The RN also operated P236, a SRN-6 radar equipped hovercraft to assist with the task, with the advantage that it could follow boats into the shallowest water or up onto a beach. Only that genuinely joint approach could be successful, as described by Sergeant Kalinski:

> One such incident occurred in the early hours (0200 hrs) on 10th July 1980 when a snake-boat was seen to be travelling at a great rate of knots north of Lantau Island towards Kowloon. We told the Marine Police who immediately deployed two Royal Marine Seariders and a hovercraft. After picking up his cargo, by illuminating him, the Seariders managed to corner him in the Yau Ma Tei typhoon shelter. In trying to get away, the snake-boat collided with a junk, the crew abandoned it and by keeping the light on them in the water, the shore police were able to arrest them and confiscate the snake-boat.[39]

In reality the task for the Scout pilot could be extremely testing and dangerous as Staff Sergeant Bonner recounted:

> The high-speed launches were sometimes faster than the RN and HK Police patrol boats so occasionally a 660 Squadron AAC aircraft was tasked to drop slow-falling flares then descend low level and chase the II launch. Before NVG, chasing a high-speed launch at 30ft AMSL [Above Mean Sea Level] that was dodging between freighters and cruise ships at night was beyond hairy.[40]

The snake-boat operators were willing to run considerable risks to escape interception and sometimes paid the price, as recalled by Lance Corporal Walker:

> On 10 August 1980 XP887 was scrambled to intercept a snake-boat north of Lantau Island. The crew were pilots Sergeants Steve Davies and Dick Kalinski with me in the back as crewman. We were vectored by the hovercraft to the last known position seen on radar. After a few minutes searching a wake was spotted. A boat travelling at 50 knots is hard to hide. We decided we were too low for a flare to be effective, so I lit up the boat with Nitesun. At this point I could only see two people in the boat as they had already dropped their passengers on Lantau and were heading for Kowloon and home. I should imagine that 30 million candle-power had somewhat of a blinding effect on the helmsman as he went into Kowloon Typhoon harbour at max speed and hit a junk broadside on. There was nothing left of the speedboat and the junk was almost cut in half. A police launch was quickly on the scene and picked up the two snake-heads, one of which had lost an arm on a propeller. Fortunately, the junk was unmanned. On a previous flight, a flare had been dropped from the helicopter and either due to height or parachute malfunction had landed in the back of the speedboat. The petrol tanks exploded and there was nothing left of the boat or the two drivers.[41]

38 *South China Morning Post,* 28 May 1980.
39 D. Kalinski, account to the authors, 18 September 2020.
40 S. Bonner, account to the authors, 28 May 2020.
41 J. Walker, account to the authors, 26 August 2020.

At first light each morning, after reports of high-speed boats around Hong Kong the previous night, 660 Squadron would be tasked to patrol the coastline of Hong Kong Island to try to locate and intercept any IIs that had made it ashore. Major Whidborne's successor as OC 660 Squadron, Major Douglas Fox AAC took a reporter and photographer on a sortie along the coast of the Sai Kung Peninsula to experience the difficulties in finding and intercepting illegal immigrants:

> Strapped tightly into our seats with the wind screaming through the open back doors, we had just rounded one headland of the Sai Kung peninsula when Maj Fox's voice sounded in the headphones. "That sampan wasn't there before!" The beach below revealed several derelict boats but the one that stood out was stranded on the waterline and had not sunk even partly into the sand. Farther along, where the tide had not yet washed them away, were footprints heading for a cleft in the face of a steep hill.
>
> Another helicopter was on the scene in moments and, out of nowhere it seemed, soldiers were running towards the cleft. In minutes the dream of the bright lights of Hong Kong had come to an abrupt end for 15 cold, surprised and somewhat frightened illegals. Though they might not have thought so at the time, they were actually the lucky ones.
>
> Had they landed only a few hundred yards away along Sai Wan beach they would have seen the decaying bodies of several of their countrymen who had paid for their dreams with their lives. That grisly spectacle was made worse by the sight of a pack of dogs tugging away at the plastic body bags that encased the remains. Some of the bags had been ripped away completely. As the helicopter came down the sandstorm whipped up by the rotor blades drove the dogs off – but not in the kind of fright that was likely to keep them away. The seas on Thursday had taken a grim toll with 24 bodies washed up on the shores of Mirs Bay alone. The remoteness of these beaches and the rough seas prevented police and the Urban Services Department from picking the bodies up quickly.
>
> "There has most decidedly been an increase in the number of bodies we've found," Maj Fox said. "I would suggest most are victims of their boats going down. There's no doubt in my mind that they don't understand the danger of a water crossing. The currents and tides are extremely dangerous"...
>
> [660 Squadron] have played a role in catching snakeboats making a quick run for Hong Kong harbour under the cover of darkness. In one perfect case of coordination the Royal Navy called in the Scouts of 660 Sqn which located the high-speed boat and, by hovering above it and spotlighting it, managed to drive it ashore where the occupants were arrested.
>
> Maj Fox is extremely proud of his aircraft. He points out they are perfect for the job and that they have a great record of serviceability – unlike some helicopters of similar size designed primarily for civilian use. "Most of the time it's hard to run from a helicopter," he says. "The helicopter effort saves a lot of manpower. Without it we would need extra battalions of troops."[42]

The rate of flow of illegal immigration began to steadily fall when the Hong Kong Government revoked the Touch Base policy in October 1980. On Christmas Day the same year 660 Squadron enjoyed its first day in 19 months without a sortie being flown. Operation Culex had been drawn down and although anti-illegal immigration sorties remained a significant part of the Squadron's workload, it was able to return to a more manageable pace of operations. For 660 Squadron and for the pilots who deployed to Hong Kong temporarily, Operation Culex had proved valuable experience:

42 *South China Morning Post,* 16 December 1979.

> Having qualified as a Scout pilot exactly a year previously, and already having been on one operational deployment on Op Agila in Rhodesia/Zimbabwe, this was a great consolidation of my flying experience and broadening of my flying knowledge. I left Hong Kong to return to 656 Squadron in Farnborough on 19 August 1980 having flown a total of 78.7 hours (72.7 operational) over 38 flying days.[43]

660 Squadron AAC

By 1981 the illegal immigration problem had reduced to a manageable level although unforeseeable events in China could still cause an isolated surge. At the very end of March 1981, just after Major David Cranston AAC had taken over as OC 660 Squadron, an earthquake in China caused a surge of illegal immigrants trying to escape into Hong Kong. On 31 March dozens of Chinese junks began arriving off the coast of Hong Kong and all available Scouts and their crews were deployed to monitor the situation. By the evening approximately 50 junks were concentrated in Joss House Bay at the southern tip of Sai Kung, east of Hong Kong Island, holding an estimated 2,500 to 3,000 potential illegal immigrants. 660 Squadron aircraft were tasked to lift infantry to the beaches and the cliffs above Joss House Bay and one Nitesun equipped aircraft was held on standby to assist as darkness fell. The aircrewman however, were not called on to make any arrests as the RHKP Marine Division kept all the junks under surveillance and guard overnight until they could be dealt with the following morning.[44]

Such incidents, however, were few and far between and Corporal P. W. S. Rooke AAC gave an idea of the range of tasking for 660 Squadron aircrew on a typical day, starting in the squadron lines for an early morning sortie:

> I find out which aircraft I am in, XV122, pick up all the kit. Picture this: 4 ALJs (life jackets for the passengers) slung over my left arm, 4 headsets and my lifejacket slung over my right arm, map pack under my left arm pit, task sheet in my right hand and helmet slung by its strap between my teeth. I walk out of the ops room, discover that XV122 is a dot on the horizon, start my forced march towards it.... Pilot is briefed and we have booked out with air traffic control (it is essential that ATC know the complete movement of an aircraft because of the hazardous terrain in Hong Kong). The aircraft is started up, I check it for leaks, aircraft is fine...
>
> [The pilot asks] "Which way are we going?" "Through the fire station gap and onto 060° Sir" I answer without even thinking. I'd done this task so often, (the Sharp Peak and Tolo II recce that is), still it is one of my favourite sorties.... The next hour and a half is spent along the north and east coast looking for overnight evidence of Illegal Immigrants (IIs) movement. With nothing to report we land at Man Yee for a refuel and a 'Bhat' breakfast with the Gurkhas.

Corporal Rooke's day continued with a liaison sortie from the HQ British Forces, HMS *Tamar*, across Hong Kong Island to Stanley Fort Barracks and back across to Sek Kong. His aircraft then conducted a casevac from a serious traffic accident at Lye Mun Barracks before spending the afternoon helicopter-abseil training at Sek Kong.[45]

---

43 D. Kalinski, account to the authors, 18 September 2020.
44 AFM 215: '660 Sqn AAC Int Book', 31 December 1981.
45 P. W. S. Rooke, 'Oh! To be an Aircrewman in Hong Kong', *Army Air Corps Newsletter,* No.114, November-December 1983, pp.10–12.

By the time Corporal Rooke was writing in 1983, the arrest rate of illegal immigration had continued to fall to an average of just eight per day. Undoubtedly the demise of the Touch Base policy and the introduction of compulsory ID cards across Hong Kong had precipitated the decline but it had been assisted by other factors. The Chinese had built a security fence along their side of the SEZ. 4,000 extra People's Armed Police had been drafted to work in the SEZ, conducting a concerted anti-crime operation which had resulted in the arrest of a large number of potential illegal immigrants. Furthermore, widespread and heavy flooding in Guangdong had persuaded thousands more potential illegal immigrants to stay at home and help rebuild their family farms.[46]

In 1984, uncertainty surrounded the future of Hong Kong, the UK and China stepped up talks on the status of the colony post 1997, on the expiry of the 99-year lease on the New Territories. The issue was, however, viewed as the distant future and was rarely raised in conversation in 660 Squadron which remained active and fully tasked. The challenges of flying in the mountainous New Territories in support of the security forces remained undiminished as the OC 660 Squadron Major Bill Burdett AAC explained.

> ...We were expected to provide 6 manned aircraft every day: 2 for HQHK [HQ Hong Kong], 2 for Gurkha Field Force and 2 for general tasking to units, visitors, parachute school, etc. The other 4 were retained for flying training and servicing. We encouraged individual units within the Field Force to ring our Ops Desk directly to request aircraft support. This contrasted with RAF tasking which was all centralised at HQHK.
>
> One of our daily tasks, sponsored by HQHK was to fly the entire HK mainland coastline at first light every day to look for, and recover, the bodies of swimmers who had drowned whilst attempting to swim across from China to HK. The phenomenon of these poor people attempting to escape from China was covered internationally in the mainstream press and I had read about it before arriving in Hong Kong. My predecessor, David Cranston, had also briefed me on handover. However, in my two years in command, my unit never found a single body or trace of a crossing. In fact, the whole illegal immigrant issue never raised its head during my time there.[47]
>
> The 660 Squadron aircraft have the 'facilities' of about 100 landing sites often just two sheets of medium stressed platform (MSP) marked by an up-turned bucket marked with reflective strips. No easy task at night often in adverse weather conditions...Finding these isolated hill top OPs quickly at night, say in the case of a casevac, presents a difficulty. There are no beacons or landing lights in these mountains, just the up-turned bucket devices...
>
> If the engine goes wrong now, we're dead and that is that! The thing about Hong Kong that is not generally appreciated is that it is virtually all mountains and that is where we operate the single-engined Scout. If the blades stop going round what are you going to do? The Scout glides like a brick. Trying to find IIs and searching through broken mountains is hazardous flying when weather conditions are at their worst.[48]

In fact, the venerable Scout continued to prove an indomitable workhorse as it approached a quarter century of service with the British Army. REME, with the assistance of the Hong Kong Aircraft Engineering Company (HAECo), who were responsible for major servicing, airframe repairs and modifications, worked ceaselessly to keep the aircraft flying. That was no mean feat as the Scouts in Hong Kong included the oldest airframe and the airframe with the most flying hours

46 AFM 215: 'Monthly SITREPs', 7 July 1983 and 4 January 1984.
47 B. Burdett, account to the authors, 31 August 2020.
48 *Soldier Magazine*, January 1985, pp.17–19.

in the fleet, operating in hot, humid, and salt-laden air conditions. Throughout the late 1980s, the older aircraft were returned to the UK and replaced with younger airframes, which improved the flying hour/maintenance ratio and helped secure the Scout's future in Hong Kong.[49] At the end of 1984, however, that future became more certain. On 19 December the leaders of the People's Republic of China (PRC) and the UK Governments signed the Sino-British Joint Declaration in Beijing. In the Declaration, the PRC Government stated that it would resume sovereignty over Hong Kong (including Hong Kong Island, Kowloon, and the New Territories) with effect from 1 July 1997 and the UK Government declared that it would restore Hong Kong to the PRC with effect from 1 July 1997. In accordance with the 'One country, two systems' principle agreed between the UK and the PRC, the socialist system of the PRC would not apply to what would become the Hong Kong Special Administrative Region, and Hong Kong's previous way of life would remain unchanged for a period of at least 50 years.

With the ratification of the Joint Declaration in 1985 Hong Kong's future became certain and most of the incentives to move from China to the colony disappeared. The flow of illegal immigrants across the border dwindled further as did the threat from an incursion or invasion by the Chinese PLA and by the end of the 1980s the emphasis of 660 Squadron's tasking had changed. The AAC in Hong Kong had always worked closely with the RHKP and that included support to the regular police and the Special Duties Unit. Routine support included traffic control, emergency liaison flights, internal security support and continued assistance with illegal immigration control. The Special Duties Unit was a counter terrorist organisation and 660 Squadron's support included providing recce and surveillance tasks, and helicopter-abseiling in urban areas and on to container vessels at sea. Aircrew also practised flying counter terrorist assault teams to oil rigs and directly on to the wings of airliners at Kai Tak Airport.[50]

From 1991 AAC support to border security was gradually withdrawn with the responsibility being handed over to the RHKP and RHKAAF, but other activity began to increase to fill the gap, or at least in part. Smuggling between Hong Kong and China had always been an issue, but it became big business in the early 1990s. As some restrictions were eased in China and a degree of affluence began to emerge, there was a growing desire for 'luxury' goods that could not be obtained under the communist regime. The Hong Kong Government established an Anti-Smuggling Task Force that coordinated complex operations involving Gurkha infantry in OPs around the coast, Marine Police vessels, RM and RE Rigid Raider assault craft, RN patrol vessels and the Scouts of 660 Squadron. WO2 Dave McGee AAC was working in the Joint Air Tasking Cell on HMS *Tamar* as the Scout Tasking WO and helped coordinate 660 Squadron's part in the operations:

> If you wanted a high-end Range Rover [in China], you would order it, the smugglers would go and nick it, put it on a boat and it was a straight line to China. The Anti-Smuggling Task Force was basically a bunch of ex-UK police who had been employed by the Hong Kong police and the Scouts ended up chasing the smuggling boats, which had up to five engines on the back, on their behalf and helped coordinate the RN ships to intercept.[51]

It was no easy task. The smugglers operated two types of boat. The first, small fishing boats with one or two small outboard motors, would carry household goods such as TVs and video recorders and were difficult to identify but relatively easy to intercept. The second type, speedboats fitted

49 *Defence Helicopter World,* December 1988-January 1989, p.42.
50 R. C. Hall, 'The Army Air Corps in support of the Royal Hong Kong Police', *Army Air corps Newsletter,* No.149, January-March 1989, pp.18–19.
51 D. McGee, interview with the authors, 16 July 2020.

A 660 Squadron Scout approaching a jetty on the north shore of Hong Kong island. The building directly beyond the Scout is HMS *Tamar*, the HQ British Forces Hong Kong. (Army Flying Museum)

with multiple engines capable of producing up to 300HP, were a more difficult proposition as they were capable of outrunning the patrol vessels and the Rigid Raiders. Many were modified with a large open hold able to carry a car. The boats were fitted with communications and scanners to listen to the RHKP frequencies. A typical smuggling run would start with the vessels illegally entering Hong Kong waters, usually at Deep Water Bay or Mirs Bay and Tolo Channel, where they would meet a local junk to cross-load the goods to be smuggled. If vehicles were being picked up the speedboat would have to tie up at a jetty. Once loaded the boats would take the fastest route back to Chinese waters. The Scouts were often the only assets tasked that could keep up with speedboats and were therefore vital in keeping track and communicating their location and route in order to coordinate an interception. The aircrew were equipped with cameras with telephoto lenses to take photographs that could be used by the Anti-Smuggling Task Force to identify boats and individual smugglers and at night the Nitesun was invaluable. The RHKAAF would also deploy their Sikorsky S-76 helicopters to drop water bombs on the speedboats in an attempt to drown their engines.[52]

The anti-smuggling operations were often complex and protracted but ultimately successful. WO2 Nick Nicholls AAC from 660 Squadron was committed to Operation Fairway in support of the Anti-Smuggling Task Force for three months in 1992. Aircrew flew in support, often at

52 M. J. Whitcombe, 'Smugglers in the South China Sea', *Army Air Corps Newsletter,* No.152, October-November 1990, pp.16–17.

short notice and in 'difficult and dangerous weather conditions', to lift men and equipment into isolated OPs and to conduct photographic sorties. The information gathered allowed intelligence to be developed that resulted in widespread arrests on land and at sea. Two shipbuilding yards in Sam Mun Tai in Tolo Harbour, producing illegal hulls and modified engines for smuggling, were put out of action and the personalities behind them arrested. Twenty-eight boats were seized and major players in the smuggling syndicates were arrested. Overall, the joint military/police effort broke the largest smuggling operation in the New Territories. The head of the Anti-Smuggling Task Force thanked the OC 660 Squadron, Major Northam, for the aircrew's 'dedication, perseverance and cooperation in the most trying of conditions.'[53]

With the handing-back of Hong Kong looming, the fate of 660 Squadron was sealed in 1991 as part of the 'Options for Change' defence review. The OC, Major Peter Roberts AAC, received a personal letter from the Chief of the General Staff, General Sir John Chapple, informing him that the Squadron would be disbanded but with no details as to date adding, 'Your Squadron has upheld the finest standards and traditions of the British Army for many years and I know that I can look to you and all who serve under you to maintain these standards and traditions in the future.'[54] Once the decision was made public however, 660 Squadron lost much of its priority for aircraft spares, equipment and personnel as recounted by Air Trooper Jason Ward AAC:

> Once the squadron was told that it was to close we became non-operational. This meant spare parts for the aircraft stopped becoming a priority. Aircraft were used as donors to keep other Scouts flying. Major servicing was stilled carried out. A scout from C Flt Brunei was taken to HAECo located at Kai Tak for the last servicing of a Scout there.[55]

By 1993, under Major Hugh Northam AAC's tenure as OC, it had become clear that 660 Squadron would disband well in advance of the 1997 British withdrawal from Hong Kong. The Squadron's tasking had become almost entirely routine liaison and training support tasking, flying range recces, familiarisation sorties and abseil and parachute training as described by the Squadron 2IC Captain Dave Roberts AAC.

> From an operational perspective, it was very much normal jogging: little or nothing with respect to border incursion/illegal immigrants coming across on lilos etc. We did see evidence of this in and around Deep Bay (lilos and detritus from those still trying to get to Hong Kong from China), but that's about it during my time for illegals. We supported the Gurkha battalions dotted around the province in the normal manner, and we flew liaison sorties. We also had a standing task to photograph the border, and as far 'over the other side' as possible. For this we used a magnificent camera supplied and fitted by the Int [Intelligence] Corps, all very hush, hush! We also conducted various border patrols both with and without interested parties.... We also supported the freefall club at times; I hated climbing to 10,000 ft in the Scout and then auto-rotating back to the airfield, something not right about being up with the 747s approaching into Kai Tak![56]

With the disbandment of 660 Squadron imminent during 1993, the opportunity to view the colony from the air became ever more popular with a myriad of 'official' visitors and the Scout

53 AFM 215: Letter, 'Senior Staff Officer, Anti-Smuggling Task Force to OC 660 Squadron AAC', 2 July 1992.

54 AFM 215: Letter, 'CGS [Chief of the General Staff] to OC 660 Squadron', 23 July 1991.

55 J. Ward, account to the authors, 28 August 2020.

56 D. Roberts, account to the authors, 3 September 2020.

pilots all became adept aerial tour guides.[57] The Squadron's days were numbered however, and a disbandment ceremony and final fly past of the Scout took place on 20 December 1993. Airtrooper Ward and Captain Roberts took part in the fly past as aircrewman and pilot respectively:

> [The fly past] consisted of a three-ship formation around the Colony followed by a ceremonial 'wake', where a half size model of a Scout (built by the REME), was escorted onto the dispersal by the 'Grim Reaper' (the OC in costume), and then it was set alight and we all had a party with various members of the Colony's military personnel. A small cohort then remained in-country for several weeks and put the Squadron to bed. This included selling the Scouts to a UK contractor who came out and bought the lot and shipped them back to the UK. Story has it they encountered a storm on the way back, and not all of them made it!...
>
> I was the last person to fly a Scout in Hong Kong, as I was the last in the formation, and I made sure I was the last 'skid-on'.... We were all very sad to go, and to leave the Hong Kong Chinese soldiers behind. We had several of those on the Squadron and our departure left them in the lurch, as it were. Also the knowledge that we were living through a momentous time in history, and that life in Hong Kong was about to change for ever. We all wanted to remain; I certainly had a great time.... It was obvious that the Wessex would be chosen to remain until the last, but a sad moment for the AAC when this was announced.[58]

Service in Hong Kong never attracted a campaign medal although the operational nature of much of the tasking for the AAC was undeniable. While the threat of a major incursion by the PLA was unlikely the constant battle against the illegal immigrants called for the highest levels of skill, dedication, and courage. The disbandment of 660 Squadron ended a 36-year AAC presence in Hong Kong. All those that had served in the colony had worked hard to live up to the motto of 660 Squadron: *E Coelis Auxilium* (Help from the Heavens).

57 J. Ward, account to the authors, 28 August 2020.
58 D. Roberts, account to the authors, 3 September 2020.

Chapter 6

Cyprus

See Map 5

The Cyprus Emergency

Cyprus became a British Protectorate in 1878, was formally annexed into the British Empire in 1915, and became a Crown Colony in 1925. The Eastern Mediterranean Island's population was traditionally divided between Turkish Cypriots and Greek Cypriots. During the first half of the twentieth century Greek Cypriots became increasingly dissatisfied with British rule and agitated for Enosis, the concept of political unification between Greece and Cyprus. In 1951 the Greek Government in Athens suggested that if Cyprus became part of Greece, part of the 'Hellenic Motherland', Britain would be given land for military bases on the island. The British Government rejected the idea. In the same year, across the Mediterranean, Egypt abrogated the 1936 Anglo-Egyptian Treaty that allowed the UK to station up to 10,000 troops in the country to protect the Suez Canal. Britain agreed to begin withdrawing its troops in 1954 including HQ Middle East Land Forces (MELF). The British intended to relocate MELF to Cyprus which further inflamed anti-British sentiment on the island, particularly within the Greek population.

In 1948 Michael Mouskos was elected Bishop of Kitium and adopted the title Makarios III. In 1950 he was elected Archbishop of Cyprus and became the figurehead for his fellow Greek Cypriots, proclaiming that they would conduct an uncompromising struggle for Enosis. In 1952 Makarios' old school friend, former Greek Army Lieutenant Colonel Georgios Grivas, returned to Cyprus and began forming ideas on how to remove the British from Cyprus through insurgency. He formed *Ethniki Organosis Kyprion Agoniston* (EOKA). Although they did not always agree, together Makarios and Grivas became the political and military representatives for Enosis respectively. Grivas was inspired by Gamal Abdel Nasser's overthrow of the established powers in Egypt from 1952 to 1954. Over the same period EOKA recruited fighters and obtained arms, developed tactics and doctrine, conducted reconnaissance and created plans. Grivas launched his insurgency on 1 April 1955, although the initial results were not dramatic.

As time went on, however, EOKA stepped up its campaign. On 10 June 1955 the insurgents conducted their first attack on a police station in Limassol and nine days later the first of many Greek Cypriot police officers was killed in the campaign of violence. Later that month British military married quarters were subjected to grenade attacks. EOKA also began attacks on military bases, patrols and convoys which increased throughout the summer. In September the first joint patrols between the Cyprus police and British Army were mounted. Nevertheless, the security situation continued to deteriorate and when the Cypriot Government appealed to London for help more troops were sent to the island. EOKA was declared an illegal organisation and cordon and search operations were launched to catch members of the group and their sympathisers. At the beginning of October 1955 Field Marshal Allan 'John' Harding arrived in Cyprus as the Governor with both full civil and military powers and responsibilities. Harding strengthened the defences of government buildings, military and police buildings, and bolstered the strength of the

Cyprus police. He established an operational HQ and reorganised the response to the insurgency along similar lines to those being employed in the Malayan Emergency. On 26 November 1955 Harding declared a state of emergency; the Cyprus Emergency had begun. By the end of the year 12 members of the British armed forces had been killed along with five police officers and seven civilians. Over a thousand buildings had been damaged by EOKA activity and the courts had sentenced 600 people deemed to be involved in the insurgency.[1]

The pattern of violence continued to escalate through 1956 and into 1957. The British Army, assisting the Cyprus police, had the twin tasks of maintaining law and order in the urban centres while tracking down EOKA insurgents in their hideouts in the rugged countryside. The latter task was made more difficult by the predominantly rural districts across the island, particularly the Troodos Mountains which reached over 6,400 feet altitude and were snow covered in the winter, along with the widespread support of the Greek Cypriot population. The troops and police on the ground were assisted by the Austers of 1910 Flight RAF, part of 651 AOP Squadron RAF and based at RAF Lakatamia, five miles southwest of Nicosia in the centre of the island. When the flights first moved to Lakatamia in April 1956 the conditions were far from ideal.

> We left our strategic position on the peri[meter] track at RAF Nicosia in April. Here at Lakatamia we occupy our own camp, for the guarding and running of which we are entirely responsible, but we share cookhouses and messes and Camp QM with various other small units which have their camps near but not contiguous to ours. The tents in our camp were erected by Capt [Owen] Lewis [GPR] and the men of 1915 Flight but the rest was not by any means ready when we moved in and morale fell to rock bottom during the first month and a half owing to intermittent water supply, no lights, appalling food (flies everywhere) and no means of entertainment or amusement.[2]

The position did steadily improve through much self-help, building recreational facilities including an improvised cinema and a tennis court. The Flight also operated its three Austers from eight other improvised airstrips across Cyprus including a golf course at Xeros (now Denizli) on the coast at Morphou Bay, a football pitch at Famagusta in the east of the island and a road near Episkopi on the south coast. 1915 Light Liaison Flight RAF was formed at Middle Wallop in April 1956 and then moved to Cyprus as the Independent Light Liaison Flight to work alongside 1910 Flight at Lakatamia.[3] The Austers provided valuable support to troops on the ground including the use of Skyshout, an airborne loudspeaker system used to broadcast information and propaganda. The system was not ideal and presented problems for the aircraft and the pilot.

> 1915 Flight have one aircraft modified for loud hailing. It has a gigantic loudspeaker on each wing and a vast amplifier inside the aircraft. We have not as yet received all the figures but according to our calculations the aircraft works out as being 60lb overweight with a normal passenger. The broadcasting is done by a tape recorder or a very small passenger strapped into a very small seat. She is a cow to fly but contrary to rumour has a normal stalling speed.[4]

1 For more information on the background and conduct of the Cyprus Emergency see D. French, *Fighting EOKA: The British Counter-Insurgency Campaign on Cyprus, 1955–1959*, (Oxford: Oxford University Press, 2015) and R. Holland, *Britain and the Revolt in Cyprus, 1954–1959*, (Oxford: Clarendon Press, 1998).
2 AFM 179: Letter, '1915 AOP Flight RAF, Newsletter for June 1956', 18 July 1956.
3 N. van der Bijl, *The Cyprus Emergency; the Divided Island, 1955–1974*, (Barnsley: Pen & Sword, 2010), p.67.
4 AFM 179: Letter, '1915 AOP Flight RAF, Newsletter for June 1956', 18 July 1956.

Despite the difficulties, the Skyshout system was used extensively and clearly had some effect, although whether positive or not was more difficult to judge.

> In early July the Sky Shouting aircraft was used extensively in an area south of the Troodos range in which a terrorist sweep took place. A tape recorded message was broadcast to a large number of villages in the mountains. On 7th July Captain [John] Tippen [RASC] covered 72 of these villages in five hours flying. The aircraft fitted with this equipment was unfortunately grounded late in the month due to mechanical failure requiring an engine change. The new engine, a Gypsy Major Mk VIII was flown out from England and the aircraft is now in service once more. In fact it was used again extensively in August when the surrender terms were broadcast all over the larger towns and some of the more remote villages. Captain Morton shouted the odds for half an hour over Nicosia and 22 villages in the mountains on August 24th. He was relieved by Sergeant [Thomas] Driver [GPR] who covered another 40 odd villages. On these occasions a Turkish Cypriot was carried in the aircraft and did the broadcasts using the microphone. His voice stood up to the endless repetitions extremely well, partly due perhaps to lubrication by Coca-Cola between flights! These broadcasts did not seem to be received too favourably. We later discovered that the jeering in Kyrenia was considerable and Captain Hespe reported a display of fists in Famagusta. We have discovered that the best height at which to circle for a broadcast is 600 feet above ground level and it is effective up to 1000 feet.[5]

On the formation of the AAC on 1 September 1957 the two RAF Auster flights were renamed as 10 Recce Flight and 15 Liaison Flight AAC. By the beginning of 1958 Makarios had been deported to the Seychelles, although Grivas was still at large at the head of EOKA. Harding had been replaced as Governor by British diplomat Sir Hugh Foot. The key to ending the Emergency at a political level increasingly appeared to lay with a move towards Cypriot independence rather than Enosis. The Turkish Cypriot community were concerned, however, that a Greek Cypriot dominated independent nation would discriminate against its interests. This added a further dimension to the Emergency as the British forces found themselves amidst inter-ethnic violence between the Greek and Turkish communities. It was against this background that, in May 1958, 10 Flight and 15 Flight were amalgamated to form 653 Light Aircraft Squadron AAC. The Squadron was based at Kermia (Metehan), immediately northwest of Nicosia where it drew its administrative support from 43 Light Anti-Aircraft Regiment RA. There were early issues when REME inspections found all six Auster AOP6 had distorted fuselages through age and extended use. All aircraft engineering had to be conducted in the open under the blazing Mediterranean sun, while the offices of the workshop HQ were constructed from aircraft crates and packing cases. As the problems were fixed or worked around, every available flying hour had to be focused on supporting operations.

Greek-Turkish intercommunal strife came to a head in June 1958. Recce cover was flown over Nicosia with aircrew providing minute by minute reports to the police and army on crowd movements, roadblocks and fires within the city centre. The Austers flew at 2,000 to 3,000 feet and the aircrew used binoculars to identify incidents in the streets before reporting them to the security forces on the ground. Counterinsurgency operations also continued. Detailed and systematic photo reconnaissance was carried out to try to uncover EOKA hideouts in the mountains. Convoys were escorted by Austers in an attempt to negate the threat posed by electronically detonated mines. Roadblock patrols were flown identifying and then tracking any vehicles that turned away from police and army checkpoints. In the latter scenario 653 Squadron aircrew worked with

5 AFM 259A: Letter, '1915 Light Liaison Flight RAF, Newsletter for July and August 1956', 1 September 1956.

the armoured cars of the Royal Horse Guards (Blues) and developed novel approaches to stopping errant vehicles.

> In one case an aircraft succeeded in stopping a van load of thugs who were trying to escape after causing mayhem in a village, by doing a sort of sheep dog act until ground forces arrived. On another occasion a pilot attacked a motor-cyclist with a Verey pistol, and caused him to drive into a ditch. Hardly copybook tactics but in this case it worked.[6]

During such operations Army aviation suffered its first fatality in Cyprus. On 8 December 1958 Captain Dom Mulady of the Queen's Royal Regiment was flying Auster VX116 on a road patrol covering armoured cars from the Blues. He descended to low level three miles east of Nicosia to identify a suspect vehicle when he was dazzled by the sun and hit a set of high-tension power cables which wrote off the aircraft and killed Captain Mulady.[7]

Alongside 653 Squadron six Westland Whirlwinds of the JEHU were operating on counterinsurgency operations. The aircraft arrived in Cyprus in early December 1958 at the request of the Director of Operations, Major General Kenneth Darling and were operational by 10 December. A platoon from 3rd Battalion, Grenadier Guards (3 Gren Gds) formed a QRF along with the Whirlwinds and practised fast roping techniques to speed up their response at the scene of any incident. Each helicopter carried an officer or NCO and four men, and the lead aircraft also had a tracker dog on board. During Operation Dovetail, an extensive cordon and search around Lefkoniko 20 miles northeast of Nicosia, the QRF responded when a scout car was blown up by a mine. The Whirlwinds 'swarmed to the scene' of the explosion and the guardsmen had the incident secure within minutes. Operation Dovetail resulted in the arrest of 27 EOKA suspects and the recovery of large quantities of arms and ammunition. During Operation Maresnest at the beginning of 1959 the Whirlwinds dropped parties of 3 Gren Gds high in the Troodos Mountains to establish OPs. Those OPs remained in place throughout the operation, resupplied and then extracted by the helicopters. The guardsmen quickly learned how to create and improve an HLS and to give an indication of wind direction for approaching aircraft using smoke grenades. JEHU's stay on Cyprus was brief but successful and the Whirlwinds were withdrawn back to the UK at the end of February 1959.[8] It was clear that the helicopters were much in demand and that if the AAC did not embrace rotary flight it would soon be left in the wake of the RAF.

> Mount Olympus roughly in the centre of the island is 6,403 feet high, about half the island is covered with well wooded mountain slopes (vertiginous). Scattered in this area is a handful of top terrorists committing tip and run crimes and commanding the allegiance of the town thugs. In the operations to eliminate these bandits we are seldom involved in direct support and unfortunately we have no airstrips in the operational area. As far as light aircraft are concerned, here helicopters occupy the prima donna role, virtually anything except pilot observation which can be done from an Auster can be better done from a helicopter. We Auster pilots are the chorus and the remainder of the cast and sometimes when the leading lady has overworked we might even provide an understudy. It is a real tragedy that this is yet another place where the army is losing valuable practical experience with helicopters.[9]

6 Anon, '653 Light Aircraft Squadron, Army Air Corps', *Army Air Corps Association Journal*, 1959, pp.18–19 & 21.
7 Anon, 'Appreciation – Captain T.D.G. Mulady, Queen's Royal Regiment' *Army Air Corps Association Journal*, 1959, p.19; Anon, Army Air Corps Aircraft Losses Since 1 September 1957, RHQ AAC, 13 June 2019.
8 P. R. Richardson, 'Whirlwind Operations in Cyprus', *Army Air Corps Association Journal*, 1959, pp.35–38.
9 AFM 179: Letter, '1915 AOP Flight RAF, Newsletter for June 1956', 18 July 1956.

> ...in this country 10 Recce Flight and 15 Liaison Flight, whatever our titles, are in the communication business together. We provide the lower grade transport for those people (or goods) important enough to go by air, but whom the firm consider need only travel tourist. If there are any operations in this country and a requirement arises for troops to be landed astride the enemy or a General Officer needs to make a reconnaissance or an operation has to be directed from the air, then Bristol Sycamore or Westland Whirlwind helicopters are used. If there are any important passengers requiring movement from one place to another, whether it has an airstrip or not, then they travel in helicopters, comfortable multiple seat helicopters and all those helicopters come from the RAF.
>
> We, the Auster pilots, are the poor man's Air Force and it is not surprising that the poor man remains poor. With our inefficient two seater aircraft (payload 232lb) and our horde of expensive pilots, ground crew, drivers, signallers, etc (annual wage bill circa £35,000) we are an expensive way of providing the services we give. With more modern equipment a tighter organisation could be produced. Two De Havilland Beavers (or Prestwick Pioneers if they have overcome their cooling problems) based on RAF Nicosia could put us out of business.[10]

JEHU was released from Cyprus in part because the political situation on the island changed at the beginning of 1959. On 18 December 1958 NATO Secretary General, Paul-Henri Spaak, met with the Foreign Ministers of Britain, Greece and Turkey in Paris. During the meeting the British Foreign Secretary agreed to Cypriot independence in exchange for British Sovereign bases being established in Dhekelia in the west and Akrotiri and Episkopi in the east of the island. All parties agreed and Greece applied pressure to Grivas who announced a cessation of EOKA operations on 24 December. The initial negotiations were followed up by further discussions and agreements in Zurich on 11 February 1959 and a final agreement on the future constitution of an independent Cyprus was signed in London on 19 February. That final agreement was also signed by Cypriot community representatives – Archbishop Makarios for the Greek Cypriots, and Dr Fazıl Küçük for the Turkish Cypriots.

Despite Grivas' declaration there were still isolated incidents including the death of two RAF junior ranks when their fuel tanker ran over a mine in north-east Cyprus. Nevertheless, the state of Emergency was formally brought to a close on 19 March 1959. Since its declaration at the end of 1955 the Emergency had claimed the lives of 371 service personnel and 21 British policemen. 653 Squadron noted that by the end of the Emergency the full value of light aircraft in the recce role had been recognised and exploited by ground troops and the police. The Squadron looked forward to a return to training but much of its flying time was taken up with liaison sorties between Nicosia and Episkopi while the details of the sovereign base arrangements were refined through an endless stream of conferences. Even so, 653 Squadron personnel did manage to conduct exercises across the island during 1959 including live AOP and NGFS sorties.[11]

During 1960, with Cypriot independence imminent, 653 Squadron was forced to move from its comfortable base at Kermia to a new site within the Eastern Sovereign Base Area (ESBA) at Dhekelia, in Larnaca Bay in the south-east of the island. Squadron personnel arrived at the site just over a mile north-east of Dhekelia harbour to find a new tarmac strip approximately 400 yards long and orientated roughly north-north-east to south-south-west, a parallel dispersal, a couple of Nissen huts, a deep trench latrine and not much else. There was a flurry of self-help activity to make the site more inhabitable. Seventeen Auster crates, corrugated iron and other scrounged material was used to create crew rooms, offices, stores and even a control tower. In the winter east to west crosswinds were likely

10 AFM 179: 'Newsletter for 1957 – 10 Indep Recce Flight Army Air Corps', 23 December 1957.

11 Anon, '653 Light Aircraft Squadron', *Army Air Corps Association Journal*, 1960, pp.28–31.

to render the new tarmac strip unusable and so, with RE and Royal Pioneer Corps assistance the Squadron cleared, levelled and rolled a 400-yard cross strip through the scrub. Cypriot independence was declared on 16 August 1960 and Makarios was elected as the country's first political leader. With the Western Sovereign Base Area (WSBA) around Episkopi and Akrotiri separated from the ESBA by more than 50 miles the Austers were in high demand for liaison flying between the two. Nevertheless, flying training and exercises continued across Cyprus and in North Africa.[12]

1961 brought changes to Army aviation in Cyprus. On 16 January 653 Squadron HQ and workshop along with 15 Flight departed the island for operations in Aden, leaving 10 Flight as an independent recce flight at Dhekelia. The slimmed down establishment brought some advantages and single personnel were able to move out of their tented accommodation at the airstrip into new, purpose-built barrack blocks 200 yards from the Mediterranean. It took longer, however, to convince the wider Army that the Flight was not capable of maintaining the same level of flying as the full Squadron had been. Later in the year Auster AOP6 were replaced by new Auster AOP9. Exercises frequently took place in North Africa with RAF Idris in Libya (now Tripoli International Airport) used as a base to conduct FAC sorties in support of the RAF's English Electric Canberra bombers. The Austers deployed to Libya via Anatalya in Turkey, Rhodes, and Heraklion in Crete before crossing the Mediterranean to El Adem and then on to Tripoli. There was an unusual task for the Austers when a banking strike in Cyprus disrupted the usual methods of cash distribution. The aircraft were tasked to fly to Nicosia Airport to collect the British Army's £300,000 pay roll in £50,000 batches carried in mail bags and to fly it to the Dhekelia, Episkopi and Akrotiri for collection and distribution. Towards the end of the year 10 Flight began to receive Sud Aviation Alouette II helicopters requiring another round of conversion training for all concerned. The new American Ambassador to Cyprus, Fraser Wilkins was an early passenger in the new aircraft.[13]

The Alouette was initially brought into British Army service as a stop-gap measure following delays to the Saunders-Roe P531, which would eventually see the light of day as the Westland Scout. It was the first production helicopter to be powered by a gas turbine engine and was capable of 130 knots with three hours endurance. The cockpit cabin could carry the pilot and up to four passengers or 500lb of internal freight and had external litters for prone casualties. Following evaluation, a total of 17 Alouette II's were ordered for the AAC. Three went to 12 Flight in Wildenwrath and three to 18 Flight in Detmold. Three went to 8 Flight at Wilson Field near Nairobi and four were allocated to 10 Flight in Dhekelia. The remainder were allotted to Middle Wallop for trials, development and training.[14] Three Alouette pilots arrived in Dhekelia to reinforce 10 Flight; Captain Paul Walters RASC, Captain John Athron RA and Captain Ewart Pavey RA. Shortly after the aircrew the aircraft arrived and were soon in use.

> The four Alouettes arrived in bits in enormous wooden boxes at RAF Akrotiri in December 1961. Our REME people built three of them XR384, XR385 and XR386 and during the last week of December we air tested our three new helicopters and flew them to Dhekelia. [The fourth remained in reserve at RAF Akrotiri]. We were immediately busy supporting 3rd Infantry Brigade commanded by Brigadier [Basil] Eugster and spent a lot of time on exercises with 1st Battalion The Royal Green Jackets and one of the Parachute Regiments then based at General Kitchener's Polimedia Barracks near Limassol. In early February Major Stuart Whitehead came up from his 8 Flight command in Kenya to check out our flying standards.[15]

12 Anon, '653 Light Aircraft Squadron', *Army Air Corps Association Journal,* 1961, pp.18–19.
13 Anon, '10 (Independent) Recce Flight', *Army Air Corps Association Journal,* 1962, pp.47–48 & 58.
14 D. Kellock, account to the authors, 20 June 2020.
15 S. P. Walters, account to the authors, 7 March 2023.

Two Scouts from 3 Flt ACC fly over Kyrenia Castle on the north coast of Cyprus in 1964. (Army Flying Museum)

Although initial experience of the aircraft was positive there were doubts over how it would fare in Cyprus' often harsh environment.

> The Alouette so far seemed to enjoy being flown hard and we hope that they continue to enjoy this treatment. But we are not so sure how they are going to like living in the open and particularly in the heavy rain we are getting now. Also, we are seriously wondering how they are going to react to being tethered out in high winds. We rather feel that we are between the devil and the deep blue sea, since the alternative is blade folding (if we had the blade folding clamps) which if done too frequently must set up wear.[16]

Despite the stop-gap nature of their procurement the Alouette II was still in service with the AAC in Cyprus until 1988. The initial sojourn of the Alouette in Cyprus was, however, brief. At the end of June 1962, the helicopters left Dhekelia with their pilots in order to form 24 Reconnaissance Flight as part of 655 Light Aircraft Squadron and then head to British Guiana. The initial transit to 655 Squadron's home in Detmold, Germany, led by Captain Walters, took nine days via Turkey, Greece, Italy and France. 10 Flight was therefore left with just three Austers in Dhekelia and was renamed the Army Air Corps Section (Cyprus) under the command of Captain Tony Boatfield RA

16 AFM 179: 'Newsletter – 10 Indep Recce Flt AAC', 4 January 1962.

who had just three pilots at his disposal.[17] Wider events, however, would soon force a reversal in the reduction of Army aviation capability on Cyprus. Since independence two years before, the British colonial influence that had gone some way to binding together the Greek and Turkish communities had disappeared and the two had moved further apart, retreating towards their own separate ethnic and cultural identities. The constitution that governed Cypriot independence further reinforced the fractures between the two communities. Many Greek Cypriots still longed for Enosis, while Turkish Cypriots continued to feel vulnerable and discriminated against. In November 1963 tension began to increase as Turkish representatives in the Cypriot Parliament began to frequently veto new legislation and taxation policies. Frustrated, Makarios abandoned the Treaty of Guarantee on Turkish rights and representation and increased the ratio of Greek Cypriots in the civil service. Many in the Greek community believed that Cyprus would never be able to progress and develop while a substantial Turkish community remained. EOKA, which had remained in the shadows immediately after independence began to stir again, this time focused on Turkish targets.

UNFICYP (UN Peacekeeping Force in Cyprus)

The inter-community tension exploded into violence in December 1963. Greek Cypriots, often posing as special constables, attacked Turkish Cypriots killing several, including women and children. Confrontations spread from Famagusta in the east to Paphos in the west. During three days of clashes in Nicosia 17 Turkish Cypriots and 11 Greek Cypriots were killed and hundreds on both sides were taken hostage. Turkey put its navy on standby and sent its air force's jets over Nicosia at low level. Cyprus appeared to be sliding inexorably towards civil war and Makarios turned to the British to restore order. The Government in London held an emergency cabinet meeting on Christmas Day and ordered British troops in Cyprus, along with reinforcements, to intervene and separate the conflicting factions. Four days later, the British commander Major General Peter Young negotiated a ceasefire and drew a demarcation line with the agreement of community leaders. The Green Line (so named as that was the colour in which obstacles were traditionally depicted on military maps) ran across the island and through the centre of Nicosia separating Turkish from Greek communities. By the end of the year 190 Turkish Cypriots had been killed with 210 missing, and 130 Greek Cypriots were dead with 40 missing. Over a hundred Turkish Cypriot villages were abandoned as their inhabitants fled to the Turkish side of the Green Line.

21 Recce Flight was recalled from Christmas leave to load its Austers and new Scout helicopters onto RAF transport aircraft bound for Nicosia. The Flight was integrated with 16th Parachute Brigade which had deployed as part of Major General Young's force. The Flight was soon engaged supporting British forces on the ground with recce, liaison and casevac sorties. They were joined by 651 Squadron's 19 Liaison Flight with its de Havilland Beavers.[18] 21 Recce Flight counted Archbishop Makarios among its passengers during the flurry of inter-community negotiations that took place during the first two months of 1964. Operating from the comforts of RAF Nicosia it was sometimes easy to overlook the urgency and danger of operations on the ground until Lieutenant Fleming found bullet holes during a post flight inspection of his Auster.

Meanwhile, far away in New York, Greek and Turkish Cypriot representatives aired their grievances at the UN. In response, on 4 March 1964 the UN Security Council passed Resolution 186 calling for a permanent cessation of hostilities on Cyprus and establishing a UN Peacekeeping Force in Cyprus. The Force was established on 27 March with a remit to prevent a recurrence

17 Anon, 'Army Air Corps Section (Cyprus)', *Army Air Corps Association Journal*, 1963, pp.76–77.
18 Anon, '19 Liaison Flight' and Anon, '21 Recce Flight', *Army Air Corps Association Journal*, 1964, pp.9–10.

of violence and to contribute to the restoration and maintenance of law and order. Australian, Austrian, Canadian, Danish, Finnish, Irish, Swedish and British troops joined UNFICYP. Major General Young's force was transferred to UN command, along with 19 and 21 Flights as its designated aviation support. The aircraft were hastily repainted in UN colours and personnel swapped their AAC berets for the brighter shade of blue of the UN.[19]

The rebranded 21 Recce Flight's first task came on UNFICYP's day of establishment, 27 March, Good Friday. Captain Derek Emery of the York and Lancaster Regiment, along with his observer, Private Jackson, landed his Scout at HQ 16th Parachute Brigade at Nicosia to pick up a Major from 1st Battalion, the Parachute Regiment (1 Para), his Sergeant and a cameraman from the *Daily Express*. He was tasked with taking the party to Ayia Irini (Akdeniz) on the northern coast of Morphou Bay in north-eastern Cyprus, where a 1 Para company was due to relieve a company from 1st Battalion, the Duke of Edinburgh's Royal Regiment the following day. On take off Captain Emery contacted Nicosia control tower and was advised that six Czechoslovak Yakovlev Yak 11 aircraft were coasting in at Morphou Bay. Soon after, he was told that one of the Yaks had made a forced landing due to running out of fuel:

> I almost immediately gained contact with five of the Yaks circling near the town of Morphou and proceeded straight to the search area. The circling Yaks departed for Nicosia before I arrived in the search area, and it took about seven minutes to actually pinpoint the Yak.
>
> We landed beside them, ascertained that they were not injured, but only badly shaken, collected their kit, and leaving my passengers to look after the Yak took off with them for Nicosia, just as the R.A.F. search and rescue boys arrived on the scene.
>
> Neither of the Czechs spoke English but understood and spoke German; they told me that had they had one gallon of extra fuel, they would have made Nicosia. They showed considerable interest in the Scout particularly the fact that it was a Turbine, they were interested in the starting cycle and impressed by 115 knots of airspeed.
>
> They were dressed in civilian clothes with lightweight flying overalls, one blue-grey, the other black, they carried seat-type chutes and a flimsy-looking type of Mae West but no dinghies, lightweight electric hat but no protective helmets.
>
> The aircraft itself was very basic in design with no aids and no crew comfort, and appeared to be at least Cat 4; so that is one less for Nasser or the Yemen.
>
> The end of the saga, and successful completion of the first mercy mission dressed as members of the UN Force but in aid of the Eastern bloc.[20]

Amid the turmoil and the establishment of UNFICYP the Flight Scheme also reached Cyprus with the associated appearance of many and varied aviation detachments across the island. The AAC Section (Cyprus) became the Air Platoon 3rd Battalion, the Green Jackets (3 GJ) and was manned by personnel of the Rifle Brigade. The Austers remained in Dhekelia where steady improvements in the infrastructure were becoming apparent.[21] In June 21 Recce Flight returned to the UK to take up a new role at Farnborough. During the same month 19 Liaison Flight flew its Beavers from Cyprus to Aden to take part in operations on the Arabian Peninsula. HQ 651 Squadron, which had briefly deployed to Nicosia to command the two flights also returned to Middle Wallop in June 1964.[22]

---

19 Anon, 'Stop Press – Cyprus', *Army Air Corps Association Journal,* 1964, p.11.

20 D. Emery, 'The Extraordinary Case of the Bounced Czechs', *Army Air Corps Association Journal,* 1964, p.21.

21 Anon, 'Army Air Corps Section (Cyprus)', *Army Air Corps Association Journal,* 1964, pp.62–63.

22 Anon, '"Pegasus Airways" – 21 Flight, Army Air Corps' and Anon, 'Cyprus – Aden', *Army Air Corps Association Journal,* 1965, pp.25–27.

In August, 3 Reconnaissance Flight of 651 Squadron AAC took over the UNFICYP commitment from its sister flight under the command of Major Jim Cullen AAC. The pace of operational sorties continued at a high rate which, along with occasional accidents took a toll on the Scouts and Austers along with the REME who repaired and maintained them. Shortly after arrival in Cyprus Lieutenant Richard Kingsley-Smith managed to flip Auster XN442 onto its back when attempting to land on the beach at Ayia Irini. The aircraft had to be recovered by road to Nicosia for repairs. A Scout suffered a tail rotor strike while on the ground at the Skouriotissa copper mine south of Morphou Bay. The small villages in the foothills of the Troodos made recovery by road impossible and so the aircraft had to be repaired in situ.[23]

Just as 3 Flight arrived in Cyprus the security situation deteriorated dramatically. Greek Cypriots believed that Turkey was landing weapons at the small port of Kokkina (Erenköy) on the western edge of Morphou Bay in order to arm the Turkish Cypriot resistance organisation *Turk Mukavemet Teshkilati* (TMT). Kokkina was essentially a Turkish exclave in a largely Greek Cypriot area and had been under effective siege for years. On 5 August 1964 Greek National Guard units disarmed the Swedish UNFICYP presence in the Tillyria district surrounding Kokkina and attacked several villages in the area. The TMT in Kokkina managed to hold out. The following day four Turkish North American F-100 Super Sabres attacked Greek fast patrol boats off the north-east coast of Cyprus; one aircraft was shot down. A Turkish naval task force then sailed towards the island while RAF aircraft from Akrotiri intercepted Greek Air Force jets over Cyprus. The UN stepped in and managed to negotiate a ceasefire before the situation descended towards a war between Turkey and Greece. UNFICYP personnel returned to Kokkina to find the port wrecked and the Turkish Cypriot population sheltering in caves in the surrounding countryside.[24]

Air Platoon 3 GJ was replaced by the Air Platoon 2nd Battalion, the Royal Anglian Regiment (2 R Anglian) under the command of Captain David Smith. Unlike most air platoons and troops 2 R Anglian's Austers were not in direct support of their parent battalion. Instead, they were in direct support of HQ Cyprus District. Their tasks were mainly liaison flying, carrying passengers and mail between the two SBAs. The Air Platoon did also conduct a successful search to find two missing canoeists in Larnaca Bay and were able to direct an air sea rescue Wessex from RAF Akrotiri to winch the pair to safety.[25]

In February 1965 6 Flight AAC deployed from Middle Wallop to replace 3 Flight in support of UNFICYP. The two flight commanders managed the briefest of handovers in the reception lounge of RAF Akrotiri as one flight landed in Cyprus and the other departed. 6 Flight soon began to tackle the complexities of working in a multinational context.

> Tactical reconnaissance using aircraft in a mixed force such as this provides interesting comparisons. Units belonging to the different nations change over as units every six months; the position is therefore in a constant state of flux. Some units are keen on the correct tactical use of aircraft for reconnaissance purposes and to maintain a current photographic record of known and suspected Greek and Turkish locations; some have their own observers who are trained in observation, map reading and signalling from aircraft and helicopters; other units are trained to a far lower degree of efficiency, where air reconnaissance does not even come within their sphere of thought.
>
> The main problem in these circumstances is to restrain the more air minded units to a quota of flying and encourage the other and less fortunate units to use aircraft effectively and

23 Anon, '3 Flight AAC, UNFICYP', *Army Air Corps Association Journal,* 1965, pp.18–22.
24 van der Bijl, *The Cyprus Emergency*, p.187.
25 Anon, 'Air Platoon 2nd Battalion Royal Anglian Regiment', *Journal of Army Aviation,* 1966, p.83–84.

> to relieve some of the strain on their troops. It was remarkable to see the change in attitude towards the Flight in some of the units after a few months.[26]

In August 1965, 21 Flight replaced 6 Flight for yet another six-month operational tour of Cyprus commanded by Captain John Coles AAC.[27] They were relieved in turn by AOP Troop, 4 Lt Regt RA in February 1966. The day after its arrival in Cyprus the AOP Troop, under the command of Major Ewart Pavey RA, was flying its Sioux under UN colours out of RAF Nicosia. The Troop was bolstered by three Austers left behind by 21 Flight along with two pilots, Captain Eric Butler of the Lancashire Regiment (LANCS) and Captain Mike Sumner of the Durham Light Infantry. As the AOP Troop got used to its new operating environment it soon became clear that the UN was viewed with a good deal of suspicion by both sides of the divided population:

> Like others before and after us we found the process of turning our coats interesting, amusing, and different! Ceasing to be "filthy Brits" (they only existed in the Sovereign Base Areas) we changed overnight into bright-eyed Peacekeepers in a force which included Canadians, Finns, Danes, Swedes, and Irish from Eire. The latter were excellent Peacekeepers, helped no doubt both by their understanding of feudin', and by their praiseworthy attitude of "everything matters, but nothing matters very much"…
>
> Sorties were varied and interesting. Of course, none of us ever put a foot wrong, but in the delicately balanced situation we could still be accused of it! Capt Canning flew an excellent and audacious dawn recce sortie and was blasted for seeing too much. The Troop Commander offended twice: a roadside landing to secure a litter strap was seen by "them," and published in the press as the delivery of a "large parcel of explosives" to the terrorists of the other side. Later, on picking up two of the Irish with their crock of gold (i.e. the Company pay) a woman with a "large suitcase" also got aboard the Sioux, "they" said. Of course, the Troop Commander denied it, saying that he might try flying with a woman on his knee, but not a suitcase as well.[28]

The Austers did not last much longer with the AOP Troop before meeting an ignominious end as described by Lance Corporal Mick Ashton REME who was serving as an air technician at the time:

> During our tour Austers were retired and were put down like old horses. But not without a final flourish, over flying us while discharging toilet rolls which unravelled like streamers as they fell. Their last flight was to RAF Akrotiri to be burnt by the fire service. While we were tasked to draw out all the spare engines and sledgehammer them, reducing them to scrap.[29]

On 7 April 1966 Major Pavey was flying the acting UNFICYP commander Brigadier Tony Wilson to Dhekelia in Sioux XT178. Having dropped off the Brigadier he decided to fly a training sortie before returning to meet the commander for lunch and then return him to Nicosia. As part of the training sortie Major Pavey conducted a practice forced landing during which the tail rotor made light contact with the ground. Major Pavey disembarked to inspect the Sioux, but on finding no

26 Anon, '6 Flight in Cyprus', *Journal of Army Aviation,* 1966, p.82.
27 *The Blue Beret*, 15 September 1965, p.3.
28 Anon, '4 Air OP Troop RA, UNFICYP', *Journal of Army Aviation*, 1967, p.70.
29 M. Ashton, account to the authors, 12 September 2022.

Three Sioux of UNFICYP Flight in the hover over the dispersal at Nicosia International Airport. (Army Flying Museum)

sign of damage he decided to continue the sortie. The tail rotor subsequently failed and XT178 crashed and burst into flames killing the Major.[30]

It was clear that the UNFICYP was likely to be a long-term commitment and that deploying flights on six-month emergency tours was not the most efficient or effective way of meeting the requirement. In August 1966, therefore, UNFICYP Flight AAC was established with five Sioux. The Flight took over tasking from AOP Troop 4 Lt Regt RA on 15 August. Although it was intended to be a permanent fixture, only two members of the Flight were resident on two-year postings; the REME Staff Sergeant Artificer and the OC, Major David Craig AAC. The rest of the Flight personnel would serve on six-month tours, replaced on a 'trickle' posting basis. This was designed to improve stability and consistency, both in the Flight and back in the donor units in the UK. Major Craig had little time in which to grasp the intricacies of the operational situation and of flying in the hot and high environment presented by Cyprus:

> The task of peacekeeping in this island is by no means as simple or straightforward as many people imagine and the Force is on a completely operational basis. The warring factions consist of the Greek and Turkish civilian communities backed up by the Cyprus Army, the National Guard, Greek National Contingent, Turkish National Contingent, Turkish Cypriot Fighters and the "police forces" of both communities. The various confrontations and incidents which take place rarely get into the news and range from the laughable to the lethal. Many of the outposts and O.P.s are difficult to access by road and some quite inaccessible except by air,

30 Anon, 'Obituaries', *Journal of Army Aviation*, 1967, p.39; Anon, Army Air Corps Aircraft Losses Since 1 September 1957, HQ AAC, 13 June 2019; *The Blue Beret*, 13 April 1966, p.1.

> and the tasks which our four Sioux are called on to carry out are both varied and interesting. There are between 80 and 90 "pads" in regular use on the island; the location and immediate surroundings of many of these make them interesting and in some cases hazardous propositions for a Sioux, especially in the high density altitude conditions obtaining in summer when the temperature hovers around 105° F, and during the periods of high winds in the mountains. At one pad it was necessary until a fortnight ago to take off underneath an HP [high power] Aerial – although not as dicey as it sounds, it always gave one the feeling of flying under a bridge.[31]

During the same summer that UNFICYP Flight was established, 2 R Anglian departed the island leaving behind its Air Platoon. As a result, it was redesignated as the Infantry Air Platoon Cyprus and continued to operate from Dhekelia providing aviation support to the SBAs.[32]

Although the British Army and UNFICYP aviation establishment in Cyprus was stabilising, the security situation was deteriorating markedly yet again. In April 1967 a group of Greek Army colonels overthrew the caretaker government in Athens through a coup, a month before scheduled national elections. They installed their own right-wing junta which was destined to govern Greece for the next seven years. In early June 1967, Israel fought the Six Day War against Egypt, Jordan and Syria. Both events caused political and social instability in Cyprus. Tension had been high for some time, particularly around the Turkish Cypriot villages of Kofinou and Agios Theodoros. The villages overlooked the important junction of the Larnaca to Limassol road along the southern coast and the road north to Nicosia. Throughout the summer months the TMT had been denying Greek Cypriot police access to the villages. On 15 November 1967 Greek National Guard units, under the personal direction of Grivas, descended on the area. There was considerable press and media interest in the intervention by Grivas and when he tried to force a way into the two villages fighting broke out. A four-hour firefight ensued, with the Greek National Guard units supported by artillery.

When the firing died down 27 Turkish Cypriots lay dead and Kofinou and Agios Theodoros were under Greek Cypriot control. Turkey warned the UN that if the Greek National Guard did not leave the villages, they would be forced out by Turkish armed forces. In order to calm the situation and avoid war, the Greek Government recalled the 7,000 strong National Guard and Grivas to the mainland. British infantry in the UNFICYP role, who were responsible for the Kofinou area, came under intense pressure during the incident but were ably supported by the Sioux of UNFICYP Flight. The Infantry Air Platoon also made a valuable contribution during the confrontation:

> The "Troubles" of 1967 were brief but caused flurries of excitement and kept everybody in an acute state of suspense. The aircraft were invaluable observation platforms and liaised like mad between the two Sovereign Base Areas, keeping to the coast to avoid unnecessarily provoking Republican reaction. At one stage the Platoon was ordered to "Stagger" both its serviceable aircraft on the dispersal to ensure that the trajectory of a stray missile could not pass through more than two aircraft at once.[33]

With the demise of the Integrated Flight Scheme the Infantry Air Platoon supporting British Sovereign Forces in Cyprus was renamed 16 Aviation Flight AAC. The Flight was under the

31 Anon, 'UNFICYP Flight Army Air Corps', *Journal of Army Aviation,* 1967, pp.66–67.
32 Anon, 'Infantry Air Platoon Cyprus', *Journal of Army Aviation,* 1968, p.77.
33 Anon, 'Infantry Air Platoon Cyprus', *Journal of Army Aviation,* 1969, p.96.

Captain Charlie Blount flies an UNFICYP Flight Sioux to help position an antenna for the UN Swedish Contingent in 1969. (Army Flying Museum)

command of HQ Near East Land Forces for policy and tasking. Major John Valenzia of the Royal Malta Artillery succeeded Major David Manktelow of the King's Own Royal Border Regiment (KORBR) in command of the Flight's pair of Sioux, XT220 and XT221, which remained based at Dhekelia in the ESBA. After the AAC finally left Aden in October 1967 it was decided that efforts should be made to retain the valuable 'hot and high' flying experience that had been gained during the four years of the Emergency. 16 Flight were designated to run month long hot and high Advanced Pilot Training courses on Cyprus for AAC aircrew visiting from the UK and Germany. The course was designed to give aircrew an opportunity to revisit their pure flying skills and to revise mountain flying techniques in an unfamiliar environment. Staff Sergeant David Carpenter AAC was one of the first to complete the course in August 1970:

> 16 Flight AAC was then commanded by Major Valenzia, Royal Maltese Artillery [*sic*], who was also the QHI. Major Valenzia had been in command of 9 Flight AAC, Soltau, BAOR, to which I had been seconded to gain experience when I first qualified in 1966. The Flight artificer was SSgt Alan Donald REME who I had also known from BAOR. Always nice to see some friendly faces…
>
> Major Valenzia gave me a QHI check and I was then launched forth into area familiarisation and general training. The flying was an excellent experience and most of the exercises possible were packed into our time there. Operating in very high temperatures was good further experience in operating the Sioux. Major Valenzia took us both through some very thorough training involving landing on Mount Troodos [Olympus] at about 6,000 feet before we carried out further practice on our own. We later practised underslung load work both by day and night and it was pleasing to see how experienced every member of the flight was in handling aircraft and loads particularly by night.
>
> Not all flights were without incident. I was tasked one day to fly to an RE detachment on a beach at the very western end of Cyprus, a flight of about 60 miles. I climbed to 2,000 feet for coolness ad set heading, due west. We flew with doors off and all was well until I unfolded the enormous map I had been given whereupon it promptly blew out the door! I was faced with returning ignominiously to Dhekelia to request another map or continuing without one; I could hardly miss the west end of Cyprus. However, I remembered that at some stage I would have to contact RAF Akrotiri and it would help if I knew where I was. Couldn't have them thinking the Army couldn't navigate!!
>
> Inspiration dawned! I realised I could see the map spiralling down, the sun glinting off of the Fablon covering. So, I simply spiralled down with it and when it finally landed in a barren area, I landed nearby, screwed down the friction, reclaimed it and flew on. And I made my original ETA [estimated time of arrival].[34]

1970 marked a period of relative political and social stability across Cyprus. Makarios, who still led Cyprus 10 years on from independence, advocated peace within a unitary state in which all citizens enjoyed equal rights. Some Greek Cypriots resented his reconciliatory sentiment and on 8 March 1970 a group attempted to assassinate the president. As he took off from the bishop's palace in Nicosia his personal helicopter was engaged and hit by machinegun fire. Makarios was not hit but his pilot, Major Zacharias Papadoyiannis, was severely wounded but skilfully managed to land the aircraft in a nearby street out of sight of the gunmen. Although it might have had severe repercussions, the assassination attempt was an isolated incident. UNFICYP continued to have its remit extended at regular intervals, but the force was reduced in size. The reduction meant that

34 D. Carpenter, account to the authors, 2 September 2022.

UN troops were thinner on the ground and aviation became even more important in plugging the gaps and extending the Peacekeepers' influence.

> Situations such as [the attempted assassination] cause some strain on the small U.N. forces and we are able to maintain the UN presence by reconnaissance flights in selected areas. To this end, the restrictions on flying have been gradually removed. Originally U.N. helicopters were not allowed to transit below 2,000 ft., now, we are able to fly at whatever height is necessary for the job, from zero upwards.
>
> Apart from the liaison flights necessary in an island of the size of Cyprus the flight has a particularly interesting task in resupplying the mountain OPs of the Finnish battalion in conditions of extreme heat in summer and normal almost English weather, with cloud and rain and adverse winds in winter.
>
> The Swedish Battalion makes great use of our abilities in the eastern half of the island to observe Turkish diggings of fortifications. In a zone of over 400 square miles the Swedes now have only 250 men. A helicopter can proclaim the UN presence effectively and direct ground troops into the area where they are required.[35]

The island remained generally stable and peaceful but there was a steadily widening rift within the Greek Cypriot community between those who supported Makarios' vision of a joint and equal pan-Cypriot society and those that still yearned for Enosis and the expulsion of the Turkish Cypriot community. On 31 August 1971 Grivas secretly returned to Cyprus and formed EOKA-B to reaffirm the commitment to Enosis. On the surface, however, all appeared calm as reported by UNFICYP Flight.

> The situation has simmered for some years without coming to the boil, even though more lately the Greek versus Greek aspect has added some interest and action… The aim of UNFICYP is threefold. In helping conditions in Cyprus to return to normal it has so far failed. In preventing a recurrence of fighting it has been almost totally successful. In contributing to world peace it is probably successful through achieving its second aim. Although the writer is biased, the flight has played and is playing its part in all this to a degree disproportionately large for its size.[36]

By 1973 life in UNFICYP seemed all but routine. The overall force, which had peaked at 7,000 personnel in its early days had been successively decreased to just 2,300 manning 53 OPs across the island. UNFICYP Flight retained its four Sioux authorised to fly up to 120 hours per aircraft per month. The Flight was established for 19 personnel: five pilots, seven REME technicians and seven ground crew. Only the Flight Commander, Major Peter Gill RA, the QHI and REME artificer were on permanent two-year postings. The UNFICYP emphasis was on maintaining a low profile and very little tactical flying was being carried out. The majority of tasking was liaison flying, carrying passengers to various conferences and providing area familiarisation sorties during the frequent arrival of new contingent commanders. Due to local complaints night flying had been restricted to essential training only. When night flying did take place there were often complaints from one side of the Cypriot community that unidentified aircraft were resupplying the opposing community. Once every six months the Turkish national contingent changed over, and an exercise was conducted to monitor the contingent's transit of Greek Cypriot areas on its way back to

35 Anon, 'UNFICYP Aviation Flight', *Journal of Army Aviation*, 1971, p.92.
36 Anon, 'UNFICYP Flight', *Journal of Army Aviation*, 1973, pp.105–106.

Ground crew of UNFICYP Flight prepare a Sioux for a casevac sortie. (Army Flying Museum)

Nicosia Airport before leaving for Turkey. Much was made of the opportunities for leisure while in Cyprus, from skiing in the Troodos to the multitude of water sports available in the inviting Mediterranean Sea. It was, however, a deceptive period of calm before what would prove to be a ferocious storm.[37]

The Turkish Invasion

On 27 January 1974 Grivas died of heart failure and was buried two days later in the garden of the house that had been his last hideout during the Emergency. EOKA-B came to an agreement with the Greek dictator Dimitrios Ioannidis and fell under the direct control of Athens. After its unexpected change of leadership, EOKA-B remained relatively quiet through the first half of 1974. Athens, however, was increasingly interfering in Cypriot affairs. Makarios sent letters to Athens and the press complaining about the Greek junta's support for EOKA-B and publicly accused Ioannidis of attempting to undermine his authority. In early July the Greek Government, in need of a diversion from the reported excesses and failures of the military regime, directed EOKA-B to seize power in Nicosia. Fighting began on 15 July and was immediately noted by UNFICYP Flight AAC.

37 Anon, 'An Attachment to the UNFICYP Flight', *Journal of Army Aviation,* 1974, pp.90–91 and Anon, 'Two Shades of Blue – The Army Air Corps and UNFICYP', *Army Air Corps Newsletter*, No.44, April-May 1974, pp.2–5.

> Monday 15 July 1974 started as a perfectly normal day, with clear skies and the promise of high temperatures to come. We had two aircraft airborne on early morning sorties, one on a recce on the North Coast and the other carrying out a routine Nicosia town reconnaissance. The first aircraft passed over the main Morphou road at 0810 hrs and having spotted what appeared to be a traffic accident carried out a few orbits to see if any help was needed. Things looked under control with quite a crowd in the area including some National Guard troops from a nearby camp. It later transpired that this in fact, was one of the many road blocks established by the anti-Makarios forces to seal off Nicosia.
>
> The sortie was continued and at approximately 0840 hrs both aircraft started to pick up some rather odd messages on the ATC net, and an indication of things to come on the Command Net from OPs in the town. The last message from Nicosia Airport Local was a somewhat peremptory and excitable "We are forced to close – no more traffic!"[38]

16 Flight AAC also had a bird's eye view of the beginning of a chain of momentous and tragic events. Major Hugo Brooke AAC had retired from the regular Army in 1967 but was persuaded to join the new AAC Territorial Army (TA) Squadron the following year as a reservist. Major Brooke had previously flown the Saunders-Roe Skeeter and so converted onto the TA Squadron's Sioux. In the early 1970s it became difficult to find candidates to attend the hot and high course in Cyprus due to the AAC's commitments in Northern Ireland. In 1972 Major Brooke was invited to join 16 Flight AAC and attend the course as part of his annual reservist training commitment. He returned in 1973 and again in the summer of 1974:

> At the start of my annual TA fortnight on Monday 15 July 1974 I was in a Sioux on my check flight with the Cyprus QHI Staff Sergeant John Scott [AAC]. We were a few miles north of the British Sovereign Base of Dhekelia, halfway through an engine failure practice, when we heard from air traffic control that Nicosia Airport was closed because the runway was blocked. No other reason was given and it was not until we landed back at Dhekelia we learned that Mr Samson's coup d'etat attempt to oust Makarios had just started.
>
> On arrival back at base we were greeted with great excitement. Staff Sergeant Scott and I were told to go and see the Flight Commander, Major Peter Child, 'immediately'. Pointing at me Peter asked the QHI "What's he like?" "Oh, he can get it up and he get it down! Why?" "Good. The control radio set for the whole island is unserviceable this morning and there's a Greenie arriving here in five minutes who wants to go to the top of Mount Olympus with a replacement – are you up for that?"
>
> So my next landing, not having flown for a year, was at 6,000 feet in a temperature of nearly 30°C. There was a heat wave on at the time. The heli-pad up near the famous 'golf ball' was fairly small, but there was no wind and I was pleased to hit it first time. We were flying without doors to reduce weight and keep us cool. However, my Royal Signals Sergeant passenger was not impressed, having discovered on the way up that I was a reservist! Poor chap was of a nervous disposition and a couple of hours later sent a message to say he was sorry but he had to stay up there to help out.
>
> Whilst waiting for that message I saw an unusual helicopter, a Fairchild Hiller, weaving its way southwards between the Troodos peaks. It later transpired it was carrying Makarios on his escape from Nicosia to the other British Sovereign Base Area at RAF Akrotiri.[39]

38 P. N. Gill, 'Island of Tragedy, Cyprus – Summer 1974', *Army Air Corps Journal,* 1975, p.29.

39 H. Brooke, account to the authors, 4 October 2022.

Fighting broke out across the island between pro Makarios Greek Cypriots and EOKA-B and those still committed to bring about Enosis. Members of the National Guard who naturally sided with EOKA-B attacked the presidential palace in Nicosia with infantry and armour. Loyal police kept them at bay while Makarios escaped. With a small escort and the assistance of loyal officers, and by changing vehicle several times, Makarios made it to Kykko Monastery in the Troodos Mountains. Once there, he broadcast a message across Cyprus to reassure all islanders that he was still alive. UNFICYP could not officially intervene in the situation as it was a purely internal Greek Cypriot affair and did not involve the Turkish community at that stage. The British Sovereign forces were also reluctant to become heavily involved, as it would put at risk the service families who lived outside the SBAs. As witnessed by Major Brooke, Makarios was evacuated from Kykko Monastery by helicopter and taken to RAF Akrotiri where he subsequently boarded the RAF Armstong Whitworth Argosy which flew him to London. The power vacuum was filled by Nicos Sampson.

Sampson had been a hardline insurgent member of EOKA throughout the 1950s and had been directly involved in the murder of at least 15 British police, servicemen and civilians mainly along Ledra Street in Nicosia. He was arrested, convicted of murder and sentenced to life imprisonment but was released under a general amnesty following the Zurich Agreement. The Turkish Government strongly suspected that Athens was behind the coup. With Sampson, a committed adherent to Enosis, now in control it was clear that the Turkish Cypriot community was under threat. With the clear risk that the Turkish Government in Ankara might decide to intervene, Britain began to reinforce the SBAs under Operation Platypus. American and Russian naval vessels also began to converge on the island. Turkey, however, was not deterred and on 18 July Ankara invoked its right under the Treaty of Guarantee to protect Cypriot independence and the lives and rights of the Turkish Cypriot community. Turkey launched Operation Atilla and during the morning of 20 July, Turkish forces began landing in the north of the island around Kyrenia (Girne). With the invasion widely anticipated, once again members of UNFICYP Flight and 16 Flight had a grandstand view of events as they unfolded, as recorded by UNFICYP Flight Commander Major Gill:

> The Sioux were flying dawn patrols observing the most obvious landing points on the North coast, and we had the unique privilege of witnessing the sea landings and the fighter ground attack assaults along the coast. Luckily the Turks discounted us as a threat and seemed to ignore our presence.
>
> Our flying during the first day was restricted to essential operational tasks and our aircraft were shot at by both sides, particularly in the Nicosia area where the local forces, the Turkish Cypriot Fighters and the Greek National Guard, were heavily engaged.
>
> On the second day RAF Nicosia was abandoned and during a negotiated ceasefire the families were evacuated to the Eastern Sovereign Base Area at Dhekelia. It was disconcerting to find that the flight was now stuck out on a limb in a very exposed position without any protection; particularly as the Airport was the focal point for most of the air attacks. We were forbidden to fly the aircraft to a less vulnerable location so we decided to move them on their ground wheels. On the third day the hangers were strafed by rocket and machine gun attacks, and there was little doubt that the aircraft would have been destroyed had they not been moved. From time to time the National Guard took up positions in the flight location but were fairly amenable to invitations to move elsewhere when it was pointed out that they were infringing UN territory.[40]

40 Gill, 'Island of Tragedy', pp.30–31.

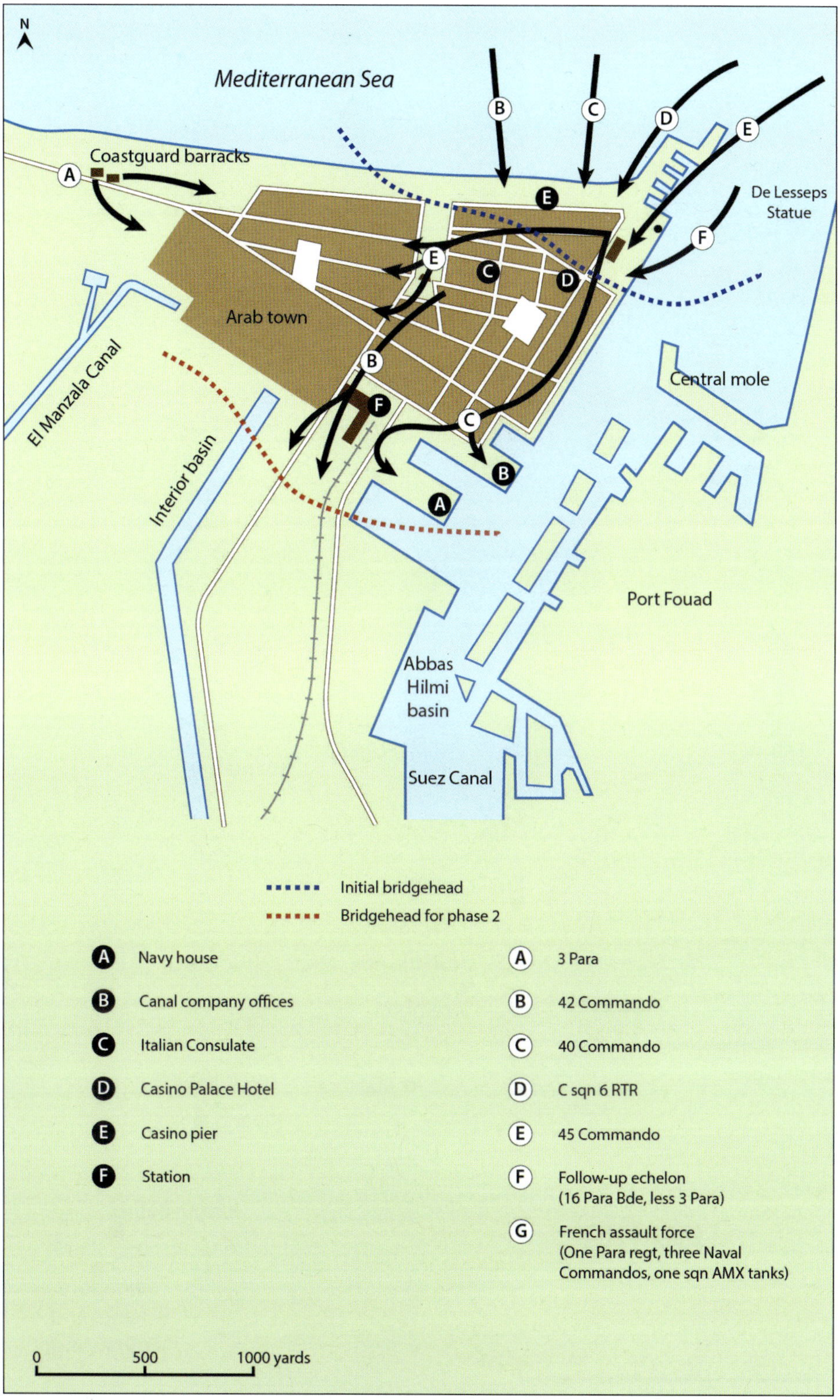

Suez – Operation Musketeer 1956.

N
South China Sea
THAILAND
PERLIS
KEDAH
Kota Bharu
Kroh
Butterworth
Grik
Cameron Highlands
Kuala Trengganu
Penang
TRENGGANU
PERAK
KELANTAN
Taiping
Perak River
Ipoh
PENANG
PAHANG
Kuantan
SELANGOR
Kuala Lumpur
NEGRI SEMBILAN
Seremban
Malacca
Malacca Straits
JOHORE
Kluang
Johore Bahru
SUMATRA
Singapore (Changi and Sembawang)
0 20 40 60 80 100 miles

Malaya.

N
Sulu Sea
Kudat
Mount Kinabalu
Jesselton
Sandakan
South China Sea
RAF Labuan
SABAH
BRUNEI
Tawau
5th
Division
Long Pa Sia
Long Semado
4th
Division
Bario
Celebes Sea
SARAWAK
Belaga
3rd
Division
Lundu
Kuching
1st
Division
Serian
2nd
Division
KALIMANTAN
(INDONESIA)
0
50
100 miles

Borneo.

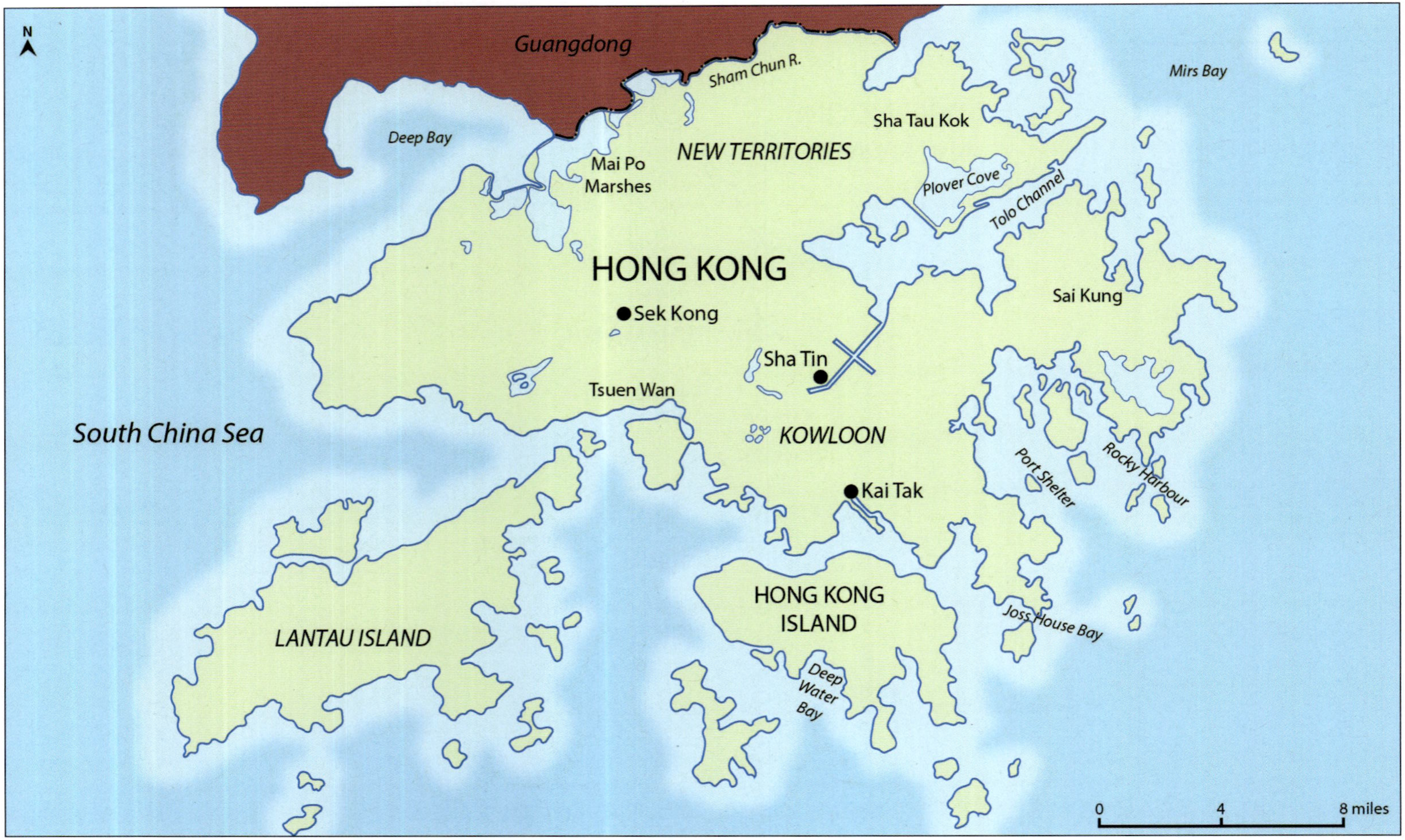
Guangdong
Sham Chun R.
Mirs Bay
Deep Bay
Sha Tau Kok
Mai Po
Marshes
NEW TERRITORIES
Plover Cove
Tolo Channel
HONG KONG
Sai Kung
Sek Kong
Sha Tin
Tsuen Wan
South China Sea
KOWLOON
Port Shelter
Rocky Harbour
Kai Tak
HONG KONG
ISLAND
Joss House Bay
LANTAU ISLAND
Deep
Water
Bay
N
0
4
8 miles

Hong Kong.

Cyprus Pre–1964.

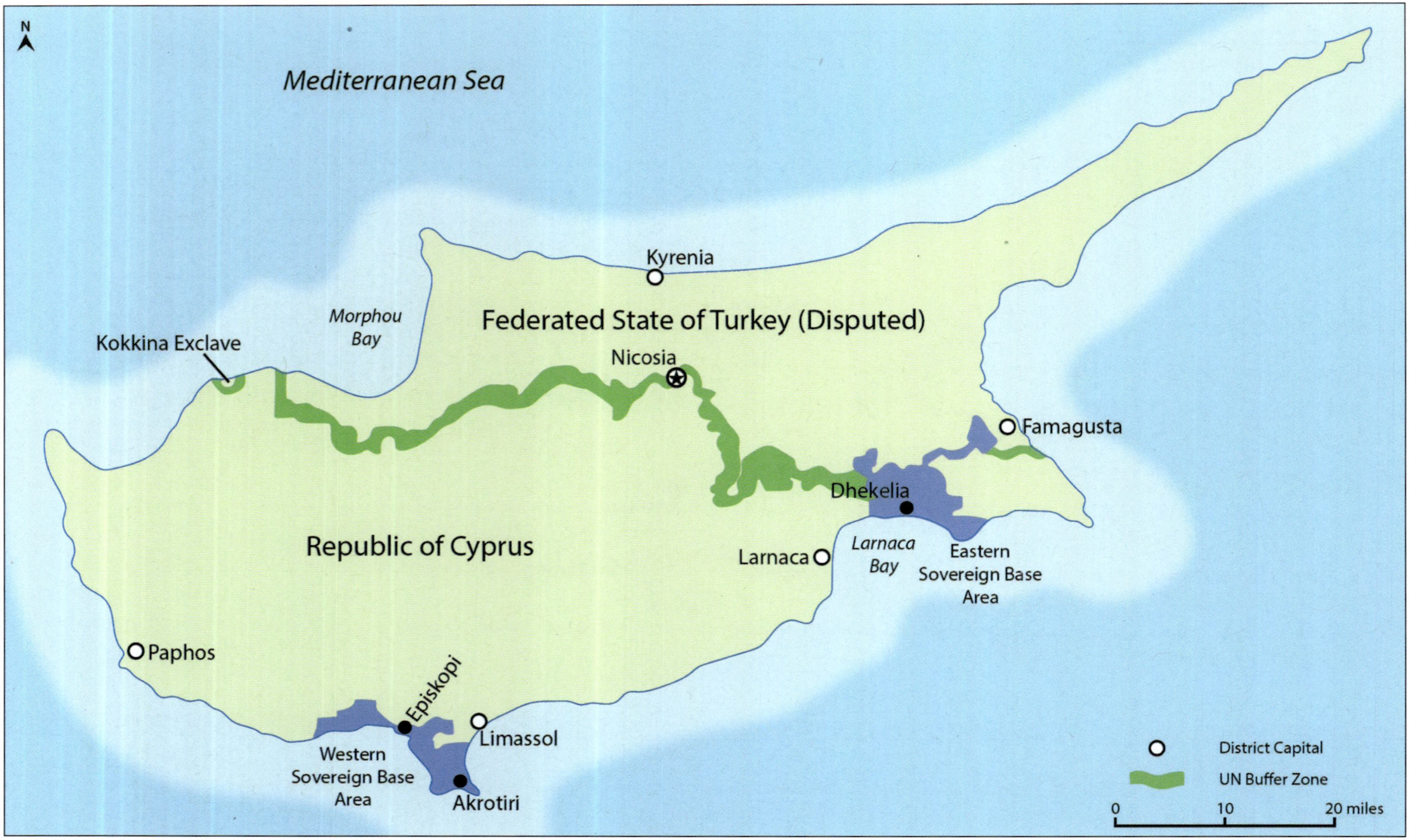

Cyprus Post Partition.

The Middle East 1960.

Kuwait Operation Vantage 1961.

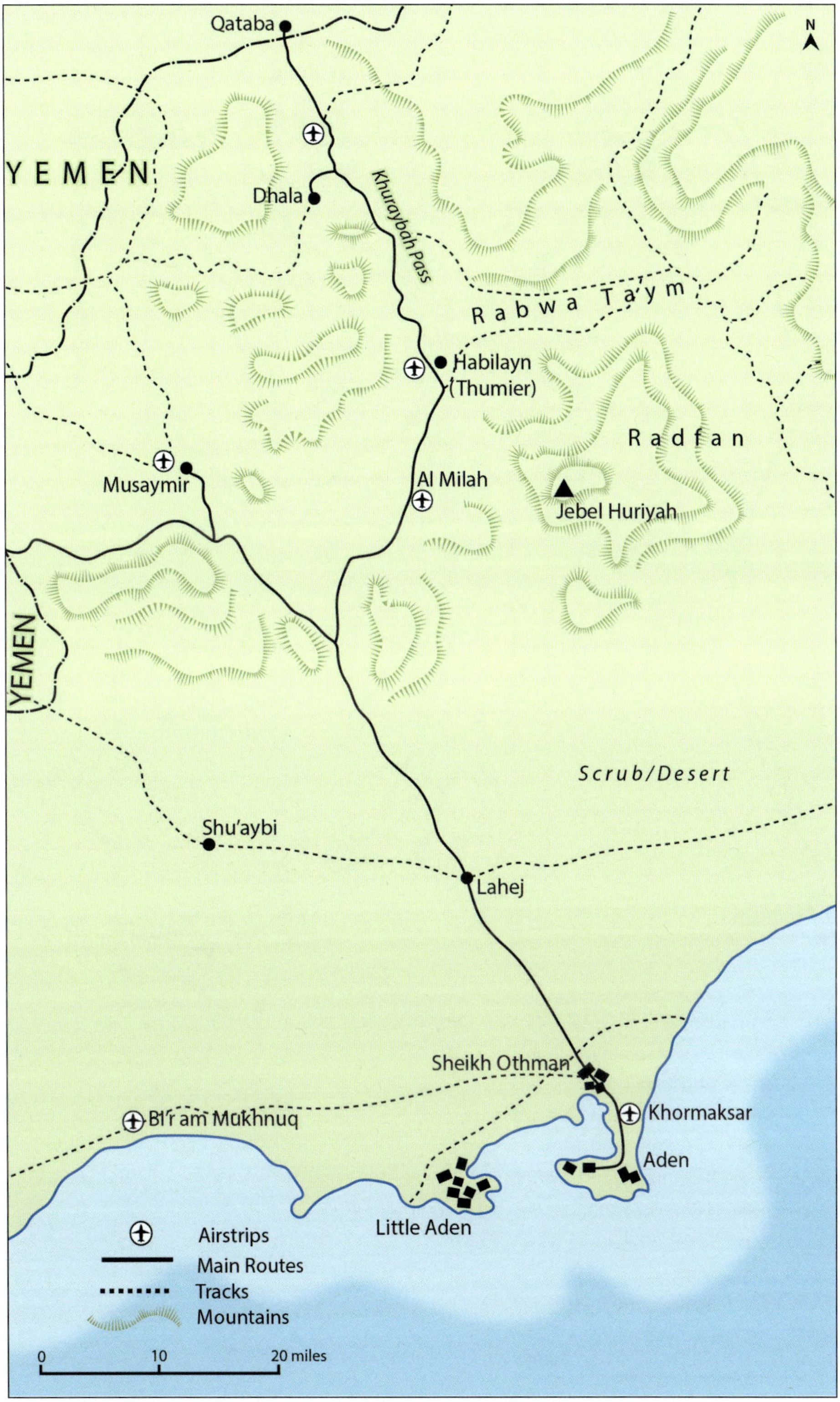

South Arabia.

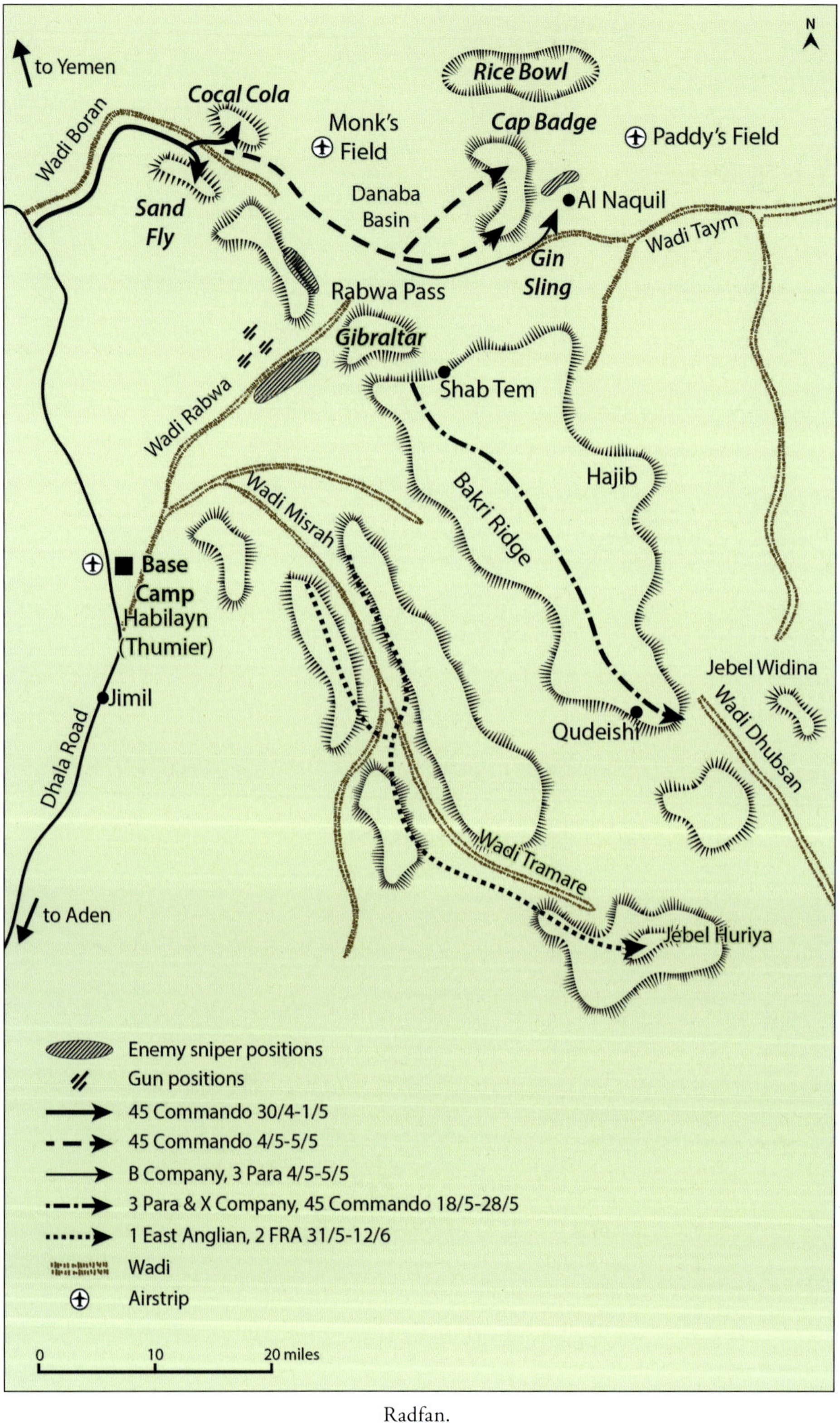
to Yemen
Cocal Cola
Rice Bowl
Monk's Field
Cap Badge
Paddy's Field
Wadi Boran
Sand Fly
Danaba Basin
Al Naquil
Wadi Taym
Gin Sling
Rabwa Pass
Gibraltar
Shab Tem
Wadi Rabwa
Hajib
Bakri Ridge
Wadi Misrah
Base Camp
Habilayn (Thumier)
Jimil
Dhala Road
Jebel Widina
Wadi Dhubsan
Qudeishi
Wadi Tramare
to Aden
Jebel Huriya
Enemy sniper positions
Gun positions
45 Commando 30/4-1/5
45 Commando 4/5-5/5
B Company, 3 Para 4/5-5/5
3 Para & X Company, 45 Commando 18/5-28/5
1 East Anglian, 2 FRA 31/5-12/6
Wadi
Airstrip
0
10
20 miles

Radfan.

Aden and Crater.

British Guiana.

N
MEXICO
MEXICO
Orange Walk
BELIZE
Belize R.
Sibun R.
Belize City
Turneffe Islands
Blue Hole
Lighthouse Reef
Belmopan
San Ignacio
Melchor de Mencos
Mountain Pine Ridge
Stann Creek
Victoria Peak
Barrier Reef
Glover Reef
GUATEMALA
Maya Mountains
Caribbean Sea
Barrier Reef
Punta Gorda
HONDURAS
0 10 20 30 40 miles

Belize.

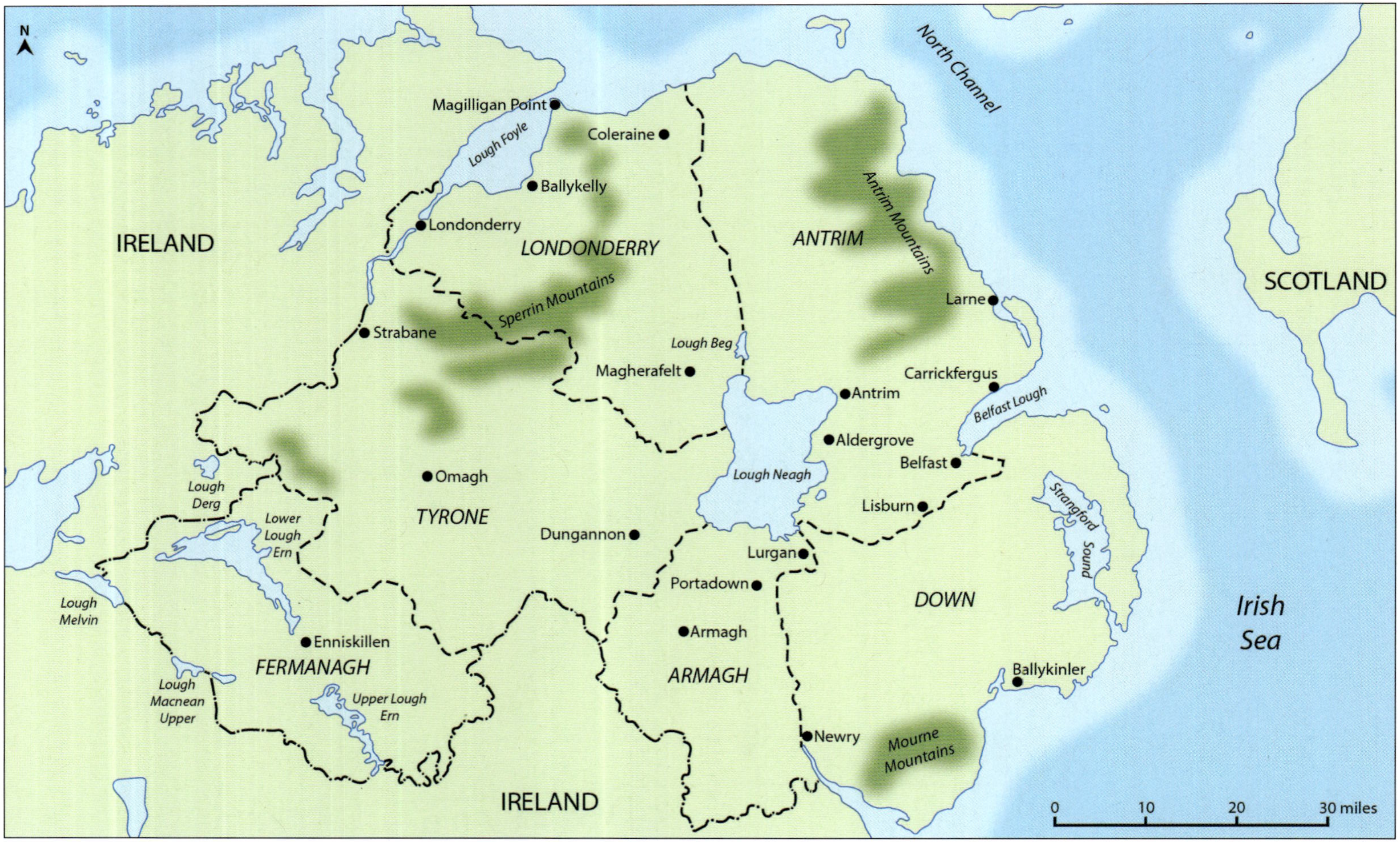

Northern Ireland.

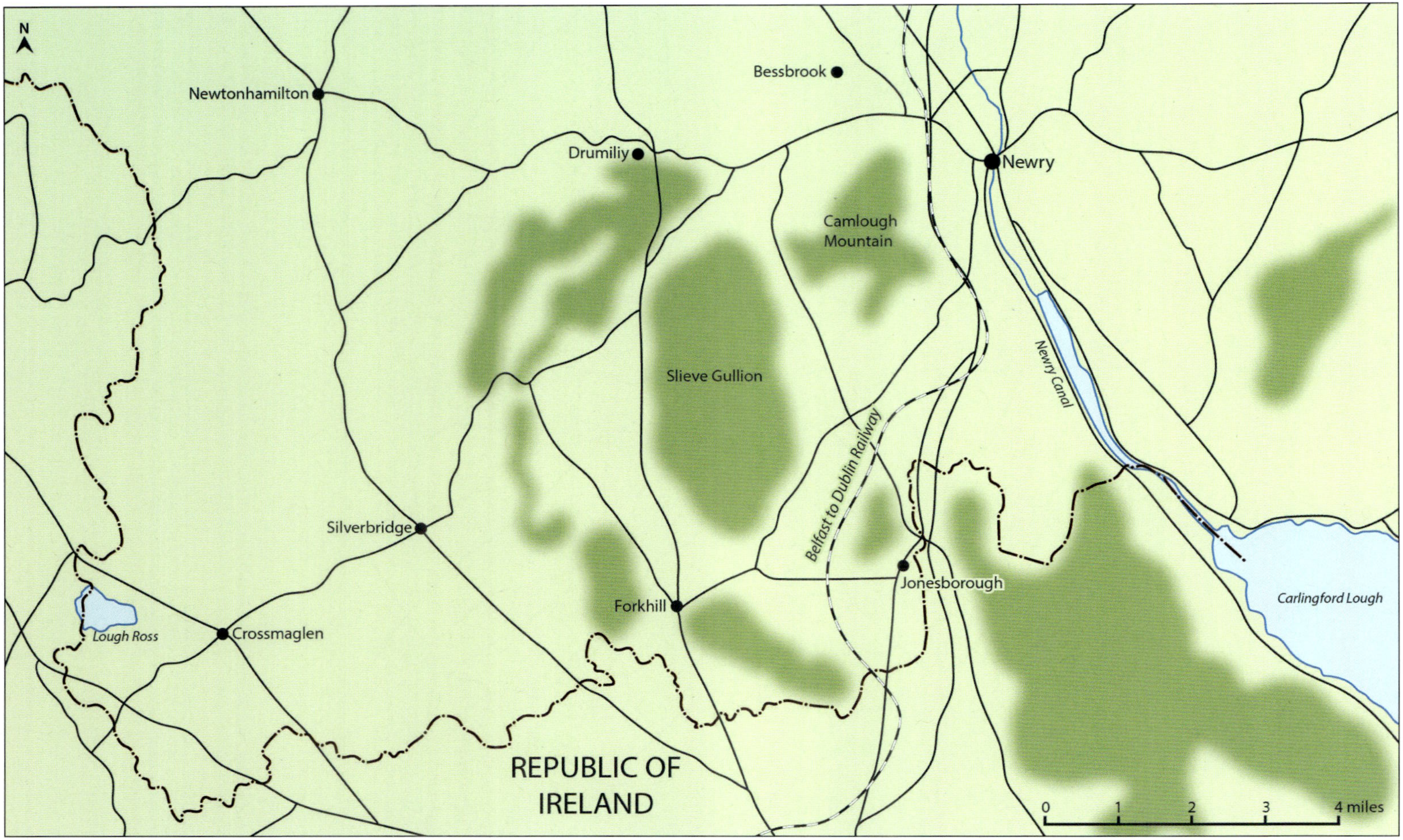

South Armagh.

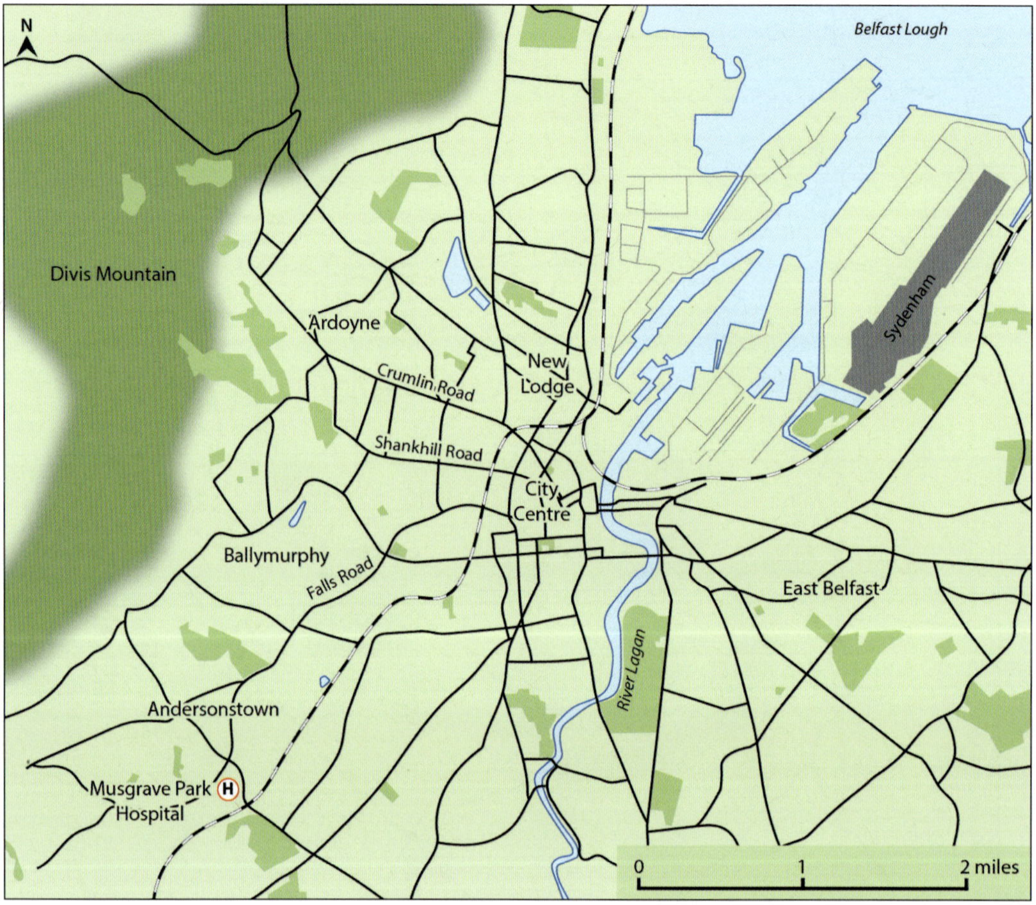
N
Belfast Lough
Divis Mountain
Ardoyne
New
Lodge
Crumlin Road
Shankhill Road
City
Centre
Sydenham
Ballymurphy
Falls Road
East Belfast
River Lagan
Andersonstown
Musgrave Park
Hospital
H
0
1
2 miles

Belfast.

Major Brooke also had a front seat view as he was sent out from 16 Flight on a reconnaissance mission as the Turkish invasion began:

> Soon after sunrise I was detailed off to go and have a look at what was happening in Famagusta. It was thought the Turks were trying to establish a second beachhead on the famous Golden Mile. If so, and the Greeks resisted, the overs and unders were likely to land in our camp. The [ESBA] Station commander wanted to know the facts.
>
> Sgt Broome of the Intelligence Corps was my observer, armed with a clipboard full of questions about Famagusta. As we climbed away eastwards from Dhekelia we had a near miss with something hurtling south. From my left-hand seat I instantly lost sight of it and asked Sgt Broome "What the hell was that?" "I don't know Sir" he said looking over his shoulder "but it's coming back!" With that a Turkish [Northrop] F-5, vapour pouring off his wings, started circling us – there was suddenly a shark in our swimming pool! I wasn't sure he would identify me correctly so I dived into a dried up riverbed. As he flew around us I did a spot turn to keep him in view, hoping to leap sideways if he lined up on us. He soon got bored and disappeared. So we went on to Famagusta, passing over the Golden Mile at about 3,000 feet, on the premise that at that height a Bell 47 would be almost invisible to anyone on the ground. Any lower and both sides would probably have a pop at us. We quickly saw that no invasion was taking place, indeed the sea was full of wakes of ships escaping in all directions. But because there was a lot of smoke at the northern end of the town it was clear that something was going on.
>
> I flew out to sea to get a better view when another F-5 appeared, this time in a shallow dive heading towards the top end of the Golden Mile. As it passed a few hundred metres north of us the F-5 let go with its rockets. I followed them down. They hit the northernmost hotel, which was promptly engulfed in a huge cloud of smoke and dust. It later transpired that Greek soldiers were firing from the hotel rooftop into the fort in the old Turkish town of Varosha, just to the north. It also showed that the Turks were well prepared for this conflict, being able to call in an airstrike well ahead of the advancing army. They had planted a Forward Air Controller on the ground.[41]

The rest of 16 Flight were also heavily involved in the repercussions of the invasion as their airstrip in Dhekelia became a focal point for the evacuation of foreign nationals, as recorded by Captain Tony Hayhurst AAC.

> At the first indication of an invasion a Sioux was sent from Dhekelia to Episkopi in the Western Sovereign Base Area (WSBA), to assist the Royal Scots in the evacuation of families from Limassol, a dormitory town. It was thought that the National Guard, in the confusion of the first war hours, might bar the legal (under the 1960 Treaty terms) convoy from entering the town. The Scots, together with the RAF Police and UN soldiers, needless to say made an impression and the evacuation was soon completed. The Sioux returned to Dhekelia along a now familiar route five miles out to sea. This first time it was greeted with the radio warning "Stay low, Turkish aircraft in vicinity." One was beginning to feel like a spectator of the war.
>
> As the roads were closed between the bases, helicopter liaison sorties were of paramount importance and this was to be the case for the next few weeks. An additional task between the bases was the carriage of documents and specialist equipment. On one occasion it was

41 H. Brooke, account to the authors, 4 October 2022.

necessary to take an underslung load by the sea route 55 miles to the WSBA, a far cry from the escorted short sea crossing from West Freugh to Aldergrove.

With the first bombs landing in Nicosia, a cry of "the Turks are coming" and a fishing boat becoming another Armada, the population was on the move. It is important to mention here that the main road from Famagusta to Larnaca passes right through Dhekelia and the Eastern Sovereign Base Area (ESBA), and it became a Flight task to assist in checking on the volume and type of movement through the base. There was a variety of National Guard and EOKA-B, together with refugees from the Larnaca area afraid of what might happen and those from Famagusta which had been bombed and strafed. It was only from a helicopter that a true picture of the refugee situation could be obtained; the numbers, whereabouts and living conditions. Cape Greco, an outpost on the south eastern tip of the Island, had to be reinforced and a daily water ration flown out to support fifty men. This was later taken over by RAF Pumas.

With the arrival of the Nicosia convoy at Dhekelia, the Officers' Club became known as the American Club, for it housed nearly 500 American nationals awaiting helicopters from the 6th Fleet to airlift them off Cyprus. 16 Flight's strip was used and all personnel found their vocation as the refugees poured into the Flight to be briefed, formed into groups and loaded together with their baggage into the aircraft. It took the [Sikorsky] CH-53 and [Boeing Vertol] CH-46 six hours to clear the 500.[42]

Once he returned from his hair-raising recce sortie to Famagusta Major Brooke became heavily involved in the management of the evacuation of US citizens from Dhekelia:

Eventually at dusk a Jolly Green Giant (CH-53) arrived. I was hatless holding a clip-board containing 150 names when it rolled to a halt just 20 metres from where I was standing. I was startled to notice I was looking down the barrel of a half-inch Browning machine gun poking out of the front side door. The rescue mission commander, a 'Bird Colonel' with antennae sticking out of everywhere jumped out, ran over to me, saluted and said "Say Captain, is this strip secure?" to which I replied "It was until you turned up!" Fortunately, his sense of humour prevailed and he said into his handset "OK Folks, you can put 'em away. The Brits have it secured."

I asked him how many passengers he could take at a time. "Fifteen and their baggage" was the reply. So, I called across to our Flight Sergeant who trying to control the 150 eager evacuees, "They can take fifteen at a time!" We had already set out a row of chairs in anticipation of this moment. When the number of chairs had been adjusted to 15, he invited the first group to sit down. They were supposed to come forward in alphabetical order, but the stampede that followed caught us by surprise. It took a while to usher all those left standing back into the hanger.

When the dust settled, we found we had 16 people on chairs – a Mum and Dad had seats but their little Johnny didn't and was standing beside them in tears. However, one of the sponge divers had got himself a chair was kind enough to offer to go on the next flight. Those fifteen then collected their baggage and boarded the helicopter.

That pattern was repeated nine times until the last flight, when all three sponge divers eventually left together. Interestingly they broke rank on their way to the helicopter, came over to me, shook me by the hand and thanked me for getting them out. They were the only people to do that.[43]

42 A. J. Hayhurst, 'The Cyprus Emergency', *Army Air Corps Journal,* 1975, pp.35–36.
43 H. Brooke, account to the authors, 4 October 2022.

A ceasefire was brokered and came into effect at 1800 hours on 22 July 1974. The timing of the invasion meant that hundreds of international tourists were by then trapped in the Turkish occupied area around Kyrenia. Their prompt evacuation became the priority to be carried out by an RN carrier task force led by HMS *Hermes* positioned off the north coast of Cyprus. The shore party to organise the evacuees was to be provided by members of UNFICYP and led by Major Gill. His multinational group, heavily equipped with various communications sets, departed at dawn on 23 July heading for the north coast. Having negotiated his way through the front lines Major Gill's priority on reaching Kyrenia was to make radio contact with HMS *Hermes.* As he approached a prospective HLS he was 'delighted, if surprised' to hear an officer from *Hermes* call up on his radio net:

> I stopped my vehicle so that I could concentrate on talking to the Commander who was booming through, and I wondered how the Navy could have managed to pick up and break into our net. We spoke for five minutes, before a Finnish officer who had been gesticulating and trying to catch my attention throughout the conversation managed to convey his message. We were speaking on the radio over a distance of roughly ten metres, the Commander having flown in by Wessex and joined the Command Post party which was only one bush away. Sheepishly I walked around the bush and met the Commander who was to be the Liaison Officer for the evacuation. I flew out to HERMES to discuss the operation and, in spite of some alarms about the beaches being mined, the evacuation was soon under way.
>
> We managed to get some of the Kyrenia tourists away by launch but, as the sea was fairly rough and we seemed to have sufficient helicopters, the bulk of the ferrying was done by air. We considered that it was essential to have a UN presence at all the pick-up points so radio vehicles moved to the main concentration areas whilst other vehicles located and gathered together the smaller and more scattered groups. Everything went very well and my only complaint was that on occasions an enthusiastic pilot would ignore or fail to see our marker panels, indicating a relatively dust free pick-up-point, and fly to a point as near to the tourists as possible, sometimes on the side of a cliff and invariably covering us all in sand.[44]

By 1700 hours on 23 July Major Gill's team had assisted in the evacuation of 1,630 civilians out to the safety of the carrier group. He was under strict instructions not to evacuate any local inhabitants or personnel, whether Greek or Turkish. However, as the day drew to a close it was clear that there were a number of Greek Cypriots and Greek tourists who were trapped behind the Turkish front line around Kyrenia. They were understandably nervous about their position and Major Gill thought it prudent if his small team stayed in the area in order to attempt to guarantee their safety. UNFICYP HQ agreed and told him to remain in position until he could be relieved by a force from the Finnish contingent. Major Gill set up his HQ in the four-star Dome Hotel, declaring it UN territory and running up the UN flag.

Major Gill's team was joined by around 600 civilians: foreign and Greek tourists, and Greek Cypriots. It was clear that among the latter group there were a number of EOKA-B members and supporters, some of whom had taken part in the fighting. Major Gill was concerned that their presence might be made an excuse for the Turkish forces to attack the hotel or that they might represent an internal threat themselves. It took eight days to negotiate and arrange the evacuation of the foreign tourists via their own embassies and consulates. The remaining 40 Greek nationals were taken into custody for interrogation. Still Major Gill's team remained in Kyrenia to try to reassure the Greek Cypriot population. Patrols were carried out and during the night intermittent

44 Gill, 'Island of Tragedy', pp.31–32.

gunfire was heard close by as the Turks conducted mopping up operations. It became increasingly difficult for the civilian population to find food and medical supplies and looting became prevalent. On 28 July the Turkish Army handed over to a civil administration who immediately told Major Gill that all UN troops were to leave the Turkish controlled area. He declined, pointing out that his orders came from New York not Ankara. The food and medical situation became worse, and the population was soon relying on supplies distributed by the International Committee of the Red Cross and the UN High Commission for Refugees.

Major Gill, still waiting for relief, resisted Turkish demands to leave the area until 14 August. During the period he had been working around Kyrenia, politicians representing the Greek and Turkish Cypriot communities had been engaged in talks in Geneva. The Turkish side demanded that a contiguous 30 percent of Cyprus should be handed over to the Turkish community. This was rejected by the Greek negotiators on the grounds it would lead to population displacement and permanent partition of the island. On 10 August the talks broke down. In response, on 14 August Turkey launched Operation Atilla 2. The north of Cyprus was reinforced from Turkey and a new offensive was launched with the aim of seizing and securing all ground north of the 'Atilla Line' which the Turks drew from Morphou in the west to Famagusta in the east. It was at this point that Major Gill's position in Kyrenia became untenable and he was forced to withdraw.

> Once again we ran the gauntlet from Kyrenia to Nicosia for this was the start of the second offensive. 30,000 men, 300 tanks and complete freedom of the air allowed the Turkish Army to secure in a few days 40% of the island. They did not have it all their own way for we were again privileged to see some very determined and courageous resistance but, as in the first phase, coordination seemed to be lacking.[45]

The renewed Turkish advance put the ESBA in a precarious position immediately south of the Atilla Line. 16 Flight AAC was given the task of monitoring the Turkish advance to ensure the integrity of the ESBA was maintained.

> The Turkish forces broke out of Nicosia and the Kyrenia salient on August 14th and it was clear that they intended to consolidate along the Atilla Line. 16 Flight was given the responsibility of reporting the Turkish forces' positions as they advanced to Famagusta and Pergamos until they were in contact with the British ground forces. Then it became a question of using helicopters to indicate to the Turkish armour and infantry the boundary of the ESBA.
>
> The observers were of great value and although they at first, even with stabilised binoculars, found reporting detail at a distance difficult, this improved rapidly. However, we still found that only within 2,000 metres could a positive identification be made in the confusion of the advance. This was indeed a unique opportunity to observe an army on the advance, with the fly line at times only a few miles away, except when they reached the SBA, when they were at a mere 100 metres distance.
>
> By August 18th the Turkish forces had reached the ESBA boundary and Famagusta. This was in fact a more southerly line than the projected Atilla Line. A daily task of checking on the Turkish positions and movements along our boundary became essential for, apart from an anticipated dash to Larnaca, they were prone to cut their journeys short by crossing the SBA at various points. Slowly this settled down, until a minefield was found being laid along the north edge of the SBA and its exact location had to be plotted. It was at this stage that we had to fly three serious casualties to BMH [British Military Hospital, Dhekelia]. The various tasks

45 Gill, 'Island of Tragedy', p.33.

> mentioned to date continued, but slowly diminished as mid-September approached and the emphasis was shifted to liaison and photographic sorties.[46]

The approach of the Turkish forces close to the ESBA northern boundary was a cause of concern and a careful eye was kept on their front line until their intentions became clear. That reconnaissance and surveillance activity was recalled by 16 Flight's QHI, Staff Sergeant John Scott AAC.

> The last phase of the Turkish advance brought Turkish forces right up to the SBA boundary. Little was known about the Turkish tactics and their positions.... We hoped that they would recognise the SBA borders and not cross into British territory. In the event the Turkish Army approached the SBA as a column of Infantry mounted on soft skinned vehicles. Very quickly the Turks were mounting patrols around the SBA boundary, never intentionally crossing into the SBA and building three man dugouts every 500m around the border. We had occasional contact with the Turkish duty officer who regularly drove around the boundary inspecting the dugouts.... Although the Turks respected the boundary, and once they had completed the capture of Famagusta and surrounding areas, they made no further advance there was no certainty at the time that this would be so. The actual advance and subsequent redeployment had to be closely and continuously monitored and this could only be achieved effectively by helicopter reconnaissance flights.
>
> Tension was high and aircraft and pilots were fully stretched and it was imperative that reconnaissance helicopters did not stray into what had become Turkish airspace. The pilots, technicians and aircraft were immediately thrown into this daunting task. We flew repeated sorties by day and night in conditions demanding a very high standard of airmanship and navigation, and with some attendant danger. Speed and accuracy of reports were vital, however that reconnaissance role was second nature to me as I had a background of reconnaissance, serving with 5th Royal Tank Regiment in the armoured reconnaissance role and subsequent to my flying course, in the regimental Air Squadron.[47]

The Turkish advance was swift if haphazard and tens of thousands of Greek Cypriots fled to the south ahead of the tanks and infantry. A large proportion of the population of Famagusta ended up in the ESBA being administered by British military forces. Late on 15 August a ceasefire was brokered by which time the Turkish forces had largely reached the Atilla Line occupying 37 percent of the island. UNFICYP Flight and 16 Flight had come through the invasion unscathed but that changed during a sortie being flown by Captain Hayhurst in the wake of the invasion.

> I was to fly early morning to the GOC's residence in Episkopi and fly him to Dhekelia. Peter Child wanted me to leave a little earlier and drop him off at RAF Akrotiri as he had to visit the hospital, and then proceed to pick up the GOC. At this stage we were now allowed to fly the direct route overland, as opposed to over the water during the crisis. What we did not realise was that the Greek National Guard had set up a small base on the high ground, halfway to Akrotiri, which happened to be below our flight path. They decided to open up with a machine gun, rather like those on a game shoot do in Hampshire. One bullet entered low and slightly left of centre of the Perspex bubble and hit Peter Child [in the thigh]. He thought it was a bird strike, but it clearly was not. I landed the aircraft to look at the injury, dress it, and check the aircraft for any damage. Just after I landed a number of the Greek National Guard

46 Hayhurst, 'The Cyprus Emergency', p.36.
47 J. Scott, account to the authors, 2 March 2023.

> turned up, probably the ones who had fired at us and asked if all was well. I said everything was fine and we took off for Akrotiri. I called up Akrotiri and reported the incident and injury. Having dropped off Peter Child I went to the GOC's residence and he was waiting by his helipad. He had already been informed. There was a fair amount of blood splattered over the cockpit, so he sent for someone to help clean it up. Then we went off to Dhekelia.[48]

Nevertheless, incidents such as the wounding of Major Child were isolated and by November the requirement for reconnaissance sorties had fallen away, although liaison flying remained intense throughout the following year. It was about 18 months before the flying rate and profile returned to normal. The initial Turkish invasion had caused the fall of the junta in Athens and put Greece back on the path to democracy. Sampson was also forced to resign and was replaced temporarily by Glafcos Clerides who filled the post until the return of Makarios on 7 December 1974. On 13 February 1975 Turkey declared the occupied area of Northern Cyprus to be a Federated State of Turkey. Thousands of Turkish Cypriots were transferred from the Cypriot south to the Turkish north. The feared partition of the island became a reality ensuring that UNFICYP would have a role in Cyprus for decades to come. Major Gill was made a MBE for his work with UNFICYP around Kyrenia and the last word on the events of 1974 goes to him:

> My own impressions and recollections vary from despair to exhilaration. The frustration, as a soldier, of trying to act without power or recognition; shock and horror at the carnage, looting, and loss of life, and wonder and admiration for the resilience and courage of some of the innocent civilians who suffered. It was a unique and rewarding experience, not unlike being a member of the Directing Staff or an umpire on a major exercise, feeling relatively secure and almost immune in spite of live ammunition being used.
>
> On the aviation side we were allowed to carry out some interesting reconnaissance and observation tasks specifically denied to us in the normal mandate. We were frequently threatened and occasionally shot at, which was not surprising since the Greek National Guard felt a great deal of animosity towards aircraft, having been bombarded and strafed with impunity throughout the hostilities…
>
> I do not believe that the partition will work, but it is difficult to be constructive and offer a workable solution whilst so many outside agencies are involved. Co-existence with integrated communities, both Greek and Turkish Cypriots having equal representation at all levels of Government, is the ideal answer, but first the Greek Cypriots must solve their own internal problems and declare a common agreed policy. The tragedy is that the majority of real Cypriots, both Greek and Turk, if allowed to express their true feelings, would probably beg for the opportunity to solve their problems without external interference. Once again human suffering is accepted in the cause of political expediency.[49]

See Map 6

Operation Tosca

16 Flight and UNFICYP Flight AAC gradually settled into the new normal situation on Cyprus. The UNBZ, which now firmly separated independent Cyprus in the south from the unrecognised Turkish federal state in the north, was extended and strengthened. British forces deploying as members of BRITCON to UNFICYP now did so under Operation Tosca. The roles and tasks of

48 A. Hayhurst, account to the authors, 1 March 2023.
49 Gill, 'Island of Tragedy', p.35.

the two flights broadly returned to those which had existed prior to the Turkish invasion. It took three years however, before British Sovereign troops were able once again to exercise outside of SBAs. 16 Flight supported what was only the second such exercise with 3 R Anglian in December 1977.

A few months previously the Alouette had made a return to Cyprus. Four Alouettes flew into Dhekelia strip for 16 Flight on 10 October 1977, 15 years after they had departed from 10 Flight AAC. Four further Alouettes also arrived for UNFICYP Flight. Two of the aircraft were actual returnees. XR385 and XR386 were originally assembled new from their packing cases at RAF Akrotiri by members of 10 Flight in November 1961. Although the Alouette was disappointing in the mountains due to its restricted payload at altitude, its usual speed, and passenger and freight capacity in comparison with the outgoing Sioux made its return to Cyprus a welcome one. It also had an improved serviceability rate over the Sioux and so it was judged that extra tasking would be possible, even taking over some of that usually directed to 84 Squadron RAF.[50] On 1 January 1978 BMH Dhekelia permanently closed its doors and 16 Flight took on the added role of medevac of seriously ill patients direct from the ESBA to the Princess Mary's Hospital (TPMH) in Akrotiri. It was estimated that events necessitating evacuation by helicopter would only occur four times a year on average. In the first 12 months of conducting the task medevac sorties were flown 38 times, 24 of them at night.[51] The task required some ingenuity and modification to the new Alouettes.

> This necessitated producing a capability of carrying a patient internally. This has been done by removing the passenger seats, putting a mattress along the port side of the cabin, modifying a Land Rover rear seat to fit behind the pilot for a medical attendant and producing oxygen bottle holders, an intravenous drip holder and a means of securing a foot operated sucker. By the end of January three successful Medevacs had been carried out, all well within the guaranteed time of 1¾ hours from the initial call out to arrival at TPMH.[52]

By 1980 UNFICYP had been operating on the island for over 15 years. At that time the force stood at approximately 2,500 personnel from eight contributing states. Over half the force was provided by BRITCON. The force looked after five sectors along the UNBZ numbered I to VI. Sector III had been covered by the Finnish contingent but when it was withdrawn in 1977 its sector was absorbed into the neighbouring sectors. There was still a Finnish contingent in 1980, but only of 11 men. BRITCON provided two-thirds of an infantry battalion, an armoured recce regiment, a support regiment, the aviation squadron, a RAF detachment operating Whirlwinds, and staff for UNFICYP HQ. At the time Major Nigel Thursby AAC was OC UNFICYP Flight AAC and described the Flight's establishment, tasks and routine:

> The UNFICYP Flight Army Air Corps forms part of the Support Regiment. Although tasked direct by HQ UNFICYP, day-to-day administration and discipline are covered by RHQ Support Regiment [Regimental Headquarters Support Regiment]; similar arrangements operate with the sappers, signals, transport, medical, ordnance and workshop detachments and squadrons, all of which are members of this unique regiment. The Flight consists of four pilots, seven REME personnel and seven ground crew/GD personnel; the GD personnel still being provided by infantry and armoured regiments...

50 AFM 259: Letter from HQ BRITCON to MOD, 'Deployment of Alouettes to UNFICYP', May 1977.

51 *The Lion; British Services Cyprus Weekly*, 16 February 1979, p.7.

52 Anon, 'The Return of the Alouette to Cyprus' and S.P. Waters, 'Alouette, Gentille Alouette', *Army Air Corps Journal*, 1978, pp.76–77 and 78–82.

An Alouette of UNFICYP Flight approaching HLS D12 on a liaison sortie to the Danish Contingent in the Troodos Mountains in 1979. (Army Flying Museum)

> ...The Flight is located on the old Nicosia International Airport, now completely disused except for 4 Alouettes and 4 Whirlwinds of 84 Sqn RAF. This combined force supplies all air requirements for UN tasks which are wide and varied and carried out daily in varying types of terrain and climatic conditions varying from 0°C in Winter to +40°C in Summer and at altitudes ranging from sea level to 6,000 feet.
>
> The task of the Alouettes is mainly that of reconnaissance in the buffer zone and the carriage of Operations, Economics and Intelligence officers carrying out their various roles. Because of distances, terrain and problems of crossing borders the helicopter provides an ideal platform from which these tasks can be carried out quickly and efficiently.
>
> The Buffer Zone, the area between the Turkish and Greek ceasefire lines is some 180km long and varies in width from 5 metres in the streets of Nicosia to 8 kilometres out in the open plains. Along the Zone are scattered some 132 observation posts, 65 of which are manned with an HLS at each site. The Zone is broken down into five sectors each manned by a different nationality with six 'air gates' for the passage of helicopters passing into the Turkish occupied northern part of the island. Flights in the north are rigidly controlled and routes flown strictly adhered to, clearance is generally required 24 hours in advance (except in emergencies) and crossings may only be made via these 'air gates' with appropriate calls to ATC stationed on Nicosia airfield...
>
> Having replaced the popular Sioux in early 1978, the Alouette quickly gained the respect of pilots, engineers and passengers alike. In spite of its nineteen years of service – some of our aircraft have over 6,000 hours on them – the Alouette II continues to prove itself to be an extremely reliable and rugged helicopter. The fact that we frequently have 100% serviceability speaks for itself.[53]

The 'new normal' of operating over the partitioned island of Cyprus continued throughout the 1980s for both UNFICYP and 16 Flights AAC. UNFICYP Flight's breadth of tasking increased as it took on some of the resupply tasking (known as Task C) from the Whirlwinds of 84 Squadron RAF. The Alouettes used their outboard Siren litters, designed for carrying casualties, to lift rations, water and mail to some of the more inaccessible UN OPs. 16 Flight's expanded medevac remit resulted in an unusual event on 23 June 1981:

> Major [Michael] Schofield AAC broke the rules of authorisation and figuratively took on an extra passenger mid sortie. Three and a half years after 16 Flt undertook the Medevac commitment, and several near misses later, the first baby was delivered en route from the MRS at Dhekelia to the TPMH at Akrotiri.[54]

A year later 16 Flight had another unexpected visitor. On 21 July 1982, much to the surprise of Lance Corporal Chambers AAC who was on duty outside the Flight's main hangar, a large twin-engine aircraft touched down at Dhekelia and hurtled down the very short runway. The 50-seat Antonov 24 from Rumanian Airlines (TAROM) continued off the end of the runway and eventually stopped just 10 metres short of the perimeter fence. Miraculously none of the 40 passengers or eight crew were injured during the landing, although one woman was taken off the aircraft and

53 N. Thursby, 'Cyprus – the Divided Isle', *Army Air Corps Journal*, 1980, pp.46–47 and Anon, 'UNFICYP Flight – March to May 1980', *Army Air Corps Newsletter*, No.95, September – October 1980, p.55.

54 Anon, 'AAC Flight UNFICYP', *Army Air Corps Newsletter*, No.99, May-June 1981, p.30 and Anon, '16 Flight AAC – Apr-Jun 81', *Army Air Corps Newsletter*, No.101, September-October 1981, p.42.

transferred by ambulance to nearby medical facilities. A conversation with the shaken crew soon revealed the cause of the incident:

> One passenger, who was taken away in an ambulance, was having difficulty breathing and required oxygen. The oxygen supply was running out and the Captain was desperate to land as soon as he could. He was in radio contact with Larnaca and approaching the edge of the SBA on a heading of 210°. Larnaca turned him on to a new heading of 216° and asked him to report "the airfield in sight". This new heading took him close to Dhekelia and in his anxiety to land as soon as possible, he mistook our airstrip for Larnaca International. We are very flattered! Neither the Captain nor the 2nd pilot had landed at Larnaca before.[55]

With disaster narrowly averted the next problem was what to do with the errant Antonov. Dhekelia's runway was just 400 yards long, not nearly long enough for the aircraft's standard take off requirements. There were three options. The Antonov could be moved to another airfield with a longer runway. That would however, entail cutting down trees and removing telegraph poles and, in any case, there was no vehicle on the island capable of transporting the aircraft. Second, the aircraft could be completely disassembled and moved elsewhere by road but understandably, TAROM was not keen to undertake that option. Therefore, the aircraft would somehow have to be flown off the short strip.

The chief test pilot for TAROM was flown to Dhekelia with a crew who conducted engine and brake tests. All non-essential fixtures and fittings were removed from the aircraft but it was still doubtful whether it could get airborne within the length of the runway. The Commander British Forces Cyprus, Air Vice Marshal Robert Davis, consulted the CAA who were reluctant to recommend any attempt to take off by the Antonov. In order to convince the CAA, TAROM conducted trials with a similarly rigged aircraft in Bucharest. When the tests proved successful the CAA did grant authority provided the temperature was below 29°C and there was 10 knot head wind component. During the late afternoon of 30 July, the conditions looked suitable and Air Vice Marshal Davis arrived at Dhekelia by Wessex to give his formal permission for the departure to be attempted. Before he could give his seal of approval however, the Antonov's engines roared into life and the aircraft hurtled down the runway, successfully lifting into the air after less than 300 of the runway's 400 yards had been used.

The following year UNFICYP Flight suffered an unfortunate and thankfully rare aircraft accident. On 11 June WO2 Tony Davies AAC, the Flight QHI, was flying an underslung load of jerry cans from a Danish OP in the Troodos Mountains in Alouette XR380. As he lifted from the HLS loose netting snagged itself on the Alouette's port skid. Aware that something was wrong WO2 Davies ejected the load which transferred all the weight of the entire 650lb load to the port skid which in turn flipped the aircraft and sent it 200 feet down the side of the mountain. XR380 was written off and WO2 Davies suffered a broken back and a broken nose but went on to make a full recovery and was commissioned the following year.[56]

International events within Cyprus and across the wider Middle East had an effect on the operations of both UNFICYP and 16 Flights. In June 1982 Israel invaded Lebanon. As the Israeli Defence Force began to withdraw three months later the UN Interim Force in Lebanon, which had been in the country since 1978, was unable to cope with the escalating violence left in the vacuum caused by the Israeli Defence Force's departure. In September 1982 a contingent of US

55 Anon, 'Air Incident at Dhekelia Airstrip, 21 July 1982', *Army Air Corps Journal,* 1983, pp.31–32.

56 Anon, Army Air Corps Aircraft Losses Since 1 September 1957, HQ AAC, 13 June 2019 and Anon, 'Army Flying Profile – Capt A.J. Davies AAC', *Army Air Corps Journal*, 1987, p.57.

The wreckage of Alouette XR380 below a Danish OP in the Troodos Mountains from which WO2 Tony Davies escaped with serious injuries on 11 June 1983. (K. Kirkaldy)

Marines landed in Beirut, ostensibly to protect US interests and infrastructure, in particular the US embassy and the international airport. Beirut was only 60 miles from Cyprus' southern coast across the Mediterranean and the ripples from Lebanon were felt in Dhekelia. The US Navy's amphibious assault ship USS *Iwo Jima* was stationed off the Lebanese coast during 1983 and when two of its CH-53 Sea Stallions required maintenance they flew into 16 Flight's Dhekelia base. The two aircraft arrived with a large maintenance team to replace blade sleeves, which it was noted, were about the same size as the Alouette's main rotor gearbox. On 23 October 1983 suicide bombers attacked the US Marines' main headquarters at Beirut Airport and the headquarters of the French Army contingent in the city. 243 US Marines and 58 French soldiers were killed in the attack and hundreds more were injured. Many of the injured were transferred to TPMH and 16 Flight assisted in the incident by flying medical staff from Dhekelia to Akrotiri.[57]

On 15 November 1983 Northern Cyprus declared its independence under the name the Turkish Republic of Northern Cyprus. The change of status was rejected by the UN through UNSCR 541. The following year, 1984, marked 20 years since the Green Line had split the island in two and UNFICYP was formed, and the tenth anniversary of the Turkish invasion. To mark the occasion many Greek Cypriots organised protest marches along the edge of the Buffer Zone. Throughout the first half of July groups such as the Kyrenia Refugee Club mounted protest marches along the southern edge of the Green Line, usually finishing in Kyrenia. Ensuring the security of the marchers and that the events remained peaceful was the responsibility of the Cypriot police. UNFICYP was stood by to provide UN civil police, military police and other troops to support the Cypriot police if necessary. UNFICYP Flight received orders to remain on standby to conduct

57 Anon, '16 Flight AAC (Oct-Dec 83)', *Army Air Corps Newsletter,* No.116, March-April 1984, p.29.

observation and surveillance and to move key personnel if necessary. In the event the protests passed off peacefully although were subject to the usual misreporting in the Greek Cypriot press.[58]

UNFICYP Flight's work continued to require the negotiation of the fragile situation that existed between the two sides of the island. Both the Greek and Turkish sides were quick to make the most of any real or perceived violation of the longstanding UN brokered ceasefire.

> A cease-fire violation does not have to be shots fired; or indeed overt violence of any sort. The introduction into a slit trench of a 12.7mm MG where only SLRs were previously deployed constitutes a violation; as does introducing concrete and reveting where only earth banks previously existed. It is permitted under the rules for example to repair storm damage but not to replace old mines with new ones.
>
> Most of the UN's task is covered by dealing with alleged violations and subsequent protests and the workload varies from Sector to Sector. In the British Sector, for example, the cease-fire lines are wide apart and the UN has made intensive efforts to persuade both sides to farm the Buffer Zone to their own mutual advantage. This has been successful to a degree, but the subsequent perambulations of tractors and ploughs have to be carefully monitored to ensure that they stick to the areas allotted and indeed stay out of the minefields. There are several rusty tractor wrecks along the Zone that bear witness to this danger.[59]

To encourage and monitor the return to agriculture, each BRITCON infantry battalion trained and deployed a small team of farming supervisors. One of the less obvious dangers that these teams faced within the Buffer Zone was the packs of feral dogs that lived and roamed there. The dogs were a collection of breeds with a high proportion of Alsatians, Dobermans and Pointers, and would frequently kill farm animals up to and including cows. They also occasionally attacked lone farmers or UN personnel out jogging. On top of the danger of serious injury, if not death, the dogs also carried a rabies-like disease called hydatid, which carried a high risk of being fatal. The Cypriot farmers could not deal with the dogs themselves as they were forbidden to carry any firearms within the UNBZ. The task of controlling the problem therefore fell to UNFICYP and the easiest way to deal with it was via UNFICYP Flight's Alouettes. British infantry officers and SNCOs would hunt the dogs from the back seat of a low flying Alouette using shotguns on a weekly basis as reported in UNFICYP's *Blue Beret* magazine:

> The dogs show considerable ingenuity in hiding from the chopper but generally several are killed during each flight. Apart from the weekly flight however, the team must be prepared to react to a 'DOGREP' (Dog Report) at any time and deal with the animals. These reports come from our OPs or by word of mouth from the farmers.
>
> Once shot the dogs are collected by the farming SNCOs and placed in 'body bags'. They are then taken for examination to the Veterinary Centre in Nicosia to test for the presence of Hydatid. So far most dogs have carried the disease.[60]

While the Alouette was eminently suitable for dog clearance sorties, by the mid–1980s it was beginning to show its age. Nevertheless, the aircraft and its idiosyncrasies continued to be appreciated by those who flew it and WO2 Tony Davies, who was flying XR380 when it crashed in the Troodos, wrote an affectionate, if tongue in cheek, tribute to the French helicopter.

58 AFM 259: 'HQ UNFICYP Operational Directives 4/84', 6 July 1984 and '5/84', 10 July 1984.
59 Anon, 'Maintaining a Ceasefire', *Army Air Corps Newsletter,* No.117, May-June 1984, pp.3–4.
60 C. H. O'Rourke, 'Dog Hunting in Sector Two', *The Blue Beret*, September 1985, p.12.

> The cockpit layout is far from conventional with enormous, strange instruments carefully mounted in the most unlikely places. Some pilots, having flown the Alouette for years, will admit to never having found the Gyro suction gauge. Air speed is shown in Noeuds, pressures in Hpz, and Tuyere, Huile and US Galls appear at random when pilots are trying to locate the clock, which is in English. Power is assessed by watching a gauge called "PAS GENERAL". This strange instrument is calibrated from seven to 20 with a red line at 15 and moves when the collective lever is raised or lowered. It is, therefore, reasonably believed to equate roughly to degrees of pitch. Pilots, however, adapt rapidly, and once they recall the adage "Instruments mean nothing" apprehension quickly vanishes.
>
> Visibility from the right-hand drive pilot position is very good indeed, and is only obscured by the instrument panel and large passengers in the left-hand seat. The middle of the three rear seat passengers is indeed privileged. He is afforded an unparalleled view of the instrument panel with its myriad switches and coloured lights which have such a soothing effect on the nervous. The intercom system is very efficient and supplies constant background engine noise and gearbox whine to assure all on board that everything is running as it should be. All radio faults can be quickly rectified by unplugging the headset lead, licking it vigorously and replacing it, preferably in another socket.
>
> The Alouette is easy to fly, and once pilots adapt to the sensation, in the cruise, of sitting on a wooden chair with two large books under its back legs flight becomes a great pleasure. Hydraulic assistance is provided on the cyclic only, but all heavy forces on the collective lever are neatly trimmed out by a very thick rubber bungee under the belly panel. Any stretching of this bungee can easily be adjusted in four hours by a thin engineer, prostrate on a trolley, working with a torch in his mouth. Excessive stretching will make it almost impossible to raise the lever for take-off giving a built-in safety precaution. Should the rubber bungee be too tight, the lever will spring violently upwards to its highest position, injuring the unwary and stoutly resisting all efforts to push it down. Massive rotor droop will accompany the resulting vertical flight, avoiding any danger of overspeeding and overstressing the main rotor head.
>
> In the hover, the aircraft has a marked right-skid-low attitude which has led to unkind remarks about pilot obesity. Any nonsense about mechanical adjustments to compensate for tail rotor drift is dismissed by Alouette aircrew, who know that this is an integral design feature to make for easier handling on tricky sloping ground.
>
> Fuel management is fully automatic and the Emergency Fuel Shut-off lever is conveniently situated approximately two inches from the front seat passenger's right hand. There is a long delay before anything actually happens when this lever is retarded, and inadvertent operation can be rectified by unhurriedly slamming it back to the original position and wagging a finger (or two) at the guilty passenger. Many pilots have their own stout Chubb padlocks and use these to pre-empt any ill feeling on the flight deck, especially when flying inquisitive senior officers...
>
> The Alouette is a joy to maintain and service, and remarks such as "It'll look nice when it's finished" are unkind and uncalled for.... To summarise, the Alouette is an amazing aircraft and will undoubtedly continue to give sterling service as long as it is required.[61]

The final comment in WO2 Davies' tribute was, however, over optimistic. By 1987 the Alouette had been in service with the AAC for over a quarter of a century and attrition had worn down the original fleet of 17 aircraft to just 9. Maintaining such a small fleet was proving uneconomic and availability was proving more difficult to maintain as systems and spares were becoming in

61 Anon [A. J. Davies], 'Flight Test No.3 – The Alouette II', *Army Air Corps Journal,* 1983, pp.79–80.

increasingly short supply. Furthermore, the AAC had a ready replacement in the Gazelle. The Gazelle outperformed the Alouette by some degree in speed, range and payload. It would remove the requirement for conversion training for aircrew, ground crew and technicians deploying to Cyprus. The Gazelle was also operationally more capable as the Gazelle Observation Aid, Nitesun and armoured seats were all options if required. The decision was made to replace each of UNFICYP and 16 Flights' four Alouettes with three Gazelles.[62] The Gazelles arrived in Cyprus in August 1988 and refresher training for aircrew was conducted through September and October before the new aircraft were declared operational and the Alouettes were withdrawn from service after a final fly past at Dhekelia.[63]

Towards the end of the 1980s UNFICYP Flight remained one of the AAC's only fully operational postings along with those sub-units in Northern Ireland and 25 Flight in Belize. Throughout 1988 UNFICYP recorded 98 shooting incidents across the island resulting in five fatalities and three wounded, the latter including two UN soldiers. When Commander Aviation UKLF, Brigadier John Hathaway, visited the Flight he was taken on a familiarisation sortie by the QHI WO2 Paul Stanton AAC. During the sortie a Turkish soldier mistook the aircraft for a Greek National Guard helicopter and fired two rounds at it. Fortunately, the soldier's aim was as good as his recognition skills and the aircraft continued the sortie unscathed. On 11 May 1988 a car bomb detonated in Nicosia killing three and wounding 15 although it was attributed to Arab terrorists targeting the Israeli embassy. A pilot from UNFICYP Flight was the first to report the incident and helped to coordinate the emergency response. The 1988 Nobel Peace Prize was awarded to UN Peacekeepers in recognition of the efforts of around half a million personnel from 53 states who had contributed to peacekeeping operations from 1948 to 1988.[64]

As well as the operational aspects of flying with UNFICYP Flight and the prospect of a UN medal to show for it there were clearly other attractions to serving in Cyprus as recalled by Major Andy Wellesley AAC who took over as Flight Commander at the end of 1989:

> Permanent staff were on accompanied tours and married quarters were available in the UNPA [Nicosia UN Protected Area]. Single and unaccompanied staff whether permanent or roulement were accommodated in their respective messes. Although the standard of all the accommodation was not the best, the location more than made up for it. There was a decent NAAFI shop and bizarrely one of the most popular food items was frozen parsnips. The Flight was based at the far eastern end of the airfield and far enough away from RHQ [Regimental Headquarters] so as not to be disturbed too much. The Flight was fortunate in having one of the few air-conditioned buildings so work conditions inside at least were comfortable in the 40 degree heat of the summer. The aircraft were hangered when not in use and this was particularly important during the summer as even with the doors off the cockpit temperature rose very quickly to over 50 degrees before the rotors began to cool it down. Despite the high temperatures the Gazelle's operational capability was excellent but could be limited to a maximum take-off weight of 1700 KGs [3,700lb] to ensure a 5% thrust margin when operating to the highest of the OPs on the Line.
>
> Life on the UNPA was always lively and fun with seven different nations making up the UN staff. All the National days were celebrated with each nation vying to out-do the others in terms of the amount of alcohol that was available. The working day started at 0700, or 0600

62 AFM 259: Army Plans and Resource Committee paper, 'The Replacement of Alouette by Gazelle', February 1987.

63 AFM 259: 'Outline Handover Programme, Alouette II to Gazelle AH1', 1 August 1988.

64 Anon, 'UNFICYP Flight AAC', *Army Air Corps Newsletter*, No.143, April-June 1989, pp.23–25.

An UNFICYP Flight Gazelle flown by WO2 Paul Stanton preparing to lift an underslung load prepared by members of the Canadian Contingent in August 1989. An AAC ground crew signaller provides communications. (Army Flying Museum)

> a few days a week for PT, and we followed the same tropical work routine as the rest of the Regiment and routine tasking ended at 1330. This allowed those who wished to top up their suntans to be on a beach on the Northern Cypriot coast soon after 1400.
>
> As UN personnel we were allowed unrestricted access to cross the border into the Turkish occupied Northern Cyprus. In the 1990s Northern Cyprus was nowhere near as well developed commercially as the South and the fishing village (as it was then) of Kyrenia, approximately 30 minutes' drive to the North provided a welcome respite to the rigours of work. The Greek area of Southern Cyprus was wonderful ranging from the cool pine forested Troodos Mountains to the lively holiday resorts of Ayia Napa and Larnaca. In winter alpine skiing was possible in the mountains and a few hardy intrepid souls managed to snow ski in the morning and water ski off the southern coast in the same afternoon. In 1990 there was more than 3 metres of snow on the pistes, more than could be found in any of the alpine resorts over the same period.[65]

For the most part the Gazelle was well received in both 16 and UNFICYP Flights and performed well within those limitations imposed by the environment. The three Gazelles in UNFICYP Flight were given a UN livery. Ten cowlings or panels were sprayed white and the battery bay

65 A. Wellesley, account to the authors, 7 September 2022.

cover, forward belly panel and port and starboard fuel tank side panels were marked 'UN' in black lettering. The panel and cowls were selected for prominence but also because they were easily removable and interchangeable so that aircraft could be exchanged between UNFICYP and 16 Flights 'to meet operational or engineering requirements.'[66]

Nevertheless, incidents could and did still occur such as that experienced by Staff Sergeant Geoff Sutcliffe AAC. On 30 November 1992 Staff Sergeant Sutcliffe was in command of Gazelle ZB668 on a recovery sortie from Akrotiri to Nicosia. While over the Troodos Mountains north of the Germasogeia dam, the handling pilot, Sergeant Colin Withey AAC, experienced increasing vibration through the flying controls. Soon after the torque meter reading became erratic, the vibration increased, and the main rotor speed began to drop. Staff Sergeant Sutcliffe took control, sent a mayday call, flared the Gazelle to regain rotor speed, entered auto-rotation and looked for a landing site. The only suitable point within range was a bend in a track running along the side of a steep valley. It was still necessary for Staff Sergeant Sutcliffe to stretch the auto-rotation, and he was forced to trade rotor speed for range. By the time ZB668 reached the selected landing site the rotor speed had fallen to such an extent that the aircraft dropped the last 5 or 10 feet vertically onto the track. The force of the landing caused the tail boom to deform, but otherwise the aircraft was undamaged and the crew were unharmed.

An 84 Squadron RAF Wessex was scrambled to respond to the incident where two firemen were winched down to assess the scene. Staff Sergeant Sutcliffe and Sergeant Withey were picked up and returned to Nicosia. The Wessex then returned with engineers from UNFICYP Flight, but the weather prevented any immediate recovery of the aircraft. Engineers returned two days later along with assistance from 16 Flight's LAD and after a nine-hour operation dismantled ZB668 to the extent that it could be moved by road through the mountains and down to RAF Akrotiri from where it was flown back to the UK. A subsequent board of enquiry diagnosed the cause of the accident as failure of the freewheel support bearing resulting in the freewheel input drive shearing. Staff Sergeant Sutcliffe's actions in dealing with the emergency and avoiding major damage to the aircraft or injury to the crew was subsequently recorded:

> Such a transmission failure is not adequately covered in Gazelle flying training as it cannot be simulated in the Gazelle helicopter. As Staff Sergeant Sutcliffe has never flown in a Lynx Simulator, this was the first time he had ever experienced such a malfunction.
>
> Because of the admirable way in which Staff Sergeant Sutcliffe handled the situation, despite the inhospitable terrain and with an emergency not previously simulated, he saved the aircraft and the lives of himself and his pilot. Due to his display of exceptional flying skill and judgement throughout this emergency, Staff Sergeant Sutcliffe is awarded the Green Endorsement and Tie of Merit.[67]

Unfortunately, despite the skill and resilience of pilots such as Staff Sergeant Sutcliffe and the dedication of ground crew and technicians, UNFICYP Flight's days were numbered. On 17 June 1994 HQ UNFICYP announced that the AAC would be replaced as the provider of aircraft and soldiers to equip and man UNFICYP Flight. The formal disbandment of UNFICYP Flight AAC took place on 30 September 1994 at the UN Flight Line at Nicosia International Airport. The disbandment parade was commanded by the OC UNFICYP, Flight Captain Steve Pidgeon AAC, and

66 AFM 259: Commander Maintenance (Aviation) UK, 'Application for approval in principle of a service engineered modification; Gazelle AH1 United Nations identification markings', 24 August 1987.

67 AFM 259: Anon, 'All in a day's work', cutting from unidentified publication, 11 December 1992 and 'Award of Green Endorsement by Director Army Air Corps – Staff Sergeant G. J. Sutcliffe AAC', undated.

A Gazelle of UNFICYP Flight hovers next to an OP in the Austrian Contingent sector along the UN Buffer Zone. (Army Flying Museum)

attended by Force Commander Brigadier Ahti Vartiainen and the UN Head of Mission, Mr Gustave Feissel. The parade concluded with a fly past by the Gazelles of both UNFICYP Flight and 16 Flight AAC. The aviation commitment was handed over to an Argentinian Flight equipped with two Hughes 500D helicopters.[68] UNFICYP Flight 2IC, Captain Andy Cash AAC remembered the feeling within the contingent as the Flight concluded its drawdown.

> I recall the flight was disappointed, obviously, not just because they were enjoying the opportunities it presented to us but because there was a feeling that the Flight did (and had for many years) add real value to the UN mission there. It was a challenging (though not arduous) role given the weather conditions and heat (for the Gazelle) and was very satisfying, giving many soldiers the opportunity to rise above their normal responsibilities of rank and to shine. I met many of those who served in the Flight through my career and many of them went on to great things and higher rank. They had something about them. Also, the integration of ACC and REME was very close given it was such a small flight and the way we worked as one team, everyone chipped in across the board. I remember a lot of detailed and hard work put in by many to prepare for the close down in good order. We even managed a small farewell parade, that the Colonel AAC, Mike O'Donohue, attended. Our relations with the incoming Argentinian Flight were excellent. We were quite used to working with the Argentinians as they had a separate ground contingent at UNFICYP at that time and we flew them regularly. The only one I took a dislike to was the young Argentinian solider who, despite there being no doors on the Gazelle, decided to throw up down the back of my helmet and flying suit! He was made to clear up the aircraft by the REME guys, while I took myself off for a shower when we got back! The Argentinian Air Force guys were great fun and we shared much in terms of our humour and professional enjoyment of flying.[69]

Although AAC personnel continued to serve on Operation Tosca within HQ UNFICYP and elsewhere in BRITCON, the final parade marked the end of just over 30 years of the AAC being responsible for the aviation commitment and its equally long and effective record of UN service in Cyprus.

The AAC maintained a presence in Cyprus through 16 Flight for almost a further decade. The Flight continued to provide aviation support to British Forces Cyprus from its base in Dhekelia. It also provided aviation support for troops exercising on the island and continued to host hot and high flying courses each year. A parade was held on 3 June 2003 in Dhekelia to mark the disbandment of 16 Flight and the Army Air Corps' and its withdrawal from the island. The parade was commanded by the Flight's final OC, Major Adrian Palmer AAC. The Colonel Commandant, General Mike Walker, attended and the Band of the Army Air Corps played during the parade. The presence of 16 Flight in Cyprus for over 33 years was commemorated by a stained glass window erected in the garrison Church. The disbandment of the Flight ended an unbroken 46 years of Army Air Corps service in Cyprus, much of it on operational duty. Throughout the Emergency, UNFICYP service, the Turkish invasion, and the contribution of 16 Flight the AAC had made a major contribution to the history of Cyprus and the Cypriot people and learned valuable lessons for future operations.[70]

68 AFM 259: Pamphlet, 'Parade for the Disbandment of UNFICYP Flight Army Air Corps', 30 September 1994 and *Army Air Corps Newsletter,* No.164, July-September 1994, p.29.

69 A. Cash, account to the authors, 21 August 2023.

70 Anon, untitled photographic article, *Army Air Corps Journal*, 2003, pp.185–186.

Chapter 7

Aden

On 19 January 1839, the Royal Navy landed an expeditionary force of British and Indian troops on the heel of the boot of the Arabian Peninsula. After an impressive naval bombardment and a brief skirmish on land, the village of Aden and its harbour were under British control. The squalid village and the 20 square miles of territory around it became the new colony of Aden with the action recorded as the first annexation of territory during the reign of Queen Victoria. The location of Aden's harbour made it the ideal location for a coaling station on the route to India and other British possessions in the Far East. The British would remain in Aden for the next 128 years, but they had intended to remain longer. Situated one hundred miles to the east of the entrance to the Red Sea, the port of Aden was not only a useful staging post to refuel on the trade route to Imperial India, but it was also an ideal jumping off point for any military expedition into Africa. It remained Britain's only colony in Arabia. The strategic importance of Aden ensured that over the following years the colony was developed, expanded and garrisoned by the British. The significance of the colony was further enhanced by the opening of the Suez Canal in 1869. The Aden garrison also acted as an imperial outpost in the path of any potential aggressive thrust by the Ottoman Empire through Arabia towards the canal.[1]

Aden was officially made a Crown Colony in 1937, at which time the boundaries of the colony encompassed just 75 square miles of land. The all-important harbour comprised two large bays on either side of the town. The uninhabited 1,700-foot Jebel Shamsan dominated the skyline over the Peninsula and the entire harbour. On the eastern side of the Peninsula was the bowl of an extinct volcano which was the site of the town's commercial district. Known as 'The Crater' the commercial area looms large in the narrative of what later became known as the Aden Emergency. On the western side of the Peninsula was the township of Ma'alla which linked to Steamer Point and the port facilities. Beyond the confines of the Crown Colony lay a huge hinterland referred to as the Protectorate. Almost the size of England and Scotland combined, the Protectorate territory covered around 112,000 square miles. The coastline of this expanse stretched for over 700 miles along the Arabian Peninsula. The British saw no reason to march into the Protectorate and expend 'blood and treasure' in futile attempts to conquer, garrison and govern the tribes that inhabited the space beyond the limit of Aden itself. Well-proven imperial methods were adopted; protective treaties that placated dominant tribal leaders and thereby secured British interests were maintained by Political Officers. In the event of a breakdown of a treaty, the colonial authorities could resort to force. After the First World War this usually entailed the use of airpower, and the successful use of what was termed Air Control by the RAF in Iraq and Somaliland suggested that these methods could also be used in the Aden Protectorate. Captain Robert Hamilton, Lord Belhaven

1 For a more in-depth history of British involvement in Aden see R. J. Gavin, *Aden Under British Rule: 1839–1967* (London: C. Hurst & Company, 1975).

and Stenton, of the Royal Scots Fusiliers, seconded to the Aden Protectorate Levies described an Air Control operation carried out in the Protectorate in retaliation for a tribal murder:

> The Air Staff would work in the closest contact with the political officer. It was my task, equipped with a portable wireless set, to camp as close to the scene of operations as I considered possible, so as to reduce the extent of operations to a minimum. Two one-day warnings were dropped on the tribe, followed by an hour's warning before the first attack, so that women and children could be taken to a place of safety and every effort was made to inflict losses in property rather than lives.[2]

The tactic of advance warning before bombing a target, referred to as 'Proscription Bombing' was widely used during the inter-war years. It also elevated the RAF to prominence in the military hierarchy of the Middle East for some years.

By the late 1950s, in the years following the ill-fated Suez Crisis British military thinking and planning in the Middle East was dominated by the age-old conundrum of striving to protect strategic commitments while simultaneously reducing the forces available to police and defend them. The post-Suez reappraisal of these commitments, many of which had existed for centuries, led to the momentous decision to withdraw from longstanding colonies and possessions as soon as was practicable. The protracted process of withdrawal steadily weakened the network of land-based installations that had historically been essential to the support of British military operations across the Middle East. This contraction highlighted the need to maintain a credible land-based foothold to reassure traditional allies and stabilise the oil-rich Gulf region. There was also a realisation that such reassurance required an equally credible capability of military intervention east of Suez.

What followed was a shift away from a headlong withdrawal towards a desire for strategic consolidation in the region. A minor contributing factor to this change of stance were successful counterinsurgency operations in Oman. British intervention in the Sultanate of Oman had helped the Sultan defeat Saudi backed rebel forces. After a difficult year in 1958, the campaign in Oman culminated in January 1959 when troops from 22 SAS mounted a spectacular assault on the previously impregnable Jebel Akhdar. The success of this operation demonstrated that airpower alone was no longer the optimum solution when dealing with rebellion or insurgency in South Arabia. 'Ground Troops, well acclimatised and with air support, were deemed to be the way forward.'[3]

The solution that eventually evolved in the early 1960s was twofold. Firstly, an enhanced land base close to the Gulf would be located in the barren, but ideally situated Aden Colony. This significantly expanded outpost would be complimented by a mobile strike force based on the RN's commando carriers, HMS *Albion* and HMS *Bulwark*. The carriers were to be supported by the recently delivered amphibious assault ships, HMS *Fearless* and HMS *Intrepid*. The new commando carriers were re-roled aircraft carriers, refitted to accommodate large numbers of RM Commandos and the helicopter squadrons required to launch them from their carrier into battle. This potent, roving naval force also included numerous RN escort vessels and support vessels from the RFA.

The selection of Aden over other potential locations for a new forward base was primarily based on the assumption that the colony could be retained in the medium to long term. Other options such as Kenya were ruled out as they had already made clear their desire for rapid decolonisation and full independence. The construction of the military infrastructure required to support the new base would not be insignificant but pouring UK taxpayers' capital into a project that could well be

---

2 Quoted in J. Walker, *Aden Insurgency: The Savage War in South Arabia, 1962–1967* (Staplehurst: Spellmount, 2005), p.7.

3 Quoted in Walker, *Aden Insurgency*, p.16.

lost to a newly independent and potentially hostile government was not an option. Therefore, the ideally located Aden Protectorate was selected as the safest option for investment and expansion. The plans for the new forward base included housing for 2,500 service personnel and a further 1,000 dependants. The embryonic concept of a regional forward base in Aden was to be put to the test almost immediately in June 1961.

See Map 7

Operation Vantage – The Kuwait Crisis 1961

The British Government's overwhelming desire to complete a rapid withdrawal from longstanding commitments east of Suez continued to evolve as a complex and drawn-out enterprise. What had been envisaged as a carefully staged and deliberate extrication from the legacy of empire had, in places, degenerated into what has often been labelled as, at best, a fighting retreat.[4] The loss of national confidence in the wake of what many described as the 'Suez debacle' was virtually palpable. Almost immediately dissident factions across British territories and colonies saw a chance to agitate for their own independence, as British influence and the appetite to govern far-flung overseas possessions waned. Old rivals on the international stage such as the USA, Russia and China also recognised this tangible world-weariness as an opportunity to flood into the vacuum left behind by the exiting British. The new superpowers ruthlessly exploited the situation and set about projecting their own economic and military power, either directly or by use of proxy forces. Nationalism and self-determination were at the fore as Britain adjusted to its new and unfamiliar position in the pecking order of global politics. At the same time, ambitious regional powers, long kept in check by Britain's Imperial muscle, also felt that they now had a golden opportunity to extend their territory and political influence over neighbours left weak after the withdrawal of British security forces. Prominent among the emerging Middle Eastern powers hungry to expand their borders was the young Republic of Iraq. The Iraqi claim over Kuwait pre-dated the Republic by centuries. Successive Iraqi monarchies had coveted the territory of its tiny Gulf neighbour to the south. The ongoing expansion and exploitation of Kuwaiti oilfields and the Kuwaiti capacity to refine and export their product, made the prospect of a swift annexation even more attractive.

Winds of Change

Kuwait had been under British protection since 1899. The agreement made at the time with the ruling Al Sabah dynasty gave Britain control of Kuwait's defence and responsibility for all Kuwaiti foreign affairs. Iraq had consistently disputed Kuwaiti independence, officially rejected the 1932 Anglo-Kuwaiti agreement and, at the same time, reaffirmed its claim that Kuwait had been Iraqi territory for centuries. The political landscape of Iraq shifted seismically in 1958 when the 14 July Revolution swept away the Hashemite monarchy and created the first Iraqi Republic. Anxious to prove itself as a strong and independent Arab nation the new Republic reaffirmed old territorial claims to its north on its border with Turkey, and over Kuwait in the south. The view that the British no longer had the political will nor the military means to contest any Iraqi annexation of territory was further encouraged in February 1960. The British Prime Minister, Harold Macmillan, gave what became known as his 'Winds of Change' speech in Cape Town, South Africa. Macmillan's speech was interpreted by nationalists across the remaining British Crown

4 See for example R. Neillands, *A Fighting Retreat; The British Empire, 1947–97* (London: Hodder & Stoughton, 1996).

Colonies and Territories as a statement that Britain was finished with empire and its role as 'world policeman':

> Ever since the break-up of the Roman Empire, one of the constant factors of political life has been the emergence of independent nations. The wind of change is blowing through the continent and whether we like it or not this growth of national consciousness is a political fact. We must all accept the fact, and our national policies must take account of it.[5]

The increasingly ambitious leaders of the Arab world were desperate to hasten decolonisation and interpreted the speech as a green light to oust the British occupiers from the Middle East. The following year during the summer of 1961 a dramatic series of events were to test the British Government's new-found will to withdraw from the burden of being an 'world policeman'. Any withdrawal had to be measured against an irreducible minimum of commitments. The reality was that Britain still considered itself to be a major power and as such had to maintain international influence and credibility. The inescapable reality was that the protection of allies and resources vital to the national interests of Britain would require an enduring military capability that could be projected into the Middle East as and when required. Nowhere was this strategic reality truer than in the oil-rich Gulf states.

Crisis in the Gulf

The planned downscaling of responsibilities included relinquishing Britain's role as diplomatic controller and military protector of Kuwait. Full Kuwaiti independence, which took effect on 19 June 1961, was welcomed across the border in Iraq. The Iraqi Prime Minister and self-appointed Defence Minister, Major-General Abd al-Karim Qasim, acted swiftly. The Iraqi Government announced the planned annexation of Kuwait six days later, on 25 June. This was a bold move based on the widely held belief that Britain no longer had the will or the nerve to project its dwindling military forces into the Middle East to protect its Gulf allies. The Amir of Kuwait, Sheik Abdullah III Al Salim Al Sabah, did not hesitate to put the brand new Kuwait Treaty to the test. On 30 June he invoked Clause 4 of the Friendship Agreement and Kuwait formally requested immediate British assistance against an external military threat. Twenty-four hours later the lead elements of a large British military force began to arrive in Kuwait. The Iraqi leader had made a massive miscalculation in underestimating British resolve to support its allies and its national interest in the region.

What became known as the Kuwait crisis was an early test for the very recently restructured Headquarters Middle East Command (MEC). The new joint headquarters established in March 1961 was located at Steamer Point in Aden. The establishment and location of the new headquarters was an indication that, although Britain intended to reduce its permanent presence east of the Suez Canal to a minimum, there was still the intent to project military force east if required. At the time the Crown Colony of Aden appeared to be the ideal location for what was intended to be a form of sovereign base area through which expeditionary forces, reinforcements and routine military traffic could stage and reconstitute as required. The HQ MEC area of responsibility sat between Near East Command in Cyprus, covering the Eastern Mediterranean, and North Africa and Far East Command in Singapore. The MEC was responsible for the Arabian Peninsula, the Persian Gulf, South Arabia and all territories down the east coast of Africa as far as Swaziland.

5 Quoted in Neillands, *A Fighting Retreat*, p.352.

Out of Africa

The threat of invasion and the anticipated call for assistance from Kuwait triggered contingency planning in units across the MEC, including in 8 Independent Recce Flight AAC which operated from Nairobi in Kenya. The flight flew the Auster Mk AOP6 and had recently deployed detachments on a diverse range of tasks to Uganda, Tanganyika and, early in June 1961, another detachment of a pair of Austers had been sent to reinforce the King's African Rifles in Zanzibar where they were involved in riot control operations. What had been planned as a six-week operation was cut short on 29 June when the flight received a warning order to be at 24 hours' notice to move for deployment to Kuwait. The flight was ordered to prepare to move with Headquarters 24 Infantry Brigade (24 Inf Bde) as part of a Theatre Reinforcement Plan, codenamed Operation Vantage. The Amir of Kuwait's official request for military assistance followed the next day.

Operation Vantage was the first real test of the MEC and the component forces it had dispersed across the region. The lead elements would arrive by a combination of strategic airlift and a large naval task force. Part of Sandys' post-Suez reforms had been the introduction of the helicopter carrier concept. The idea was based on the use of converted aircraft carriers packed with RM Commandos and the integral helicopter lift required to project them ashore, as tested and proved during Operation Musketeer. The concept of a large and potent floating reserve, able to operate offshore where it remained free from interference, had obvious military and political appeal. The Kuwait crisis offered the perfect opportunity to test the theory. At the same time as 8 Flight began its deployment to Kuwait a large RN task force was steaming towards the Kuwaiti coast. The naval force included the helicopter carrier HMS *Bulwark* carrying 42 Cdo RM. The lead elements of 42 Cdo were landed by Fleet Air Arm Whirlwind helicopters at the newly constructed Kuwait Airport on 1 July. The landing was supported by Hawker Hunter jets from 8 Squadron RAF and later in the day Centurion tanks from C Squadron, 3rd Carabiniers landed by sea. The initial landing was reinforced by more Royal Marines from 45 Cdo and troops from 2nd Battalion, the Coldstream Guards (2 Coldstream Gds). Further reinforcements poured in by air and sea over the coming days. Once ashore the land component of Op Vantage was under the command of HQ 24 Inf Bde.

Officers' Flying Club

The 8 Flight detachment of four pilots and eight air mechanics commanded by Major Tom Lacey RA landed at Kuwait Airport at 0800 hours on 2 July. The detachment had travelled on a RAF Bristol Britannia transport aircraft with the Brigade Commander and his staff. The Britannia had flown from RAF Eastleigh (Nairobi) into Kuwait via RAF Khormaksar in Aden. The 12-man detachment deployed with just 200lbs of freight, consisting mainly of tools and Auster spares. Unusually, 8 Flight had deployed without their own aircraft. Existing commitments in Africa and the need to deploy quickly had resulted in an unconventional plan. The Brigade Commander, Brigadier Derek Horsford, was anxious to have an aviation flight under his direct command. He had directed Major Lacey that, if necessary, he was to commandeer civilian aircraft known to be housed at the Kuwait Aero Club. On arrival in Kuwait the 8 Flight group set off immediately to the Aero Club hangar where they were helpfully provided with five Auster J5 Aiglet civilian aircraft. Just six hours later Major Lacey took off on the first of many operational sorties flown by 8 Flight during Op Vantage.

The Kuwaiti Aero Club Auster J5 Aiglets were inferior in performance to the Auster AOP6 normally operated by 8 Flight AAC. (Army Flying Museum)

The J5 Aiglet was an older and less capable aircraft than the Auster AOP6 that the 8 Flight pilots were familiar with. The conversion to the older type was therefore simple and the assembled pilots were all 'checked out' on type quickly and available for tasking.[6]

The newly assimilated Austers were given Kuwaiti national markings and began operating in support of the numerous ground elements now occupying defensive positions overlooking the Iraqi border from the vital high ground of the Mutla Ridge. The joint British and Kuwaiti defence line effectively blocked Highway 80, the main road from the Iraqi border town of Safwan to Kuwait City, which was the only viable route for an invasion force. As long as the Mutla Ridge was held by determined troops Highway 80 was closed to the Iraqi Army. Holding the ridge was not easy, however, and living conditions were extremely challenging for the troops dug in on its rocky slopes. A total absence of shade and extreme, relentless, inescapable heat had the potential to degrade poorly trained units. The first weeks of the operation were also plagued by high winds and sandstorms that reduced visibility to an average of 350 yards. Temperatures in the daytime averaged around 54°C in the shade and navigation over the featureless desert was difficult. The 8 Flight Austers were employed on daily border patrols, liaison tasking, and numerous training tasks with Brigade troops. Training reflected anticipated tasking should the Iraqis decide to cross the border, with sorties practising pilots in the AOP role, direction of NGFS, and FAC with RAF and RN FGA aircraft of numerous types.

See Map 8 As the operation progressed environmental conditions took their toll on personnel and aircraft alike. Pilot fatigue was a constant consideration when planning and allocating tasking. Pilots were rotated periodically between Kuwait and Kenya for rest and recuperation. The original

6 Anon, 'Flood Relief in Kenya or an Unusual Task', *Army Air Corps Journal*, 1962, p.43.

Some of the 8 Flight AAC Pilots: Captains Nigel Barry-Taylor, Tony Brown and Paul Courtenay pose in Kuwaiti headgear. (Army Flying Museum)

detachment under Major Lacey comprised Captain C. N. Barry-Taylor RA, Lieutenant Bryan Wright, Lieutenant John Turnbull and Sergeant Goddard QRIH. The relief pilots were Captains Paul Courtney Royal Sussex, Tony Brown and James Knott RE, Sergeants David Burman RSigs, and Neville Law KRRC. The austere conditions of the Kuwait deployment presented significant serviceability problems for the air mechanics maintaining the borrowed Aiglets. The spares brought across from Kenya were not all compatible with the much older civilian specification Auster. The scarcity of spares was a constant problem and the logistic difficulties of maintaining obsolescent, unsupported aircraft featured prominently in the POR.

The large-scale British presence in Kuwait ended officially on 19 October 1961 when troops from the Arab League took over responsibility for guaranteeing the Kuwaiti border. The Kuwait crisis had lasted for three months, during which time 8 Flight had flown over 500 hours in the most testing conditions. Over the 12 months of 1961, the nine pilots of 8 Flight accumulated an impressive total of 3,500 operational flying hours. The year was not quite over for 8 Flight, however, as they returned to Kenya where conversion to the Beaver was planned for the majority of the pilots. As elsewhere, technical problems with the engines of the new and much anticipated Westland Scout helicopter had resulted in an urgent requirement for a stop-gap aircraft, and so the Alouette was also deployed to Kenya. The first of four Alouettes were delivered to 8 Independent Recce Flight AAC at RAF Eastleigh, Nairobi in September 1961. The Alouette was flying in African skies with 8 Flight within a few weeks.

The AAC's brief operational foray to Kuwait in 1961 would prove to be the precursor for a much sterner test a full 30 years later when Iraqi threats to invade its smaller neighbour were carried through.

Operations in Southern Arabia

The success of the Kuwait deployment seemed to vindicate the expansion and enhancement of Aden as the lynchpin of the British military presence in the Middle East and beyond. The planning and construction of defence infrastructure to support this key strategic presence carried on at pace. Diplomatic efforts to stabilise Southern Arabia ran in parallel to the military plan.

The British Government's intention to eventually relinquish its responsibilities in South-West Arabia was based on the idea of establishing a federation of Arab states. This fledgling federation would be governed by a newly formed Arab Government based in Aden. The granting of sovereign independence was entirely dependent on the caveat that the UK retained its strategic military presence in the new state. Colonial Secretary, Duncan Sandys, successfully reassured the Aden Legislative Council of British intentions regarding its sovereignty. On 26 September 1962 the Council narrowly voted in favour of merging what would be Aden State with the Protectorate states to create the Federation of South Arabia. The merging of what were two very disparate groups was formalised in January 1963.

This was an apparently logical solution to the British problem of creating a stable buffer zone around what would be a vital cornerstone of UK strategic policy in the Middle East and beyond. The plan for the new federation was however desperately flawed. The two factions that created the new Federation of South Arabia were diametrically opposed in every aspect of their demographic composition, religion, and culture.

A milestone in the ongoing expansion of the UK military presence in Aden and the build-up of the new Middle East Command was the arrival of Headquarters 24th Infantry Brigade from Kenya. The brigade was part of the UK Strategic Reserve, and it was also the permanent headquarters for the land component of the MEC. HQ 24 Inf Bde arrived in Aden in December 1963, the arrival of this much-needed command and control capacity was appreciated by the British officers that were seconded to Aden's Federal Regular Army. The main body and fighting units of the Brigade would begin to arrive in Aden in March 1964 and from that point onwards a phased redeployment plan moved 24 Inf Bde units from Kenya into Aden right through to December of the same year.

The RAF had maintained an enduring presence on the Arabian Peninsula since its earliest days and since 1928 British Forces South Arabia had been commanded by a senior RAF Officer. The formation of MEC on 1 March 1961 did not affect the custodianship of the command appointment and the RAF retained the post. Airpower had played a major part in British Middle East doctrine, which in the aftermath of Suez leaned even more heavily on the reach and firepower of the RAF. The development of Aden included a huge construction programme at the main RAF airfield at Khormaksar. The station at Khormaksar was already the main staging post for all RAF traffic between the UK and Far East Command and became the scene of extensive works in the early 1960s. The airfield was home to 75 percent of the large RAF aircraft fleet allocated to the MEC and also provided a staging post for civil air traffic transiting through Aden. Ultimately the expanded airfield would provide fuel, ATC and servicing for nine RAF squadrons and an additional three flights. A large proportion of what was a diverse mix of aircraft types were 46 Hawker Hunters. The Hunter was a new addition to the RAF's Orbat and AAC pilots would work closely with reconnaissance and ground attack Hunters throughout the Aden Crisis.

The reinforcement of the Aden Garrison with much-needed aviation assets began with the transfer of 653 Squadron AAC under command of Major Bill McNinch AAC from Cyprus. The Squadron arrived in Aden by sea on board HMT *Oxford*. It initially sailed without any helicopter capability as 8 Recce Flight AAC, which was equipped with Alouette II helicopters, had been retained to support ongoing 24 Inf Bde operations in Kenya. The French built Alouettes had already proved to be a more than capable stop-gap replacement for their much anticipated, but long overdue Westland Scouts.

An eye-catching image of a 653 Squadron Auster AOP9 flying over a Centurion Tank of the Royal Scot's Greys. (Army Flying Museum)

The Squadron's two fixed wing flights moved onto Falaise Airfield in Little Aden on 9 March 1961, initially a mix of Beaver and Auster AOP 9 deployed on aviation operations across Aden and the Protectorate. The five Beavers of 15 Liaison Flight AAC were an instant success and extremely popular with pilots and user units alike. The rugged Beaver seemed almost purpose-built for Aden, and it proved itself to be an ideal airframe in all conditions. The pilots of 13 Recce Flight AAC soon discovered that the Auster AOP 9 was much less suited to the heat, humidity and dust of the desert. Three Austers were written off in a cluster of crashes between May and July 1963. No fatalities were reported other than the airframes. In a further incident Lieutenant Edward Hardman RTR, an Auster pilot serving in Aden with the Air Troop of 4 RTR lost an eye. Hardman's aircraft suffered a bird strike when a large vulture smashed into his cockpit. Although severely injured, Hardman managed to land his aircraft with its passengers safely. He was subsequently awarded a well-deserved QVCS for his actions. The Auster's 180HP Blackburn Cirrus Bombardier engine proved to be vastly underpowered at altitude. Consequently, take offs and landings were restricted to below 3,000 feet AMSL (above mean sea level). This lack of power prevented the carriage of passengers, 'much to the passenger's relief.'[7]

7 AFM 204c: 653 Sqn AAC, Scout AH Mk1, Radfan Operations 27 Apr–30 Jun 1964, p.2.

The reality of flying the Auster in a hot and high environment; XP240 after a forced landing which ended in a 60 feet deep Wadi near Amriga. (Army Flying Museum)

The Radfan Rebellion

The Radfan region lay some 60 miles to the north of Aden and 30 miles from the Yemen border. The Radfan takes its name from the Jebel Radfan that dominates the central area of the 400 square miles of the region. Although barren, Radfan's volcanic mountains were of strategic importance. The ancient pilgrimage trail to Mecca passed through the foothills, it was also a long-established trade route. The British had replaced the ancient camel route with the modern Aden-Dhala Highway cutting through the Radfan at the Kuraybah Pass. Beyond the pass at 6,000 feet is the fertile Dhala Plateau. The terrain of the Radfan could be characterised as complex in nature. An almost lunar landscape broken by steep hills (jebels) and divided by interconnecting riverbeds (wadis). The entire area is littered with fields of boulders scattered by frequent flash floods (seils). Centuries of flooding and the subsequent evaporation of standing water had hardened the jebels and the wadi beds. This hardened, impervious surface acted as a channel for the rainfall, forcing it into the interconnected wadis generating powerful flash floods that poured out of the wadis onto the desert plains. Its energy expended, the torrent of water then formed standing pools that stagnated and became infested with several unpleasant parasitic diseases. What vegetation there was consisted of scrub or camel thorn bushes. This austere environment was the battleground over which the AAC would mount its first major operation beyond the Aden Colony. The enemy that they faced in Radfan were Queteibi tribesmen who had called the mountains and hills of the Radfan region their home for centuries. Queteibi pride in their lands was matched by almost inbred martial skills, a violent nature, and a reputation for duplicity.[8] Queteibi villages consisted of clusters of stone-built dwellings built into a hillside and they usually included ready-made defensive positions in the form of stone sangers.

At the start of what would become known as the Aden Emergency, the local government forces charged with policing the Radfan region were organised as the Aden Protectorate Levies. This colonial force had operated under British command and was led by a mixture of Arab officers and seconded British officers and SNCOs. The Aden Protectorate Levies was reformed as the FRA on 30 November 1961. The new force was better equipped and therefore more capable and was designed to support both the new Federal Government and British strategic policy in Aden. The FRA was to be a national army that would operate in Aden and across the Protectorate. The FRA recruited to a strength of 4,500 Arab Soldiers who would be trained, supported and commanded by 400 British Officers and SNCOs. Four infantry battalions formed the core of the FRA, supported by an armoured car Squadron and a small logistic echelon. Artillery and air support was provided from resident British forces in the colony.

The first FRA commander was Brigadier James Lunt OBE, an experienced soldier who had served in India and fought the Japanese in Burma during the Second World War. More importantly, Lunt had recently served in the Suez Canal Zone and latterly in Jordan with the Arab Legion. Lunt arrived in Aden to take command of the Federal National Guard (formerly the Government Guard) in early December 1961 and made this succinct assessment of the Queteibi:

> Its inhabitants dwell close-cooped in their stinking hovels, the accumulated manure of generations rotting beneath their noses and their animals stabled immediately beneath their living quarters. Riddled with disease, suspicious of every foreigner, their withered and stunted physique bears witness to their under-nourishment. Instead of growing food in their fields, they choose instead to cultivate qat, which they chew from noon until sundown, by which time their minds are sufficiently stupefied to commit any crime.[9]

Historically, the tribesmen had been a reluctant but compliant partner to the British colonial authorities. This compliance was ensured by regular shipments of weapons and timely payments of hard cash. The status quo began to break down in the wake of Suez. As British influence and authority began to wane in the region an increasingly disruptive Egypt looked to exploit uncertainty and exploit any power vacuum in the South of Arabia. Importantly at a local level, any British withdrawal meant no more weapons or cash. The presence of aggressive Egyptian forces across the border in North Yemen quickly developed into a direct threat to the longstanding balance of power in the hills of the Radfan. In September 1962, the Egyptians backed a republican coup in Yemen. The result was the communist Yemen Arab Republic and a festering civil war that would rage until 1970.

See Map 9

The Yemeni civil war brought an influx of Egyptian military advisers, modern weapons, and cash into Yemen. Newly established guerrilla training camps produced numerous well-trained cells of communist insurgents schooled in the art of revolution. All through the summer of 1963 the guerrillas of the Yemeni backed NLF imported their Marxist ideology into the Radfan. Most importantly from the Queteibi tribes' perspective, they also brought Soviet heavy weapons and the money that they desperately wanted. The NLF was also generous in its distribution of cheap transistor radios among the tribesmen. The small hand-held radios offered a direct conduit for Egyptian propaganda, spreading the NLF message into the scattered villages via the Nasser-funded 'Voice of the Arabs' radio station. The disgruntled Queteibi of the Wadi Misrah area were

8 N. Van Der Bijl, *British Military Operations in Aden & Radfan* (Barnsley: Pen & Sword, 2014) p.71.
9 van Der Bijl, *British Military Operations in Aden & Radfan*, p.71.

proving to be particularly fertile ground for Egyptian intelligence agents and the NLF. Promising tribesmen were either sent to Cairo or across the Yemeni border to remote training camps outside Taiz, run by the Yemen National Guard. During their time at the camps, they mixed with Marxist activists from the Aden Trades Union Conference and learnt the latest guerrilla tactics developed by fellow revolutionaries in Cuba and Algeria. The politicised tribesmen returned to the Radfan as fully fledged insurgents.

The development of a strategy for rebellion in the Radfan was simple. The Aden-Dhala Highway was the main artery of the Federal Government and the FRA garrisons in the region and was the traditional target and source of revenue for the tribesmen who bore no allegiance to the Federal Government. Generations of Queteibi had raided merchant caravans, convoys and pilgrims on the same route for centuries. Spurred on by their NLF comrades and armed with modern weapons they enthusiastically returned to their old bandit practices under the banner of revolution. They began mounting successful ambushes on civilian and military traffic moving along the highway. The ambushes were well executed with the NLF making use of Second World War vintage German Teller mines, Soviet small arms, and an intimate knowledge of the terrain held among the dissident tribesmen. FRA casualties quickly mounted to an almost unacceptable level and the road became impassable without a well-armed military escort. Eventually, indoctrination by the NLF and a dispute with the Amir of Dhala over the levy being taken from the Dhala Highway traffic, triggered a dramatic change in the region. The five Radfani tribes declared their independence from federal rule. The dissident tribes named themselves 'The Red Wolves of Radfan' and declared their intent to follow a path of armed resistance. The combined fighting strength of the five tribes was estimated to be in the region of 7,000 men. This was a serious escalation of tensions in the Radfan that could no longer be ignored.

The autumn of 1963 saw the challenge to federal rule become more aggressive and more overt. The FRA maintained a garrisoned fort at Thumier, which was boldly attacked by a force of 200 dissidents. Despite the scale of operations and known Egyptian involvement, British intelligence still viewed the attacks as an indication of tribal discontent rather than a cohesive insurgent campaign. There was, however, no alternative for the British other than to mount a military expeditionary operation to restore governance in the Radfan. The armoured cars of 4 RTR had already arrived in Aden from Edinburgh in late September 1963. The Regiment's Saladin and Ferret armoured cars were deployed in squadron strength to support the FRA. A and B Squadrons were in Dhala and Beihen, C Squadron went to Sharjah, with the headquarters located in Little Aden, six miles across the Gulf of Aden from the main town. This was a fraction of the force required to have any real effect on what was a steadily deteriorating situation. With the Integrated Flight Scheme in effect the RTR brought their own Air Troop with them, still equipped with the Auster AOP9. The Austers were gainfully employed on all manner of tasking in support of the patrols of their parent regiment.

The flashpoint that really ignited the Aden Emergency was an NLF grenade attack on the British High Commissioner in Aden, Sir Kennedy Trevaskis, at Khormaksar Airport on 10 December 1963. The grenade killed a woman and wounded 50 other people at the airport. A state of emergency was declared on the same day, tensions rose rapidly, and further grenade attacks were made in the Crater area and against off-duty service personnel. The most notorious attack was made against a children's party at RAF Khormaksar killing one girl and wounding four other children. The Christmas and New Year celebrations of 1963 carried on against a backdrop of training and preparations for an expeditionary operation into the Radfan. The operation, scheduled for the first days of 1964, was codenamed Operation Nutcracker.

Operation Nutcracker

Operation Nutcracker was a clear step change in British policy in Aden and the Protectorate. This was purely a military operation that signalled an abandonment of the traditional, more subtle methods of British colonial governance. There was no intent to apply any form of hearts and minds doctrine to the dissident Radfan tribes. The aim of the operation was to support the federal authorities and the FRA with enough British force to quell the current rebellion and reopen the Dhala Highway. The secondary aim was to then construct new roads that would allow more wide-ranging patrolling and federal reinforcements to enter the Radfan quickly when required. It was hoped that this twofold plan would deter any future rebellion and set stable conditions for Federal Government.

The land component of Nutcracker was built around the three infantry battalions of the FRA commanded by Brigadier Lunt. British involvement in the operation included armoured support in the form of a single troop of Centurion tanks from the 16th/5th Lancers (16/5 L). Dedicated artillery fire support was added to the expedition by J Battery, 3rd Regiment Royal Horse Artillery (3 RHA). The RHA gunners were equipped with new 105mm Pack Howitzers. In addition, 2 Troop, 12 Field Squadron RE joined the Nutcracker Orbat. Helicopter support was considered critical to the operation's success and the RN's aircraft carrier, HMS *Centaur,* arrived offshore adding four Wessex helicopters from 815 NAS (Naval air squadron) to the operation. Further helicopter lift capacity was available from the Belvedere helicopters of 26 Squadron RAF. The twin-rotor Belvedere was capable of carrying 18 troops, and it could lift small vehicles or artillery pieces underslung. The Austers and Beavers of 653 Squadron would also play a part in the operation. The AAC aircraft were to be deployed in the liaison role, assisting in the maintenance of command and control. The RAF Hunter squadrons were also on call to provide CAS as required, with an impressive 36 Hunters allocated to fly on Operation Nutcracker. Avro Shackleton MR2 Bombers were also on hand to carry out night bombing raids on insurgent camps and fortified positions. This temporary task force under Brigadier Lunt would be directed by Headquarters 24 Inf Bde whose units at this point remained in Kenya dealing with unrest there.

The planning of Operation Nutcracker followed the tried and tested methods of previous operations in the region. The age-old military principle of dominance of high ground remained all-important. The execution of this operation differed, however, from past ventures in its employment of modern military technology. The FRA troops would be lifted onto the highest key mountain peaks by helicopter. Once established they were to be resupplied by helicopters, while ground attack and photo-recce would be provided by RAF Hunters. The AAC Auster and Beaver flights would operate closely with the FRA providing AOP to direct artillery fire support and airborne forward air control for the Hunters. Tasking for 653 Squadron was heavy with liaison and reconnaissance sorties carrying British and FRA commanders.

The operation began on 4 January 1964 as intended; the first objective was to secure the Rabwa Pass. The high ridges and peaks overlooking the pass had to be secured by picquets, a forward defensive screen, from 2 FRA. The use of helicopters for this task was considered innovative and risky at the time. Nevertheless, despite the lack of any experience of operations at the altitudes required, the lifting of 2 FRA got underway. The half dozen Belvedere and Wessex helicopters were an integral part of the plan to position picquets on all the key features that dominated the pass. Performance at altitude was an obvious constraint, as was the temperature; even that early in the year temperatures could reach more than 54° Celsius. The large, ungainly Belvedere was of particular concern to Brigadier Lunt:

> There was of course, some risk attached to the employment of these machines, since no one was certain how they would perform at that height (4,000 feet) and in the blazing temperatures

The DHC Beaver proved ideally suited to operations in Aden and the Radfan. (Army Flying Museum)

> of midsummer. There was also the risk that the tribesmen might regard their appearance as a heaven-sent opportunity to try out their marksmanship and even the blunderbuss can do a deal of damage to a helicopter when it is coming into the 'hover'.[10]

In spite of the Brigadier's concerns and those of other staff officers who had little or no experience of working with helicopters, the new aircraft remained on the operation. The success of the Operation Musketeer heliborne assault, coupled with positive reports from Malaya, Cyprus and Kenya outweighed any scepticism. So, after the usual round of New Year parties the force elements began to move into the assembly area at Thumier. H Hour for Operation Nutcracker was scheduled for first light on 4 January 1964. This was delayed until later in the morning due to low mist. It was roughly 1000 hours when the first Belvedere lift dropped its FRA piquet off on their objective on the northern side of the pass. This was Number 1 Piquet, a 16-man FRA platoon commanded by an Arab Lieutenant. As the Belvedere lifted out and turned to fly back to the assembly area a second Belvedere carrying Number 2 Picquet approached its designated drop-off point. The aircraft commander was on the radio talking to the BASO six miles away in the Thumier Assembly Area. As the second FRA picquet party were disembarking from the Belvedere it was hit by small arms fire. The pilot reported hearing a loud crack and holes appearing behind the co-pilot seat, with more rounds hitting the engine bay and the front fuel tank. The insertion of Number 2 Picquet was completed, and the Belvedere lifted out of the HLG. At this crucial point

10 Liddell Hart Centre for Military Archives (LHCMA), LH1/464, Papers of Brigadier J. Lunt.

in the plan the RAF BASO ordered all helicopters to return to base immediately. Brigadier Lunt was surprised by the sudden change of plan.

> I looked over my shoulder and saw all six helicopters turning round and disappearing, and there we were stuck in the Rabwa Pass, with only two picquets established, and we had to wait until the British Air Staff Officer obtained permission from the Air Officer Commanding for the helicopters to complete their mission. This naturally took some time, and I could not wait. We were being heavily fired on at the time, and there was no alternative but to send up picquets on their feet. This is obviously a difficult situation for a commander who is carrying out an operation. If he has not got full control over all the resources on which his plan is based, he is not a free agent.[11]

The suspension of helicopter operations effectively hamstrung Lunt's plan and immediately inhibited the deployment of the FRA battalions through the pass. The aircraft were also scheduled to resupply the FRA units with ammunition, water, and rations. While Lunt adapted his plan the BASO was consulting the Air Officer Commanding, Air Vice Marshal 'Johnny' Johnson, the fighter ace of Second World War fame. Johnson considered the risks to the aircraft and the impact of their withdrawal; he then authorised the return of the helicopters to the operation. By now, the exposed picquets were in contact with large numbers of dissidents. Number 1 Picquet's position was assaulted by a force of more than 50 dissidents and forced to make a lengthy withdrawal in contact. They eventually reached Thumier at 0100 hours on 5 January 1964. On the south side of the pass, Number 2 Picquet was also forced out of their OP position and withdrew off the high ground.

The fighting on the high ground had, however, distracted the dissidents from defending the Rabwa Pass itself. This allowed 3 FRA to drive forward and to reach the head of Wadi Rabwa. At dusk the Belvederes returned to the combat area carrying two 105mm guns from 3 RHA onto the 3 FRA position. Another section of RHA guns arrived a few days later. The gun position was overlooked by a 3,700-foot mountain feature codenamed 'Cap Badge' which presented technical problems with crest clearance for the gunners of the RHA. The gunners, like most of the FRA, would remain in the field for approximately a month. After the initial problems with the use of the RAF Belvederes, the 26 Squadron crews began to develop an understanding of the RHA requirements. Two weeks later 4 FRA moved through 3 FRA positions to advance along the terraced slopes of Bakri Ridge. The movement was supported by the guns of 3 RHA. The guns had been underslung by Belvedere and landed onto their new gun position with their barrels facing in the planned direction of fire. This reduced the manhandling required to re-position the guns in the debilitating heat and brought them into action sooner. A combination of artillery, rocket firing Hunters and FRA movement pushed the dissidents away from the Wadi Rabwa. In the wake of the advance the sappers of 12 Field Squadron RE set about their task of building a sand and gravel road. A fort was also built to control future access to the Wadi.

Operation Nutcracker carried on for some weeks until 28 February 1964. The casualty rate was considered light; the FRA lost five men killed and 12 wounded. The operation had achieved its aim, the NLF and associated dissident tribesmen had been driven back and the new road running from Thumier, through Wadi Rabwa to Wadi Taym was opened on 31 January 1964. The problem now was how to retain the ground that had been gained at some human and financial cost. The FRA was fully committed to the Radfan; this left the borders and the interior of Aden itself lightly manned. The Federal National Guard was struggling to maintain internal security without

11 LHCMA, LH1/464, Papers of Brigadier J. Lunt.

support from the FRA. By mid-February the bulk of the FRA force was withdrawn to Thumier, and helicopter support was also pulled back to RAF Khormaksar. That left the single remaining FRA battalion unable to dominate the high peaks overlooking the valleys below. The 'Wolves' returned to their mountain peaks which made patrolling an increasingly hazardous proposition for the FRA. Despite attempts to use airpower and artillery to suppress dissident activities and maintain the new road, the decision to withdraw ground forces from the Radfan was inevitable.

Early in March the last force elements of Operation Nutcracker withdrew from the Radfan. Brigadier Lunt, who had been sceptical about the cost and value of the expedition from the outset was vindicated. The operation had also confirmed that the FRA lacked the numbers, resources and reach to maintain the security of the new federation without substantial British support. The NLF and their Egyptian backers made much of the withdrawal and the propaganda value of the 'Wolves of the Radfan' driving out imperialist aggressors was a priceless opportunity. To make the propaganda victory even more tangible the Queteibi returned to destroy the new patrol road and level the now abandoned fort. Traffic on the Dhala Highway was again subject to disruption and ambush. The strength of the dissident tribesmen along the highway route was estimated at five hundred, with an additional two hundred Egyptian-trained NLF insurgents.

There was now an enforced lull in operations while the staff at MEC and the Federal Government decided what to do next. Clearly the Radfan could not be abandoned to the NLF and the insurgents, the Dhala Highway was just too important to both the British and the fledgling Federation. A new operation would have to be mounted and to avoid the extreme heat of the summer it would have to begin before the end of April that year. One thing that was certain about the next operation was that there would be a much more significant British involvement. It was also clear that after a shaky start, correctly handled helicopters had proved to be ideally suited to this type of counterinsurgency operation. From this point forward in Aden there would be a constant demand for more helicopters of all types. Up until that time the AAC had been unable to offer an in theatre helicopter capability; the Skeeter was hopelessly underpowered and totally unsuited for operations outside Northwest Europe. The Westland Scout had been plagued by delays and there were very few Alouettes available. Nevertheless, this deficit in capability was soon to be rectified.

On 24 March 1964 the capability of 653 Squadron AAC improved significantly when the Squadron was reinforced with the arrival of two brand new Scout AH1 helicopters. The Scouts had originally been despatched to 8 Independent Recce Flight AAC in Kenya. The Flight was however already trialling two Sud Aviation Alouette II helicopters. This offered the timely opportunity to get much-needed AAC helicopters into Aden, and the Scouts were diverted en route and allocated to 13 Flight AAC. The reinforcement was announced locally in a press release to the Aden Garrison Newspaper, *The Dhow* on 2 April 1964:

1. In the last few days, helicopters painted with Army camouflage colours have been seen flying over the Little Aden Area.
2. These are the first two Scout helicopters to be received by 653 Light Aircraft Squadron, Army Air Corps. They arrived on HMS *Bulwark* and were flown direct from the carrier to the Squadron Airfield.
3. The Scout helicopter is a five seater aircraft powered by a free turbine, it will be used for liaison and reconnaissance purposes both upcountry and in Aden State.[12]

12 *The Dhow*, Press Release, 653 Sqn AAC to the Editor, HQ Aden Garrison, 2 April 1964.

Suez JHU veteran, Lieutenant Colonel Francis Graham-Bell was in command of 653 Sqn when the first Scouts arrived in theatre. (Samantha von Huetz)

The arrival of additional aircraft was welcomed by 653 Squadron, but the Scout was very much an unknown quantity. Lieutenant Colonel Francis Graham-Bell AAC was commanding the Squadron.

> We had very little idea of how its performance would be affected by the temperature and humidity or what effect the sand and the dust would have on engine life and efficiency. Furthermore, for these two SCOUTS we had only three pilots, myself, one Theatre Helicopter Instructor and the Flight Commander.[13]

The introduction of the Scout was fortuitous because as the new helicopters arrived in Aden the nature of counterinsurgency operations was becoming more complex, and the tempo and scale of future operations were set to increase. There was much for 653 Squadron and their RN and RAF counterparts to learn from the limited success of Operation Nutcracker. It was generally accepted that the use of air and aviation assets should be an integral component of any future operation in the Radfan. The principal lesson was that clarification was needed on the ownership (command) of helicopters during future operations. Experience from operations in Malaya and Borneo had shown that the RAF preference for central command and control of helicopters and their tasking had proven to be much less effective than allocation of helicopters down to Army brigade, battalion or even company level. The resistance to this proven doctrine and inter-service argument on the subject rumbled on throughout the Aden campaign.

The demand for aviation support in Aden Colony itself was easily met from the Squadron's main operating base at Falaise Airfield in Little Aden. During Nutcracker and for ongoing operations the Squadron forward based an aircraft of each type at Thumier. This was a logical method of providing Auster, Beaver and Scout over the Dhala Highway and other trouble spots at a reduced notice to move. The sceptics that had initially doubted the value of helicopters were, in the main, now enthusiastic converts as Lieutenant Colonel Graham-Bell commented.

> Aircraft at Thumier were tasked by Brigade HQ as they saw fit. The Scout rapidly became a status symbol and all Recces 'had to be carried out by Scout'...even though Beaver was more suitable. The absurdity of the situation was well brought out on one day having flown three recce sorties in a SCOUT, I wanted to do a Recce of the Wadi Taym for myself and chose a BEAVER as the more suitable aircraft. Obviously, control had to be exercised if precious SCOUT flying hours were not to be wasted and the system was introduced where the unit said what they wanted done and we allotted the appropriate type of aircraft. Thus, the users could no longer stipulate a SCOUT and many flying hours were saved. At this time the ban on Austers carrying passengers was lifted so that they could be used for passenger recce. It was interesting to note how quickly a recce party grew to Beaver size as soon as an AUSTER was allotted to a task![14]

The month of March 1964 saw renewed levels of dissident activity. Reports from the Radfan indicated the increasing sophistication of NLF tactics and training. Individual dissident units were making greater use of radio communication to coordinate multiple attacks on commercial and military traffic on the Dhala Highway. A steady rise in the number of land mine attacks and associated casualties was also a cause for concern. The Federal Government was convinced that a second operation must be mounted in the Radfan and this time it must be British led. The window

13 AFM 204c: 653 Sqn AAC, Scout AH Mk1, Radfan Operations 27 Apr–30 Jun 1964, p.2.
14 AFM 204c: 653 Sqn AAC, Scout AH Mk1, Radfan Operations 27 Apr–30 Jun 1964, p.4.

of opportunity for such an operation was, however, reducing by the day. The approach of the summer and climbing temperatures meant that any operation would have to be launched before the end of April.

The problem in the Radfan was clear, the solution equally self-evident. Any venture into the troubled region required a larger and more capable force than had been available for Operation Nutcracker. That force would be mainly British with a British commander. With time marching on and all of its brigade headquarters committed to ongoing operations, an unorthodox solution was required. The first thing to do was to find a commander for the new operation. Major General John Cubbon, GOC Middle East Land Forces, wasted no time resolving the problem. On 14 April 1964 he appointed the newly arrived Commander Aden Garrison, Brigadier Louis Hargroves to run the imminent Radfan operation. Both men were veterans of previous operations in the Middle East and Hargroves also had considerable experience of mountain warfare. A temporary headquarters was established by Hargroves at Thumier.

A composite force designated as 'Radforce' was then assembled under Hargroves to mount what had been codenamed Operation Cap Badge. The Radforce planners grappled with numerous problems, not least a lack of quality intelligence and unclear direction as to how long the operation was to last. It was abundantly clear that even with the arrival of the 13 Flight AAC Scouts, and also making use of the Westland Whirlwinds of the local RAF Search and Rescue Flight, there would still be a desperate shortage of helicopters. By 17 April, Hargroves had constructed an orbat for Radforce and settled on an operational plan. In spite of its commitments in East Africa, HQ 24 Inf Bde managed to release 45 Cdo RM to bolster Radforce. The marines arrived to join other force elements gathered from the FRA and British units drawn from the Aden garrison and the MEC. In a matter of days Radforce had grown from a tiny headquarters to a brigade sized formation of 2,500 troops.

The Radforce Orbat consisted of 45 Cdo RM, two battalions of the FRA, B Company 3 Para, A Company from the 1st Battalion, The East Anglian Regiment (1 E Anglian), a troop of armoured cars from D Squadron, 4 RTR, the guns of J Battery, 3 RHA and an Engineer Troop from 12 Squadron RE. This impressive feat of military organisation was almost undone on 25 April. The Radforce Command Group was travelling in two Land Rovers when the lead vehicle hit a landmine. One officer was killed instantly, a second was mortally wounded and a third died later in hospital. Brigadier Hargroves was travelling in the second Land Rover and narrowly escaped being killed. Undeterred by this close call, Hargroves ensured that planning for the operation carried on at pace. The first phase of Cap Badge began in the area of Wadi Taym. The opening stage involved a night assault to secure the Danaba Basin and the nearby Rice Bowl Ridge on the evening of 30 April/1 May. As with all operations in the Radfan it was essential to secure the high ground and dominate the area of operations. The 3,700 feet feature known as Cap Badge completely dominated both the Danaba Basin and the Rice Bowl. It was therefore vital ground and had to be occupied by Radforce by dawn on 1 May 1964. A night infantry assault was settled on for a number of reasons, not least the lack of enough helicopters to consider a daylight operation. A night attack also robbed the tribesmen of the advantage of good observation. With the daytime temperatures peaking at 49°C, movement at night offered obvious advantages.

The Marines of 45 Cdo RM were tasked with making a night approach to Cap Badge on foot. Three companies would move through the Wadi Boran as the route was considered a more difficult but a less likely route for an assault. It was hoped that the less obvious approach would be undefended. The plan also included a diversionary thrust by a mixed British/FRA force along the obvious route through the Wadi Rabwa. The approach march through the Wadi Boran could be covered from a hill codenamed Sand Fly. The task of occupying Sand Fly and denying it to the dissidents was given to Z Company, 45 Cdo RM. With Sand Fly secured, X and Y Companies

RM were to pass through Wadi Boran and then on through the Danaba Basin to secure Rice Bowl Ridge. While this was happening, 120 men of B Company, 3 Para were to parachute onto Cap Badge. The DZ for B Company was to be marked by 3 (Mobility) Troop, 22 SAS. The SAS pathfinder patrol was to be inserted in advance of the main operation by two Scout helicopters from 653 Squadron.

See Map 10

Operation Cap Badge

During the few weeks leading up to Operation Cap Badge, 653 Squadron had integrated the new Scouts alongside their Beavers and Austers. The Scouts were allocated to 13 Flight, and it was hoped that the Scout would initially operate alongside and eventually completely replace the Auster in Aden. Also, newly arrived in Aden from Hereford were A Squadron, 22 SAS. The code-word to activate an A Squadron emergency call out in April 1964 was 'Free Beer'. The callout was initiated on 22 April by the SAS squadron commander, Major Peter de la Billière. Under the cover of a practice callout for a training exercise on Salisbury Plain, the 75-man Squadron had flown out to Aden. The men of A Squadron were almost all veterans of operations in Borneo and Malaya, many had also served in Oman. They were also experienced with helicopters, having used them extensively for the insertion, extraction and resupply of jungle patrols.

The Squadron were given orders for their task to mark out the DZ atop the Cap Badge feature for the drop by B Company, 3 Para at midnight 30 April 1964. The plan required an approach march to Cap Badge under the cover of darkness on the night of 29 April. The A Squadron planners requested insertion by Scout. It was decided to use both Scouts to lift the nine men of 3 Troop, commanded by Captain Robin Edwards, to an HLG some 5,000 yards up the Wadi Rabwa. An appreciation of maps, photographs and local knowledge suggested that the HLG could be easily overlooked by dissidents. The Scouts would therefore be vulnerable to small arms fire during the drop off. It was decided to send a troop of armoured cars from 4 RTR onto the HLG during the morning of 29 April to secure the area. Once this armoured cordon was in place the Scouts would fly a dummy run and prove the HLG was viable. In the event, the armoured cars were under almost constant fire from the hills above from the moment they arrived. It was therefore decided to abandon any thought of flying a route recce.

The Scouts would have to carry out the insertion of the nine-man SAS patrol based on a map recce and an appreciation made from photographs. This was a first and there was little time to develop any standard drill; time on the HLG had to be reduced to an absolute minimum. The plan required three lifts to get all nine men forwards, along with their 60lbs of equipment each. The aircraft was stripped of doors and safety straps to reduce the vulnerable offloading period and the SAS were instructed to return fire from the rear cabin if any opposition was encountered. The Scout was much easier to get out of than the much higher Belvedere and Wessex; troops could be out of the aircraft and clear of the rotor disc relatively quickly.

The insertion was scheduled for dusk, with the first run to the HLG lifting from Thumier at 1820 hours. In the fading light, the Scouts began a series of low level, 80 knot runs following the bed of the Wadi Rabwa. As the Scouts approached the HLG, the armoured cars lifted their fire to allow them into the HLG, and the RHA also switched fire to an alternate target. Once each drop was complete and the Scouts were clear of the HLG, both fires were turned back onto the dissident positions. The dust created by the rotor downwash and the artillery strikes reduced visibility considerably for the pilots. It was later described as being like flying through porridge. Each of the drops took less than a minute on the HLG and all three lifts were completed in 20 minutes. Despite the dissident fire and the dust, the mission to insert the SAS patrol was a complete success. The Scout had proved itself as the ideal aircraft for this type of small-scale troop lifting task.

Operations in Aden and the Radfan were the beginning of a long and productive association between AAC Scout crews and the SAS. (Army Flying Museum)

Once the Scouts had departed from the HLG for the last time, Captain Edwards and his two four-man teams waited for dark before moving off. The Troop had a six-mile march ahead of them as they moved off in silence, heading for the DZ on Cap Badge. Shortly after leaving the HLG, as they began to climb the Jebel Ashgab (4,000 feet), things began to go wrong. Trooper Nick Warburton, the patrol signaller, started to feel sick and then became violently ill. The patrol slowed and eventually they were forced to halt. Edwards ordered his men to redistribute Warburton's radio equipment and kit among themselves and they set off again at a much slower pace. As they moved in the darkness members of the patrol reported that they thought that they were being observed or followed. Progress had been painfully slow; Warburton's condition was not improving, and dawn was approaching. If his men's suspicions were correct and they were being watched, a helicopter casevac was out of the question. At 0200 hours Edwards made the decision to lay up through the day and hope that Warburton's condition would improve enough to allow the mission to continue the next night. He informed A Squadron Headquarters at Thumier of his decision by radio. The patrol established itself in a good defensive position in two hillside sangars. They had settled three miles from the DZ on the slopes of the Bakri Ridge. The following morning the rising sun brought the realisation that their position was just 3,000 feet above the rebel village of Shab Tem. From their position the patrol could see 25 uniformed dissidents moving around in the village.

The situation worsened when a goat herder began moving his animals close to the patrol's position. In spite of their efforts to remain undetected, the goat herder spotted the SAS men and raised the alarm before he was silenced by one of the snipers in the patrol. The shouting of the goat herder and the sniper shot had compromised the lay up position, and tribesmen gathered and began an assault on the SAS position. The initial assault and a subsequent attempt to outflank Edwards and his men were beaten back by accurate fire. As the morning moved on more dissident tribesmen arrived to reinforce their comrades. The dissidents managed to get behind the SAS and fire down onto the patrol from above. This made a breakout in daylight unthinkable; it would be suicide. Edwards called for air support and for most of the day pairs of Hunters from 8 and 43 Squadrons RAF blasted the surrounding ridges and hills with their 30mm cannon and 3-inch rockets. Later in the afternoon the 105mm pack howitzers of J Battery, 3 RHA also began to bring fire down around the patrol position. This gave Edwards and his men breathing space and helped them to conserve their ammunition. In between the Hunter strikes the firefight continued. Corporal Paddy Baker was hit in the knee by a .303 round. The round passed through Baker and killed Trooper Warburton. Later, Trooper Hamilton was lightly wounded and the patrol's radio was knocked out. The battle intensified and a charge of 70 tribesmen was only stopped by the rapid and accurate fire of Sergeant Geordie Tasker and his Bren LMG. As dusk approached, and with it any prospect of further air support, Edwards ordered his men to prepare to break out of their besieged position. The patrol would fight their way out in two fire teams using well-rehearsed fire and manoeuvre drills to support each other's withdrawal.

Edwards gave the order to move and was immediately killed by an incoming round. His body and that of Trooper Warburton had to be left behind on the position. The remaining seven members of the patrol fought their way out, stopping twice to ambush their pursuers. All seven reached Thumier at first light on 30 April 1964. While the battle had raged through the afternoon at Shab Tem, it was obvious to de la Billière and his command team that 3 Troop were not going to make it onto the DZ. A contingency plan was worked up and another SAS patrol would have to be sent in to mark the DZ. The new plan involved 1 (Mountain) Troop led by Captain Ray England. The two 13 Flight Scouts were called forward again, but this time the SAS intended to drop from the Scouts by parachute. England and the two pilots recced the route and overflew the DZ in a 15 Flight Beaver. During the recce the three identified a stone building that would make a good defensive strongpoint if required. England was happy with what he had seen and set off to brief his Troop on the plan. At the same time, for reasons unrecorded, the 13 Flight pilots changed over. OC 653 Squadron, Graham-Bell, would now be leading the Scout pair with Captain Claude Surgeon AAC flying the second. England was far from happy with the last-minute change of pilots and made his views clear to Major de la Billière. The availability of the Scouts was also constrained by the requirement to return to Thumier before last light. There was little to be done at such a late stage, daylight was fading, and 3 Troop must by now be low on ammunition and water. The RAF Hunters were due to withdraw, having flown 18 sorties, fired 127 rockets and expended 7,131 rounds of 30mm cannon ammunition.[15]

Prior to launching the insertion of 1 Troop, Major de la Billière made a blind radio call to Edwards ordering him to break out to the east and make for the strongpoint identified during the Beaver recce flight. All communication with 3 Troop had been lost, so there was no way of knowing whether the message had been received. The two Scouts set off with England and some of his patrol on board. Flying tactically at low level the two aircraft deviated from the planned route. Both aircraft came under LMG fire and the lead Scout was hit in the fuel tank and began leaking fuel. The tail rotor of the second aircraft was hit and damaged by tracer fire. With both Scouts damaged, Graham-Bell was forced to abandon the mission and limp back to Thumier. Graham-Bell would later report on the two SAS insertion missions:

> These two attempts, one successful, and one unsuccessful, showed the potential of the Scout for offensive action and dictated the pattern for future operations. They also emphasised the vulnerability of the helicopter – the expected result of every sortie in close proximity to dissident areas had to be weighed against the possible loss of the helicopter.[16]

15 van der Bijl, *British Military Operations in Aden & Radfan*, pp.93–94.
16 AFM 204c: 653 Sqn AAC, Scout AH Mk1, Radfan Operations 27 Apr–30 Jun 1964, p.3.

Unaware that 3 Troop were already fighting their way out of Shab Tem, England suggested a parachute drop from an RAF Short Pioneer onto the DZ. This last-ditch attempt was vetoed by Major General Cubbon, Commander MELF. A few days after the 3 Troop survivors reached Thumier, Yemeni radio reported that the heads of two British SAS soldiers had been displayed on stakes in the Yemeni town of Taiz. On 11 May 1964 troops from 1 FRA were involved in a firefight with 25 NLF insurgents at Shab Tem. During the fighting, the FRA called in artillery support and drove the insurgents back. In the process, the decapitated bodies of Captain Edwards and Trooper Warburton were discovered in a shallow grave. The bodies of the two SAS men were picked up by a 13 Flight Scout from what was described as an almost impossible HLG and flown back to Aden for burial.

Phase one of Operation Cap Badge began for the remainder of Radforce on the night of 29 April 1964. As planned, the three companies of 45 Cdo RM struck out into the darkness heading into the Wadi Boran. The marines of Z Company left the column at the mouth of the wadi to occupy Sand Fly hill. The advance continued with X Company leading Y Company further into the wadi. The advancing marines encountered problems; the terrain was much harder going than had been anticipated. At midnight CO, 45 Cdo RM Lieutenant Colonel Stevens RM, informed HQ Radforce that his companies were three miles inside enemy territory at the base of the Coca Cola hill feature. Radforce in turn informed Stevens that the drop by 3 Para was cancelled. Brigadier Hargroves ordered Stevens to occupy and consolidate defensive positions on Coca Cola and Sand Fly. The occupation of Coca Cola in darkness was not a straightforward proposition. The climb was 1,600 feet of steep, rocky slopes laced with treacherous ravines. The Recce Troop of 45 Cdo RM, led by Major Mike Banks RM, found a way up onto the peak and four hours later 45 Cdo RM reported that Coca Cola was in their hands. The dissident tribesmen awoke on 30 April to find 250 Royal Marines with their Vickers Medium machine guns and 81mm mortars on the two peaks dominating the Danaba Basin.

The failure to reach Cap Badge on the first day presented HQ Radforce with the challenge of securing the peak without the element of surprise. An alternate plan would have to be developed and executed quickly. The logistic plan would also require some adjustments. The logistic estimate for sustaining Radforce in the field on Operation Cap Badge had concluded that nearly 14 tons of combat supplies a day were required. Two-thirds of that total was water. Prior to H Hour on 29 April 1964, 35 helicopter lifts per day were needed to bring those supplies into the Brigade concentration area at Thumier. The resupply of the diversionary force and the Royal Marines on Sand Fly and Coca Cola would require every available aircraft. The vulnerability of the large Belvedere helicopters and their limited operational window due to heat and altitude were significant constraints on any heliborne resupply programme. Nevertheless, the newly occupied mountain-top positions had to be resupplied and helicopters were the best method available. So, with little or no experience of mountain flying, the pilots of 13 Flight AAC were tasked to carry out the resupply of 45 Cdo RM on top of Coca Cola. The POR described the debut of the new aircraft type and the unforgiving flying conditions encountered in the Radfan mountains:

> Our first task was the Resupply of Objective Coca Cola – first experience of flying SCOUT in mountainous conditions. Coca Cola – razor backed peak a quarter of a mile long and just 30 feet wide at its widest point. Standing 4,000 feet above Sea Level …. 1,500 feet sides dropping down vertically to Wadi beds on either side. From mid-morning wind speed and turbulence became extreme. Landings and take offs were hazardous undertakings. On one occasion while taking off and attempting to dive the aircraft over the side of the cliff in order to gain airspeed, the aircraft actually rose vertically for 500 feet in a right-hand spiral. However, it was very good practice and stood us in good stead as we advanced into the Radfan.[17]

17 AFM 204c: 653 Sqn AAC, Scout AH Mk1, Radfan Operations 27 Apr–30 Jun 1964, p.5.

The brief pause to consolidate allowed Brigadier Hargroves to modify the plan to capture Cap Badge. He ordered B Company, 3 Para to move up by road to Thumier. The paratroopers would join X Company, 45 Cdo and prepare to take Cap Badge from the southwest. At the same time, Y Company would move off Coca Cola and capture the peak of Gin Sling. This would allow Y Company to observe and dominate the Danaba Basin and Wadi Taym. The FRA would continue to dominate the Wadi Boran and draw away dissidents from the Danaba.

On 2 May, the UK Spearhead Battalion, 1st Battalion King's Own Scottish Borderers were flown in from their barracks in Folkestone. The arrival of a reinforcement battalion in Aden allowed Hargroves to order the already acclimatised troops of 1 E Anglian to move up from Aden and reinforce Radforce. The East Anglians relieved the Royal Marines on Sand Fly and Coca Cola on 4 May. The now dehydrated marines then joined B Company, 3 Para in the valley below. The combined force set off in darkness on the five-mile march across the Danaba Basin. At dawn on 5 May 1964, the men of Y Company successfully climbed 1,000 feet and captured Gin Sling, X Company made a similar 1,200-foot climb and finally took the peak of Cap Badge. The dissidents who had occupied the peak could be seen scrambling down the far side after hastily abandoning their sangar positions.

Below and to the east of Cap Badge, B Company 3 Para were close to the village of Al Naqil. The paratroopers advanced on the village and became embroiled in a protracted firefight. The company second in command, Captain Edward Jewkes, was killed while treating a badly wounded sergeant in the open. Six other B Company men were wounded. Eventually, B Company took control of the village which was then renamed Pegasus Village. The dissidents were not prepared to give up the village completely and continued to mount counterattacks against B Company. Artillery fire support was available, although the gunners of 3 RHA were firing at extreme range. Air support was also called in once the sun rose, but it was difficult to identify the dissidents in the rocky terrain and the shadow of the hills. Water and ammunition were desperately needed if the village was to be held and resupply was required at all the new Radforce positions. After a long night march and some hard climbing with heavy loads, water bottles were nearing empty. As the sun rose the RAF Hunters reappeared in the skies over Cap Badge and 653 Squadron's Beavers and Scouts returned to the battle area:

> There was an urgent requirement for resupply of water and ammunition at CAP BADGE and a SCOUT was sent in at 0700hrs to ascertain if the area was safe for helicopter resupply. Although it was fired on during the first sortie it was not hit, and subsequent sorties proved that it would be safe for the more unwieldy BELVEDERES to carry on with the resupply. At the same time a resupply was called for on to PEGASUS VILLAGE and two Beavers were detailed for this sortie. Flying in low they dropped their loads by parachute and the drop was completely successful. During this drop both aircraft were under fire from the dissidents who were covering PEGASUS VILLAGE and one of the BEAVERS was hit.[18]

Twenty-four hours later, Z Company RM had been flown forward by Belvedere from Coca Cola to Cap Badge. The marines then moved down off the peak to drive the dissidents onto the guns of B Company in Pegasus Village. The dissidents were driven off, and B Company were relieved. This was the end of the first phase of Operation Cap Badge. Radforce had achieved all its objectives, casualties were relatively light, Radforce had lost just two dead and eight wounded. The tribal elders of the Hujari sought to negotiate terms with Hargroves. The approach was rejected as none of the dissidents were willing to negotiate. Phase one had secured the north and west sectors

18 AFM 204c: 653 Sqn AAC, Scout AH Mk1, Radfan Operations 27 Apr–30 Jun 1964, p.6.

Undeterred by ground fire, Beavers of 15 Flight AAC supported the assault on Cap Badge with vital parachute resupply drops. (Army Flying Museum)

of Radfan, although the NLF and dissident tribesmen remained active in the mountains of the south and east of the region. The FRA moved up onto Coca Cola and Sand Fly to relieve the 1 E Anglian. HQ MEC ordered a pause to allow Radforce to reconstitute and to carry out reconnaissance of the next series of objectives.

There was much for 653 Squadron to reflect on at this stage. The SAS patrol insertions and support to Operation Cap Badge itself had enhanced the credibility of the AAC in Aden. The introduction of a rotary capability in the form of the Scout had been well-managed and a host of new lessons had been learnt in a matter of weeks. The theory and practice of mountain flying on operations had been tested to the full although trial by combat did of course expose some shortcomings in the peacetime training syllabus.

> …it was necessary to hastily modify the mountain flying technique taught at Middle Wallop. It was quite obvious that a 360-degree orbit of a landing site held little future if 270 of those degrees was covered by enemy fire. Thus, a straight in approach technique was developed, checking sink with power and aborting the approach if sink could not be arrested. The direction of approach was dictated by the arc of ground held by our troops and by the terrain – wind direction had little bearing on it, and we were lucky if the approach was into wind.[19]

The full potential of the Scout was by now being realised, and a regular shuttle service was started to move men and equipment from Coca Cola onto Cap Badge. Initial suspicion and wariness of the Scout had been replaced with blind faith in its powers. At this stage of the operation little attempt was being made by troops to calculate the weight of stores being loaded onto the aircraft. It was impossible for pilots to question the troops loading the stores as most of the activity took place with rotors running. In any case such questioning often received unreliable answers. One notable incident saw two soldiers staggering to carry a box under the rotor disc and onto a waiting

19 AFM 204c: 653 Sqn AAC, Scout AH Mk1, Radfan Operations 27 Apr–30 Jun 1964, p.6.

Scout. When challenged by the pilot about the weight of the box they stated that it weighed 40lbs. Lessons were learnt after Operation Cap Badge and user units were educated about the lift capacity of the new helicopter; until then numerous incidents of dangerous overloading occurred, where the 1,500-foot drop-off was the only thing that saved the Scout. The technique adopted was to carefully hover to the edge of the Coca Cola feature before diving the aircraft over the edge and then flying onto Cap Badge.

A significant planning variable that had affected 653 Squadron's ability to provide direct support to the operation was the 60 mile transit from Falaise Camp up to Thumier. Pilots were lifting from Falaise at dawn and then transiting to Thumier. After an intensive day of flying in extreme temperatures they were returning to Falaise Camp at dusk to allow for aircraft servicing and pilot recuperation. This daily transit also consumed a valuable flying hour every day. The consistently intense tempo of operations, coupled with extreme heat meant that pilot fatigue was a constant concern. The effects of fatigue became more and more evident as the days passed and symptoms included excessive yawning, a harassed expression, depression and a tendency towards irritability. The average flying day began at 0430 hours with the first take off. The working day usually lasted through to completion of the evening brief at about 1950 hours. Constant noise, dust and oppressive heat ensured that there was little chance of snatching sleep during any lull in tasking. Pilots were generally in bed asleep by 2030 hours having grabbed a quick evening meal.

The nature of the flying itself also sapped energy. Tasking frequently involved long waits in the cockpit exposed to the sun, constant thirst, irregular meals and being continually bathed in sweat. Pilots were consistently required to fly to the limits of the aircraft, at altitude, in turbulent air. Landings on some of the exposed and difficult HLGs required great accuracy and imposed a higher degree of strain than would be expected when landing on a more conventional HLG. The employment of the Scouts in the intimate support role generated some interesting statistics. Analysis of landings per flying hour during their first 610 flying hours in theatre revealed that pilots were completing a landing on average every 8.5 minutes.[20]

Rest periods at Falaise were programmed at the end of the first week and pilots were sent back from Thumier for 48 hours rest and recuperation. This short break was found to be inadequate and fatigue levels quickly rose again, so a further 48 hours was built into the programme after 14 days. Eventually a policy of seven-day rest periods alternated with seven days of mountain flying was adopted. This was not always achievable, but it was considered to be the optimum solution. The requirement to rotate pilots out of the line eventually led to a policy of three pilots per airframe. During the initial months after the Scout's arrival in theatre the pilot to aircraft ratio was just one to one. While the obvious waste of flying hours and the risks associated with aircrew fatigue were an important planning constraint, the impracticability of operating from a rear base was having a negative impact on the Squadron's all-important operational output. The daily withdrawal of the aircraft from the area of operations created a feeling that was described as a lack of 'touch' and not being 'with it'. The general feeling was that a 60 mile gap did not facilitate intimate support. The Squadron Commander, Graham-Bell, had strong views on the subject and this statement contained in his POR captures the ethos of Army aviation to this day:

> In order to give intimate air support, it is essential to be part of the force being supported and this cannot be when you are living going back sixty miles every night to home comforts and civilisation. You must live on the spot, sharing the discomforts, really getting to know the people that you are working with, building a sense of camaraderie, and keeping in the picture

20 AFM 204c: 653 Sqn AAC, Scout AH Mk1, Radfan Operations 27 Apr–30 Jun 1964, Annex A.

> with every minute change in the situation on the ground. It is useless to try and support troops on the ground if you are doubtful or uncertain of their exact location.[21]

On 7 May, just as phase one of Operation Cap Badge was drawing to a close, the 653 Squadron Tactical Command Post (Tac CP) was deployed from Falaise Airfield to Thumier. The Squadron also increased its numbers of forward deployed aircraft to two Scouts, two Beavers and two Austers. The Tac CP maintained a radio rear link back to the Squadron Headquarters at Falaise and was connected to brigade HQ by field telephone. The Squadron now had an Operations Officer working directly from the Tac CP to Brigade. In this way the pilots were kept up-to-date with the current battle picture and aircraft were allocated to tasks as required. In addition, a small REME detachment moved with the aircraft and were able to carry out first line Battle Damage Repairs and maintenance. Any more complex servicing that involved stripping the aircraft down and exposing the engines to the elements had to be carried out overnight back in the hangars at Falaise. The REME had a difficult task as the Radfan operations were asking far more flying hours from the aircraft than the prescribed rates. The Scout was flying 60 hours a month on average in the Radfan, compared to the usual average of 35 hours. For the Beaver the rate was 49 hours against the prescribed 45 hours and for the Auster 53 hours compared to the normal 30 hours.[22]

There may have been a lull in offensive operations on the ground, but the resupply task load remained constant. The Beaver pilots of 15 Flight AAC continued to fly supplies forward from Aden to Thumier. The Scouts of 13 Flight AAC were extremely busy working between the Radfan peaks. The absence of airstrips in Radfan constrained any wider use of the Beaver and placed greater load on the available Scouts. At this stage, the transition from Auster AOP9 to Scout was still in progress and the Austers still had a part to play. In spite of its inadequacies, the Auster was far from useless and they were committed daily to the first and last light 'Camel Spotting Sorties'. These tasks usually lasted about an hour and involved patrolling caravan routes searching for camel trains carrying supplies to the dissidents. The Austers were also on call for AOP tasking, working with the gunners of 3 RHA. Tasking also included photo-recce sorties, light resupply flights and ABFAC.

The reinforcement of the Squadron continued with the arrival of three new Scouts by RAF Beverley transport on 12 May 1964. Consolidation of the position in Radfan also progressed with a recce to identify locations for new fixed wing airstrips in the area of Wadi Taym. Two sites were identified and the RE set about constructing 'Paddy's Field' and 'Monk's Field', both of which were suitable for Beaver operations. The new airstrips allowed Beaver and Auster to operate in the Danaba Basin and over Wadi Taym which reduced the amount of tasking being placed on the hard-pressed Scout Flight. The main runway and facilities at Thumier were also enhanced to sustain RAF medium transport aircraft. Most accounts refer to the airfield as Habilayn from that point onwards. The 15 Flight Beavers had proven effective in the parachute resupply role but the availability of this useful form of resupply had been hampered by a shortage of parachutes. The expansion of airfield infrastructure, and reinforcement of the RASC in theatre, brought with it an RASC Air Supply Troop and more parachutes. As the campaign progressed Beavers would regularly be employed in close support, making para-drops to ground forces. The pilot to Scout ratio had been improved with the arrival of four Scout trained pilots, including Major Jimmy Nunn AAC from 8 Flight in Kenya, Major Jake Jackson AAC from Boscombe Down, Captain Claude Surgeon AAC, a QHI from Middle Wallop and Captain Hardie AAC from BAOR.

21 AFM 204c: 653 Sqn AAC, Scout AH Mk1, Radfan Operations 27 Apr–30 Jun 1964, p.6.

22 B. M. Kennett and J. A. Tatman, *Craftsmen of The Army, REME Official History* (London: Leo Cooper, 1970) p.351.

'Keeping in the picture with every minute change in the situation on the ground.' The 653 Squadron ops room forward at Thumier. (Army Flying Museum)

The Auster AOP9 maintained a credible presence in Aden operating in the ABFAC and light liaison roles. (Army Flying Museum)

The establishment of a FOB at Thumier saved vital flying hours, increased loiter time and reinforced the need for dedicated signallers, drivers and refuellers. (Army Flying Museum)

There were other changes underway as Brigadier Hargroves returned to his post in Aden Garrison. Radforce was only created as a temporary command and Hargroves was replaced by Brigadier Cecil 'Monkey' Blacker MC, Commander 39 Infantry Brigade (39 Inf Bde) from Northern Ireland. Hargroves had achieved wonders with a tiny headquarters staff and a hastily gathered composite force; the next phase of the Radfan campaign needed a fully functioning brigade headquarters. The appointment of Blacker, and the arrival of his brigade headquarters, along with the construction of forward airstrips signalled an intent to retain the initiative in Radfan. Blacker was determined to reinforce the initial success of phase one of Cap Badge and press on with phase two of the operation.

The 39 Inf Bde plan was to seize the Bakri Ridge on or around 22 May 1964. This date was dependent on the arrival of the Aircraft Carrier, HMS *Centaur*. The Centaur's Air Group included six Wessex helicopters from 815 NAS. It was intended that the Wessex would augment the Belvederes and provide much-needed additional troop lifting capacity for the assaults on Bakri Ridge and beyond. With the planned date in mind, the Belvederes of 26 Squadron RAF had been taken offline for a week of repairs and major servicing. In hindsight this total lack of any contingency for heavy-lift helicopters appears to show a lack of awareness of joint operations on both sides of the arrangement. During the operational pause, the new brigade was reinforced with additional armoured cars, an increased number of RA guns and additional RE strength. The SAS also increased the range and frequency of their patrolling and intelligence gathering activity. The main body of 3 Para also arrived in Aden from Bahrain to join B Company and reinforce the Brigade. All that was needed, as ever, were more helicopters.

The 653 Squadron report on the operation gives some indication of the mood at the time and the working relationship that the Squadron developed with the new arrivals:

> The comparative peace and quiet was rudely shattered by the arrival of 3 PARA who decided to take Bakri Ridge at a hard gallop. Their enthusiasm and verve were very catching, and the result was a magnificent spirit of cooperation built up between the Regiment and the Squadron.[23]

Brigadier Blacker carried out a recce of the phase two area by Scout with OC 653 Squadron flying the sortie himself. The Brigade Commander was accompanied by CO 3 Para, Lieutenant Colonel Anthony Farrah-Hockley and CO, 1st Battalion King's Own Scottish Borderers, Lieutenant Colonel Alastair Thorburn. As a result of the recce Blacker changed his mind and decided that he could not allow the dissidents to gain further strength and improve their defences, therefore he could not wait for helicopters. HMS *Centaur* and her Wessex were still five days steaming away and there was no guarantee that the ridge would not be heavily reinforced in that time. Blacker decided to go without the Belvederes and the Wessex and new orders were issued. The plan envisaged a three-phase operation: first, the Bakri Ridge must be occupied; this would be followed by 3 Para driving through the Wadi Dhubsan. Finally, 1 E Anglian was to take Wadi Misrah and clear Jebel Huriyah. This plan thrust deep into the fertile heartlands of the dissident tribes, thereby demonstrating that the British, and therefore the Federal Government, could penetrate any area of the Radfan at will.

Operation Cap Badge – Phase 2

Orders were given for 3 Para to prepare to advance to the Bakri Ridge on the night of 16 May 1964. Farrar-Hockley had estimated that due to the terrain and the paucity of helicopter support, the initial advance would take two nights. The 3 Para plan requested two 13 Flight Scouts to act as dedicated aviation support with the Flight tasked to provide troop-lift, resupply and casevac. The first task for 13 Flight in the 3 Para plan was to lift water, ammunition and radio batteries onto the battalion start line at Shab Tem on 16 May. Heavy rainfall and hailstones that day grounded both aircraft and washed away the track through the Wadi Rabwa making it impassable to vehicle convoys. The 3 Para operation was delayed by 24 hours and with no helicopter support and the road impassable, the battalion simply carried its entire stock of combat supplies on foot to their start line. This gave 653 Squadron enough time to repair the storm-damaged Scouts and get them in the air later that day.

> It all started quietly enough with a Scout Recce to find a line of advance. A camel track was spotted, and a couple of sorties were carried out so that Company Commanders could have a look. And then, the road up the Rabwa Pass was blocked! An urgent request was received to ferry supplies, water, and ammo to Shab Tem, the Start Point, so that the advance would not be delayed. All the stops were pulled out and by nightfall sufficient stocks had been built up at Shab Tem.[24]

23 AFM 204c: 653 Sqn AAC, Scout AH Mk1, Radfan Operations 27 Apr–30 Jun 1964, p.8.
24 AFM 204c: 653 Sqn AAC, Scout AH Mk1, Radfan Operations 27 Apr–30 Jun 1964, p.8.

The Scout resupply lift on 17 May had further developed the mutual understanding and respect between 653 Squadron and 3 Para as each was aware of the other's capabilities and requirements. The advance got underway on the night of 17 May, each man carrying 180lbs for six miles over extremely rough terrain. The next morning a 15 Flight Beaver carried out a parachute resupply, dropping supplies onto the 3 Para position at Lethoom. The march carried on the following night and the 13 Flight Scouts were also active in support of the battalion advance. The battle rhythm of the operation can be gauged from the POR:

> Recce by Scout – Advance by 3 Para – ferry everything forward from last position and continue in gigantic leaps. The tempo was such that as soon as a Scout landed, whether forward or back, it was immediately filled up with men, water, ammo, anything. No time was wasted on the ground and maximum use was made of every Scout minute. It is impossible to assess accurately what was carried up and down the Bakri Ridge, but it is safe to say that whatever was asked to be carried was carried.[25]

The 3 Para advance continued until at dawn on 18 May, after a six-mile approach march, the Battalion were in control of the 2,000-foot Hajib Escarpment. Next day the anti-tank platoon was leading C Company at the head of the advance. The route took the Battalion down the Al Dhuhira Ridge when the lead platoon encountered 12 dissidents who were leaving the village of Ar Zuqm. The platoon commander, WO2 Company Sergeant Major Nobby Arnold ordered pursuit and his men captured three of the fleeing dissidents. Subsequent interrogation revealed that there was a previously unknown path down from the escarpment. The Battalion and supporting units had successfully established a position on the western end of Bakri Ridge and taken the first prisoners of the entire operation. The 3,300-foot spur on which the incident took place was nicknamed 'Arnold's Spur'. That night, as 3 Para consolidated, their patrols observed what were thought to be dissident light signals on the hillsides below. At dawn, aware of heavy mist on the peaks, 13 Flight were back on task, carrying out the daily resupply of 3 Para combat supplies.

Over the next three days, 3 Para maintained pressure on the dissidents on the ridge. An intense programme of patrolling cleared the western slopes and numerous wadis. A number of dissident sangars were discovered, numerous firefights took place, and several dissidents were killed or wounded. There was an exotic haul of captured weapons, ammunition, tons of flour, coffee and kerosene, all of which was destroyed in line with Brigadier Blacker's intent to demonstrate the reach of British forces. The advance from Shab Tem continued and the Scouts continued to carry out resupply runs, doing whatever possible to lighten the load of the 3 Para troops. On 22 May the Scouts flew a series of supply runs from Shab Tem to the Battalion position on the Hajib Escarpment. During the day Farrah-Hockley conferred with Graham-Bell about helicopter support to the next phase of the operation. Farrah-Hockley said, 'I know that you've been busy, and I have refrained from asking for any helicopters….but when I do ask for them, I shall want them quickly. I will not ask for them unless I mean business.'[26]

The next stage of the Bakri Ridge operation began on 23 May as C Company carried out an advance to contact along the Bakri Ridge towards Wadi Dhubsan. They came into contact with about 40 dissidents in the village of Al Qudeisha. The dissidents were armed with automatic weapons and five LMGs and were determined to hold the village. Their position was well chosen, built on the highest point of the ridge the stone buildings were linked by underground tunnels. The 3 Para attack was supported by RAF Hunters, 3 RHA guns and integral fire support from the

25 AFM 204c: 653 Sqn AAC, Scout AH Mk1, Radfan Operations 27 Apr–30 Jun 1964, p.8.
26 F. Graham-Bell, *Helicopter on the Heights*, (Unpublished manuscript, 1968)

Battalion's own heavy weapons. The battle was hard-fought, and each time C Company thought that they had cleared a sector of the village and moved on, dissidents would appear behind them and counter-attack. The dissidents made use of two dry stone-built watch towers that dominated the village and the surrounding area. The towers sat just outside the range of 3 RHA's guns and were too tough for the Battalion mortars to neutralise. Consecutive RAF Hunter strikes were called in to attack the towers by the 3 Para FAC. The Hunters targeted the towers with bombs and 30mm cannon, empty shell cases raining down on the lead elements of C Company in the village.

The village was eventually secured, and A Company moved around to the left onto higher ground, from where they could observe the Battalion's final objective 3,200 feet below, Wadi Dhubsan. After a week of hard marching and an equally fierce fight to clear Al Qudeisha, the 3 Para seizure of the Bakri Ridge was complete. The operation had cleared dissidents from the 200-mile length of the Bakri Ridge and 653 Squadron had played a significant part in that success. The next phase of the operation, the seizure of Wadi Dhubsan, was imminent. There was a short pause on 25 May to allow 3 Para to reconstitute and prepare for the next battle. 653 Squadron were tasked with another resupply serial, lifting further combat supplies forward from Shab Tem.

On 25 May the marines of X Company RM were flown in to reinforce 3 Para. Meanwhile, on the same day the remainder of 45 Cdo RM were preparing to work alongside 2 FRA to clear the Wadi Misrah of dissidents and raid Jebel Huriyah. With X Company under command, 3 Para were tasked to carry out what was classified as a raid into Wadi Dhubsan. This was a critical part of the operation as the Wadi Dhubsan was considered by the dissident tribes to be impregnable; no European had ever entered it. The raid was intended to demonstrate, in no uncertain terms, that the British, and therefore the Federal Government, could project force anywhere it chose in the Radfan. D Day for the Wadi Dhubsan operation was set for 26 May. The Jebel Huriyah operation on 25 May was successful and notable for the use of Centurion tanks. Brigadier Blacker then ordered 3 Para to commence their raid into Wadi Dhubsan on 26 May. One of the preliminary objectives of the plan was to seize the 5,000-foot Jebel Haqla, the top of which was flat, and therefore the ideal location for a Scout HLG. Farrar-Hockley opted to make a direct approach to the jebel. His preferred route involved descending at night directly from Arnold's Spur into the village of Bayn Al Gidr, a route that could be observed by the dissidents. The last 30 feet of the route down into the wadi was a sheer drop and the paratroopers would have to lower themselves and their equipment down by rope in darkness. The descent into the wadi was not without hazard for 13 Flight either, as Graham-Bell recorded:

> The final move of this phase was to be the descent from Arnold's Spur to the Wadi Dhubsan. The descent was made during the night of 25 May and by first light 26 May they had established an HLG on the wadi bed at 2,000 feet, immediately below Arnold's Spur. The fly-in entailed spiralling down within the narrow confines of the wadi...rather like flying down a chimney. The exit was even more hairy and required a certain amount of prayer as well. On the first casevac sortie that required a third person to watch the patient, I made sure that the Padre was included so that he did the praying while I concentrated on the flying.[27]

The initial descent began at 2200 hours on 25 May and Farrar-Hockley led his Tac Group and the machine gun platoon undetected down into the village. At first light 26 May A Company 3 Para were on high ground overlooking the village. The surprised tribesmen had abandoned the village. A Company had reported the position as being clear of dissidents by 0600 hours. X Company RM then moved unopposed down and forward of the village along the right-hand side of the wadi. The marines

27 AFM 204c: 653 Sqn AAC, Scout AH Mk1, Radfan Operations 27 Apr–30 Jun 1964, p.9.

had been tasked to clear the wadi as far as Hafwi. By this stage, the dissidents who had been guarding the more obvious routes into the wadi were alert to the British presence. Groups of uniformed dissidents were observed to the south, rushing back towards the X Company advance. The dissidents, moving in groups of 7 to 12, numbered around 50 in strength with a half a dozen LMG. Having closed the range, they then opened fire and began a determined counter-attack towards Bayn Al Gidr.

Earlier, at 0600 hours, the first Scout sortie of the day had flown onto Arnold's Spur with a water resupply. Also on board was Brigadier Blacker who met with Farrar-Hockley to gain an update on the Battalion's progress. While the two commanders conferred, Graham-Bell ferried combat supplies into 3 Para at Bayn Al Gidr. While the meeting went on mist obscured Arnold's Spur, temporarily trapping the Scout on the high ground and leaving Blacker below in the village. The sighting of the advancing dissidents triggered a reaction from Farrah-Hockley and '…as a result of the reports concerning the accretion of dissident groups, the CO had begun to ferry his headquarters forward by Scout.'[28]

Once the mist lifted, a second Scout (XR 628) flown by Major Jake Jackson AAC was tasked to fly Farrar-Hockley and his Intelligence Officer, Captain Ian MacLeod, forward to join X Company in the wadi. The grid reference for the drop-off point at the battalion Tac CP was passed to Jackson by radio but it was incorrect. The location that Jackson was flying to, was in fact roughly 500 metres in front of X Company's most forward position. Down in the wadi, 3 Troop RM had occupied a house burnt-out after a Hunter strike and were preparing to move further forward to another burning house. As the marines moved forward, they came under accurate fire from the second house. It was at this point that XR628 appeared in the wadi and the incident that followed figured in the 45 Cdo RM POR:

> The Troop advanced towards this house but came under fire from its left front. Fire was returned. At this juncture a Scout helicopter flew down the wadi just beyond the forward troops and was hit by enemy fire; however, it managed to turn and land just behind the leading sections.[29]

Jackson had flown low along the bed of the wadi, passing over X Company's rear echelon and then the forward troops. He was flying close into the bank of the wadi, using the shadow of the cliffs for cover. Then, below him he saw dissident tribesmen shouting, shaking their fists, and shooting at the wayward Scout. The noise of the Rolls Royce Nimbus engine drowned out most noise, but the men in the Scout 'heard a sound like the opening of beer cans'.[30] The Scout had been hit a total of 11 times and Farrah-Hockley remembered that, 'Fuel began to flood down over the visor of the chopper.'[31] MacLeod was wounded in the wrist but the conversation between Farrah-Hockley (TFH) and Jackson remained calm:

> "Can you keep flying?" asked TFH, "This is the last place to put her down."
>
> "I've got power," said Jackson. Coolly he turned the Scout towards the head of the wadi, but then either the engine or the rotors started to make a clattering noise. As they swept forward, the advanced element of Battalion Headquarters on the valley floor swung into sight.

28 Airborne Assault Museum, Report on Operations in the Radfan by 3 PARA Group, 19–27 May 1964.

29 Royal Anglian Museum, HQ Middle East Land Forces, Operations in Radfan, 14 Apr – 30 Jun 64, p.159. http://royalanglianmuseum.org.uk/RamCollection/Collection/documents/Operations%20in%20Radfan.pdf accessed 28 November 2023.

30 Quoted in J. Riley, *Oft in Danger The Life and Campaigns of General Sir Anthony Farrar-Hockley* (Solihull: Helion Books,2015), p.262.

31 M. Arthur, *Men of the Red Beret* (London: Hutchinson, 1990), p.339.

"Down there," said TFH.

They landed safely, and Jackson closed down the engine. Around them a firefight was in progress. Fifty tribesmen had moved in position and now waited behind them – no doubt those seen from the Scout earlier. Later, TFH rebuked McLeod for suggesting that they had been 'shot down'. "Not shot down …. forced to land, old chap", he said with a grin.[32]

As the Scout shut down, WO1 (Regimental Sergeant Major) Arthur Channon of 3 Para led a group of marines from the Tac CP forward to the damaged aircraft. The dissidents continued to fire on the men around the downed Scout. The RSM was wounded on the approach, and the Battalion Padre, Captain Fred Preston, had the heels of his boots clipped by rifle fire as he carried wounded back from the frontline. Although the dissidents could easily target the Scout, no fire was brought to bear on it. The aircraft sat abandoned in no man's land. The Scout presented Blacker and Farrah-Hockley with a problem. If the aircraft were destroyed or allow it to fall into dissident hands it would create a huge propaganda opportunity for the NLF. The incident had already featured on the front pages of almost every UK newspaper. If at all possible, XR628 had to be recovered intact.

The wadi battle continued around the forlorn Scout. A Company, 3 Para joined X Company RM clearing the north of the wadi. Dissident resistance was determined, and a steady rate of fire was maintained from well-concealed positions overlooking the wadi bed. Farrah-Hockley committed C Company to outflank the dissident positions, hitting their left flank. The manoeuvre was supported with mortar fire and five Hunter strikes. As dusk approached the 3 Para battlegroup established a defensive perimeter and settled into night routine. The damaged Scout was secure inside the perimeter. Throughout the battle, 13 Flight had maintained a continuous stream of resupply lifts of combat supplies into 3 Para and lifted the wounded out of the wadi for onward casevac back to Thumier. Graham-Bell came forward during the day accompanied by a REME Air Mechanic to assess the problem first-hand. His first task was to meet with Farrah-Hockley and develop a plan to extract the valuable Scout. Carrying a bottle of whisky for Farrah-Hockley he worked his way forward on foot to locate him. The two men met where Farrah-Hockley was sheltering from accurate dissident fire. Graham-Bell dashed forward and dived into cover next to Farrah-Hockley:

> …between gasps I handed over the Whisky…. I noticed that Tony had stood up, so I did likewise and carried on talking. It suddenly dawned on me that this was a trifle unwise and I asked: "Aren't we rather exposed here?"
>
> "Oh yes" replied Tony, "I thought that, if they were still in positions, our standing up would draw their fire."[33]

During the course of the day, Graham-Bell had arranged for two REME aircraft mechanics to be flown forward to repair the Scout that night. The mechanics had assessed the damage to XR628, and Graham-Bell recounted his thoughts:

> The obvious solution was to patch it up enough to fly out and with this in mind, the aircraft was examined as soon as it was possible to reach it. Most of the bullets had gone through the canopy roof and had somehow missed the rotor head and blades – luckily it was flying with

32 Lieutenant General H. Pike, Eulogy to General Sir Anthony Farrah-Hockley, Royal Military Academy Sandhurst, 28 Jun 2006. Quoted in J. Riley, *Oft in Danger*, pp.262–263.

33 F. Graham-Bell, *Helicopter on the Heights*, (Unpublished manuscript, 1968).

The successful repair and recovery of Scout XR628 made the UK national press. (Army Flying Museum)

> the doors off and the passengers in the rear, sitting on the floor, which undoubtedly saved serious injury, if not life. Extensive damage had been sustained but it was decided that if the non-return valve and low-pressure fuel pipe were replaced, the aircraft could be refuelled and flown out in manual, it being too difficult to repair the hydraulic system in the wadi.[34]

While the fighting in the wadi and on Jebel Haqla continued, 653 Squadron raced against time to secure the parts required for the repairs. The spares were located at Falaise and flown forward to Thumier by Auster. A Scout then brought the parts forward to Wadi Dhubsan on one of the resupply runs.

When darkness eventually fell, two aircraft mechanics from 653 Squadron Workshop moved forward with toolboxes and the parts to get the Scout into a flyable state. The two men, Corporal Hustwith and Corporal Carcary, worked by torchlight under a blanket. A 44-gallon drum of fuel was also brought forward to refill the aircraft after the fuel tank holes were patched up. There was an urgency to the repairs; if the aircraft was not made flyable by first light it would have to be destroyed, or 3 Para would have to remain in Wadi Dhubsan until it was repaired. The prospect of either of these options held little appeal. The repair team were fired at during the night but continued with their work unabated.[35] At dawn, after four hours of work, the repairs were completed and the Scout was refuelled. Major Jackson flew in with the first resupply sortie and went straight out to his aircraft, 'As the Radfani grain stores were set alight, Jackson climbed into the Scout and tried the starter motor. The engine fired, the rotors turned, and when the helicopter

34 AFM 204c: 653 Sqn AAC, Scout AH Mk1, Radfan Operations 27 Apr–30 Jun 1964, p.10.
35 Kennett and Tatman, *Craftsmen of The Army, REME Official History*, p.352.

lifted, hovered and rose rapidly into the air, there was a hearty cheer from the paratroopers, who knew that heralded their own extraction.'[36]

Jackson disappeared through the billowing smoke and the departure of the Scout signalled the beginning of the withdrawal of 3 Para. The Battalion, minus X Company RM, collapsed their perimeter and climbed up onto the Alfara Ridge. The Wessex helicopters of 815 NAS then airlifted 3 Para back to Falaise. X Company RM remained in Radfan and were given the task of holding the Bakri Ridge. The battle for the ridge and the occupation of Wadi Dhubsan was regarded as a success and had killed six dissidents, captured 11 weapons, destroyed two and a half tons of food and 18 cans of fuel. Casualties were lighter than expected; one marine was killed and six others were wounded. Most importantly, Brigadier Blacker's intent to demonstrate to the dissidents the reach of the British garrison had been realised. Phase two had also reinforced the reputation of the Scout as the ideal aircraft for close support at the tactical level. The data captured for flying operations in support of 3 Para on 26 May 1964 was:

Pilots	Lt Col Graham-Bell
	Major Jackson
	Major Surgeon
	Capt Hardie
Aircraft	XR628
	XR 29
	XR630
Hours Flown	8.40
Landings	63

The three Scouts were primarily employed on resupply and casevac tasking, with the casualties being flown back to Thumier for transfer to larger helicopters. Staff Sergeant Ian Scott and Captain Homeshaw operated out of Falaise on the same day in XR600, completing 2.15 hours and four landings towards Homeshaw's conversion to type.[37] This success, coming so soon after its arrival in Aden, did overshadow the contribution of the Beavers and Austers but both had successfully supported the operation from the new airstrips.

The battle was widely reported in the national press with a detailed account featured in the *Daily Mail*; Jackson and the REME air mechanics were singled out. The *Mail* referred to Jackson as,

> ...a notable Test Pilot. While at the Boscombe Down Experimental Station, he flew light helicopters and spotter airplanes which the Army Air Corps is now flying with such dash in South Arabia. Most of the airplanes in 653 Squadron have been hit by anti-aircraft fire in the past three weeks, but their pilots still combine the kind of gaiety and gallantry associated with the Royal Flying Corps pilots who faced similar risks nearly fifty years ago.
>
> The cheerful mood of the squadron's mess tent is set by the CO, Lieutenant Colonel Francis Graham-Bell, who has carried a reputation for successfully dodging rebel machine gun fire and singing *My Boy Lollipop* less tunefully but no less vigorously than Millie.[38]

The relief after the successful recovery of the relatively new XR628 from the ignominy of denial to the enemy reached the top of the chain of command. 653 Squadron received an immediate signal

36 Quoted in Riley, *Oft in Danger*, p.263.

37 AFM 204c: 653 Sqn AAC, Scout AH Mk1, Radfan Operations 27 Apr–30 Jun 1964, Return for 26 May 1964.

38 P. Young-Husband, Radfan Report, *Daily Mail*, 27 May 1964.

from the GOC MEC: 'Recovered Scout helicopter. GOC sends his personal congratulations to 653 Sqn and all others who helped. We are full of admiration for recovery from the battle area. Well done.'[39]

Operation Test Match

The Radfan operations had gone well so far. 39 Inf Bde was now deep inside dissident territory. The third and final phase of the operation was the seizure of the Jebel Radfan itself and the Jebel Huriyah. The 5,500-foot Huriyah dominated southern Radfan and had never been climbed by a European. The approach to this new objective involved a six-mile march along a track followed by a further four miles on a camel path. The path ran through the steep-sided Wadi Misrah, a route ideally suited to ambush. The task of clearing the wadi route and seizing Jebel Huriyah was given to 1 E Anglian. The battalion was commanded by Lieutenant Colonel Jack Dye and the East Anglians were supported by D Squadron, 4 RTR and 2 FRA. The new advance was intended to maintain the momentum of success from Bakri Ridge and Dye gave his orders on the night of 29 May 1964. The four-phase plan began on 4 June and progressed through to its third phase by 7 June, at which point 2 FRA were at the base of Jebel Huriyah.

Throughout the advance 653 Squadron had been fully occupied in the positioning and resupply of picquet posts along the heights covering the approach route. The Beavers and Austers were employed in carrying out photo-recce sorties over the *jebel*. The 15 Flight Beavers also carried out a parachute resupply to the East Anglians during their approach to the *jebel*. The assault on Jebel Huriyah was made at night on 10 June while 13 Flight waited, poised to resupply the assault force on the summit at first light. Lessons had been learnt from the previous operation with 3 Para.

> It had been decided to use Scouts to move the water, ammunition, supplies forward from Lima 2 to the top of HURIYAH as soon as 'Position Firm' could be given at dawn. With our previous experience of loading difficulties in mind we got all the equipment pre-positioned in 500lb sticks. The first Scout, with a load of water, arrived at the top of Jebel Huriyah 15 minutes after first light and thereafter two Scouts shuttled backwards and forwards with the minimum amount of time wasted on loading and unloading. The fact that the whole operation went like clockwork again demonstrated the high standard of efficiency that can be achieved by very close liaison between ground forces and their supporting helicopters.[40]

The day after the occupation of Jebel Huriyah a pair of Scouts flew a recce sortie to the Jebel Radfan with Dye on board. It was decided to establish an OP on the peak, and the next morning at 0600 hours the Scouts returned carrying a flagpole and an East Anglian OP party. A sniper was silenced and the OP was established overlooking almost the entire Radfan. The flags of the East Anglian Regiment and the FRA flew on the highest point in Radfan; the message had been delivered, Blacker's demonstration was almost complete. The bulk of the troop-lift tasking during this period was carried out by the Wessex of 815 NAS. A Wessex crashed in the Wadi Misrah due to ground resonance on 12 June 1964 and a 13 Flight Scout was scrambled to act as a rescue aircraft. It arrived at the crash site within 15 minutes to find there was one fatality and 10 injured survivors to recover from the site.

39 Quoted in F. Graham-Bell, Helicopter on the Heights (Unpublished manuscript, 1968).
40 AFM 204c: 653 Sqn AAC, Scout AH Mk1, Radfan Operations 27 Apr–30 Jun 1964, p.13.

The third phase of the Radfan operation was almost complete. On 14 June Brigadier Blacker handed over command of 39 Inf Bde to Brigadier Chandos Blair. The new commander set about consolidating the gains made in previous months. Intelligence reports indicated a large surge in the number of Egyptian-trained insurgents crossing the border into Radfan from Yemen. Blair wanted to disrupt the inflow of uniformed insurgents from outside and thereby create the opportunity to pacify the tribesmen inside Radfan. He divided the province into battalion Areas of Responsibility and sited platoon and company patrol bases all over Radfan. It was clear from the outset that 653 Squadron's mixed fleet of aircraft would have an important role to play in supporting the new strategy. The pacification programme inevitably met resistance, notably accurate sniping from the Jebel Widina, a 6,000-foot peak that overlooked the Wadi Dhubsan. Patrols over the *jebel* found traces of dissident occupation, confirming the source of the sniping. Brigadier Blair gave orders to occupy the Jebel Widina and the nearby Jebel Sababah. The operation was codenamed Operation Test Match and A Company, 1 E Anglian were given the task of seizing the *jebel* and clearing out any dissidents that may have infiltrated onto the peak.

Any approach to Widina in daylight would be exposed to observation and fire. Speed and surprise were key to success and a plan developed to use helicopters to seize the peak. An advance guard of 36 East Anglians was to assault the *jebel* and hold it until the main body of A Company reinforced them on the objective. If there was to be any chance of surprise the helicopter assault would have to take place in darkness. Patrols had been inserted at night by helicopter before, however this would be the first heliborne night assault in history. The operation was scheduled for midnight on 23/24 August and 653 Squadron allocated two Scouts to Operation Test Match. The plan was simple, the advance guard would be flown in waves by Scout from Arnold's Spur and once the *jebel* was occupied, Captain Richard Abbott and his men of the advance guard would hold it and await the arrival of the remainder of A company by RAF Belvedere at first light. The Scout crews were by now well versed in working with their infantry peers:

> Preparations were begun two weeks beforehand and plans were made to use Arnold's Spur at the end of Bakri Ridge, as the jump off position. High Recces were done to clear the route and confirm that the planned landing site was all clear. Two days before the operation the Scouts simulated the take offs and landings by doing dummy runs at night. Aircraft loads were made up and enplaning and deplaning were rehearsed to perfection.[41]

On the evening of the operation the only two serviceable Scouts available were positioned at the HLG on Arnold's Spur. Pre-flight checks revealed a fuel flow problem on one aircraft and instrument panel lighting faults on the second. The instrument panel fault was repaired but only a single Scout remained serviceable. The operation continued:

> At the appointed hour, the 81mm mortars started their bombardment, lighting the mountain peaks with sheets of flame. As soon as they stopped, the helicopter took off by the light of the full moon with its first load. On coming into land, a large cloud of dust was raised, and the pilot's vision was limited to the tip of the starboard skid. As soon as the helicopter touched down, the pilot shouted 'Out', and the troops leapt out.
>
> …the passengers carried 4 glim lights, 2 red which were placed in front of the helicopter, one at 6 paces and one at 12 paces to give the wind direction, while one white light was placed

41 AFM 204c: 653 Sqn AAC, Scout AH Mk1, Radfan Operations 27 Apr–30 Jun 1964, p.14.

> at six paces on each side of the helicopter. This established the HLG and the helicopter shuttled to and fro, stopping only to refuel.[42]

Staff Sergeant Ian Scott flew nine consecutive troop lifts onto Jebel Widina that night. The landing was unopposed, the night uneventful, and at first light the Belvederes arrived as planned with the rest of A Company. Captain Richard Abbott, the Company Commander later went on to become an AAC Pilot and eventually to command 3 Regt AAC. He remembered Operation Test Match:

> I was acting A Company Commander, when we carried out the first night helicopter assault on Jebel Widina, 5,000 feet above sea level. At the last moment, Captain Russell Wright's Scout became unserviceable and we relied on Staff Sergeant Scott to shuttle us in the moonlight to the top of the *jebel*. He was so calm and obliging and we did not realise at the time what safe hands we were in.[43]

The East Anglians occupied their mountain-top positions for five days after their historic night mission. With the demonstration complete and the point made, it was time to withdraw. The completion of Test Match, and with it the occupation of the last of the major features of the Radfan, was viewed by the MEC as the successful culmination of the Radfan campaign. The campaign had been a military success in terms of clearing out the Radfan. The NLF, however, retained its safe havens across the border, and with them the ability to supply dissidents inside Aden and into Radfan. The Egyptians remained committed to financing the political and military campaign to oust the British and their federal allies. The Aden Insurgency would continue.

Over the preceding two months of intense operations, 39 Inf Bde had further reinforced the importance of heliborne mobility in counterinsurgency operations. Helicopters had demonstrated both their versatility and utility on each of the successive Radfan operations, undermining the dissidents' traditional advantage gained from the occupation of height and the dominance of Aden's rugged terrain. Helicopters had, however, become such a key capability that they were now considered too valuable to be risked in an assault role. The Radfan operations had reinforced the absolute need for good communications between the combat arms, artillery, aviation and air support. The integral support provided to 39 Inf Bde's operations by 653 Squadron had proven to be a vital component of the newly successful operational formula in Radfan.

Post operational analysis during the summer and autumn of 1964 unsurprisingly deduced that the arrival of the Westland Scout in theatre had allowed 653 Squadron to develop a new range of capabilities. The new helicopter had almost instantly transformed the traditional spectrum of support beyond that previously available from the fixed wing types operated in theatre. The need for some form of airstrip to extend fixed wing operations by Beaver and Auster was a real constraint on the reach and sustainment of fixed wing operations. The arrival of the first Scouts and their ability to operate from austere aviation FOBs away from airstrips created a welcome new dimension to the aviation support available. The 13 Flight Scouts had immediately proven to be robust and agile delivery platforms that could be operated at mountain-top altitude. The size of the Scout had also allowed it to land on what had previously been considered inaccessible mountain-top HLG. This new reach, and its obvious suitability for special forces tasking and conventional infantry work, had given the AAC a new and welcome prominence in MEC thinking. The aircraft's suitability for casevac and the positive morale effect of that on supported ground forces could not be overestimated. The prominence of the requirement for casevac in Farrah-Hockley's

42 AFM 204c: 653 Sqn AAC, Scout AH Mk1, Radfan Operations 27 Apr–30 Jun 1964, p.15.

43 R. Abbott, 'A Journey to Army Flying' in R. Mallock, *The Likes of Leicester* (Self-Published, 2013), pp.84–6.

A Scout casevac of 3 KOSB casualties from Wadi Taym in 1964. (Army Flying Museum)

planning for the 3 Para advance along Bakri Ridge and Wadi Dhubsan gave an indication of its importance.

The dramatic operational debut of the Scout had somewhat overshadowed the other success story, that of the Beaver. The Beaver was destined to become synonymous with Aden and desert operations. The new type had transformed the scope of fixed wing operations way beyond what the smaller, underpowered Auster AOP9 could offer. The durability of the Beaver, its superior performance and endurance beyond the Auster were significant increases in capability. The Beaver's obvious all-round suitability coupled with its useful parachute resupply capability had won it many friends in Aden. The performance of the AAC through the Aden Emergency over the period from 1963 to 1964 is best summarised by Farrah-Hockley when he stated that, 'It earned 653 Squadron the best of reputations, the high confidence of those it supported.'[44]

The Aden Insurgency was, however, not defeated. The Soviets and Egyptians continued to fund and arm the NLF insurgents and the Radfan tribesmen remained disaffected. As with many revolutionary groups, the NLF had learned the hard way that conventional warfare against a sophisticated and well-armed opponent was bound to fail. Victory lay along the now well-trodden route of unconventional warfare. The adoption of the guerrilla tactics used by the *Front de Libération Nationale* against the French in Algeria and the Viet Cong against the US Military in South Vietnam were what was required. The nature of the insurgency was about to change; the result would lead to an increased requirement for AAC support. The Aden Emergency was far from over.

44 A. Farrah-Hockley, *The Army in the Air* (Stroud: Alan Sutton Books, 1994), p.198.

Reorientation

The aftermath of the 1964 campaign in the Radfan saw a marked change in the tempo of dissident activities in both Aden and the Protectorate. The NLF had realised that it would always be hopelessly overmatched by British forces in open battle. It quickly adapted its tactics to more proven covert terrorist methods. While there was less appetite for pitched battles on the mountain peaks, terrorist activity increased in Aden itself. Beyond Aden the NLF concentrated on rebuilding its resources in the Radfan and the other mountainous parts of the Protectorate. As a result of the Radfan operations, the disaffected tribesmen were now even more willing to support the NLF's aims. British political initiatives aimed at stabilising Aden had achieved little. A conference held in London in June 1964 had concluded that the Federation should become independent of the UK by 1968. However, the British garrison would remain in place to guarantee British interests in the region and to protect the newly independent state from external interference.

The intent of the London conference was to assure the population of Aden of a safe transition to independence, a national constitution, and long-term stability. The prospect of independence in such a short time had the opposite effect on Aden however, and chaos erupted almost immediately. The existing political fault lines were further exposed, and the various factions attacked each other in a bid to secure their primacy in the new order. Those who had been loyal to the Federal Government and the British authorities now felt exposed and vulnerable to the NLF and whatever regime was to follow. The Adenis serving in the Civil Service, Police, National Guard, and the FRA felt that they had to back a side. As with previous withdrawals from colonial responsibility, the public statement of the intent to do so triggered a decline in influence and authority. The election of a Labour government led by Harold Wilson in October 1964 shifted the British position significantly. The Wilson Government did not support the idea of a federation. Many Labour MPs sympathised with their fellow trade unionists among the opposition groups in Aden, principally those who rallied under the banner of the People's Socialist Party (PSP). The PSP and the Sheikhs who had historically ruled Aden were diametrically opposed; peaceful coexistence was almost unthinkable for both. Meanwhile, the security situation continued to stretch British and federal forces.

The Nasser Government continued to interfere in Aden and Nasser backed the newly formed Front for the Liberation of Occupied South Yemen (FLOSY). The Egyptians had inherited a plentiful supply of abandoned military equipment when the British had withdrawn from Egypt. This surplus was recycled and smuggled into Aden by FLOSY and the Yemeni backed NLF. Hand grenades were the weapon of choice in the urban environs of Aden itself. The NLF persuaded dissident tribesmen to carry out hit and run grenade attacks against soft targets in Aden. There was an active campaign targeting British Special Branch officers and high-ranking Arab police officers. Service families were also considered a legitimate target and a number of British schoolchildren were killed or maimed by NLF grenade attacks. There was little that AAC units could do to assist directly in the prevention of this type of attack but they could, however, support those elements of the security forces employed in preventative operations and in post-attack cordon operations. This type of early internal security tasking would soon become very familiar to AAC flights and squadrons. The versatility and agility of the helicopter offered commanders on the ground a highly effective new dimension to their hitherto limited options in a close urban environment. It was a dimension that would develop and expand at a rapid rate as the Emergency in Aden continued. As well as providing overwatch and an airborne command post, AAC crews also dropped tear gas grenades into rioting crowds, although this was not without risk. There were reports of kites being flown over NLF demonstrations to deter low flying helicopters. While the battle for control of the streets and the Crater district escalated, the fight for the open spaces and mountains of the Protectorate also continued.

Towards the end of 1964 and into 1965 the dissidents had also increased their use of mines on the dirt roads and tracks that carried traffic through the mountains. Ironically, the majority of the landmines used by the dissidents were of British design and manufacture. The Mark 5 anti-tank mine was of Second World War vintage and the more modern Mark 7 had replaced the Mark 5 after the war. Both mines packed a powerful punch and could destroy an unarmoured Land Rover or Bedford three-ton truck with ease. The advent of the landmine threat initiated a requirement to fit armoured floors to Land Rovers and Bedford trucks. The threat also tied down valuable armoured cars that were required for other tasks. The heavyweight Alvis Saladin armoured car was often sent ahead of a convoy to test the ground, which was not the most popular duty among the crews. As the campaign progressed, the six-wheel Alvis Stalwart High Mobility Load Carrier, with its 'V' shaped chassis was found to have a high survival rate. The 'Stally' became the vehicle of choice on convoy runs along the high risk routes.

Keeping the NLF contained and the tribes pacified required an active British military presence in the mountains; this carried risks. The pattern of air operations changed to counter the new dissident tactics and the requirement for helicopter support remained constant with the landmine threat ever-present and effective. The dissidents could place them at night with impunity and the resulting incidents wrecked British vehicles, disrupted road movement and created casualties. The need for a permanent, on call casevac capability could only be met by helicopter. The Scouts of 13 Flight AAC took on an enduring casevac standby role from late summer 1964. The average transit from a mine incident back to the British Military Hospital at Khormaksar was 60 miles. The Scout could carry three stretchers, one prone casualty across the rear cabin seat and two in external pods on either side of the aircraft. The casevac role was shared with the RAF Wessex, although the smaller Scout could get into many places that the Wessex could not. One of the grim realities of war was the volume of blood from the blast injuries. Pilots often had to fly back to Falaise Airfield after a casevac run as they needed to hose blood out of the rear compartment before returning on task.[45]

There was some good news in October 1964, reinforcements were on their way. HQ 24 Inf Bde finally moved into Aden from Kenya. The much-awaited arrival of the HQ brought with it 653 Squadron's missing element, 8 Flight AAC.

8 Flight AAC Arrive in Aden

The men of 8 Flight were no strangers to operating in desert conditions thanks to their experience from supporting 24 Inf Bde training exercises in the African bush. Conditions in the north Kenyan bush were semi-desert and therefore ideal preparation for flying in the Aden hinterland. Prior to deployment to Aden, the Flight had operated the Beaver and the Alouette and found both to be a generally reliable aircraft, but the latter was underpowered at altitude and, therefore, not going to be suitable for Aden. 8 Flight's pilots had kept their AOP skills honed during exercises, notably supporting the batteries from 3 RHA on rotation from operations in Aden. The Flight was also tasked to support the RE who were involved in an ongoing road building programme in Kenya. The road ran from Dadaab, for over 100 miles north to Wajir, following the Somali border. The bulk of the tasking was in support of RE recce and survey teams operating in advance of the road construction. It was not a benign environment as not everybody in the region wanted the road built.

45 S. Holtom, account to the authors, 1 June 2020.

The local Shifta tribesmen were unhappy about the intrusion into their tribal lands and harassed the road builders. The supporting Alouettes were a prime target on which to vent their rage. This was particularly true at night when the aircraft were at their most vulnerable on the ground. To protect the Alouettes from incoming small arms fire during the hours of darkness they were placed in the centre of the RE leaguer position. The sappers would settle into a position of all-round defence, surrounding the aircraft with their heavier, and much more robust bulldozers and heavy plant vehicles. On two occasions, the Shifta's nocturnal activities came a little too close for comfort and, in response, 8 Flight pilots returned the compliment by throwing hand grenades at them when passing overhead. Other routine tasking when in Kenya included maintaining casevac support. Alouettes were employed to extract injured personnel out of the bush and fly them back to Nairobi for emergency medical treatment, although direct flights to the Kenyan capital were the exception. A rendezvous for a hasty casualty exchange in the field with a Beaver or an RAF Twin Pioneer was a better and more efficient option for all concerned. All of this experience gave 8 Flight a good grounding for later operations in Aden.

8 Flight arrived in Aden on 14 October 1964. Personnel travelled together with families and dependants on a chartered British Overseas Airways Corporation Douglas DC-4 Argonaut aircraft, while heavy stores and vehicles followed on by sea from Mombasa. Two 8 Flight Beavers were flown across to Aden by three pilots. Captain David Swan RA and Lieutenant Allan Blaxland AAC (the first subaltern to be commissioned directly into the AAC) flew XP777 while Staff Sergeant Ken Barber AAC flew XP819. Two further aircraft remained in Kenya as a detached section under command of 653 Squadron AAC. A third Beaver (XP 827) was flown from Middle Wallop to Aden by Major Nick Nicholls AAC. Fixed wing reinforcements were welcomed, as the last Auster flight prior to the type's withdrawal from Aden had taken place a month earlier on 12 September 1964. 8 Flight's Alouettes were stripped down and packed up for shipment as freight back to Europe to be replaced by the Scout.

Initially, only the Flight Commander, Major James 'Jimmy' Nunn AAC, was trained on the Scout. Jimmy Nunn had deployed to Aden earlier in the year and had flown as a line pilot with 13 Flight throughout the Radfan campaign. The Flight 2IC, Captain Spencer Holtom RA, and Sergeant John Hobday of the Queen's Own Buffs, Royal Kent Regiment, would require conversion from Alouette to Scout in theatre. Additional Scout trained pilots were to be posted direct from training at Middle Wallop to augment 653 Squadron. Holtom recalls the conversion process on the only dual control Scout in theatre:

> The Scout conversion course took about a month, part of it was dual flying in the mountains, flying operationally as part of the conversion. Captain Tom Ogston AAC was the QHI.... By about New Year 1965 all six 8 Flight pilots were operational on the Scout. Initially, a single 8 Flight Scout operated daily up country using technical and groundcrew support from whichever of the two flights was resident at Thumier.[46]

Pilot numbers increased with the arrival of WO2 Tom Driver AAC and Sergeant Norman 'Bill' Bailey RA. The Flight had received its first two Scouts from 13 Flight, a further two factory fresh airframes were flown in from the UK by RAF Blackburn Beverley transport in January 1965. The Flight's Beaver pilots and their aircraft were taken on strength by 15 Flight, and assumed tasking duties after their in theatre familiarisation. 8 Flight's ground crew and REME technicians also converted onto the Scout, working alongside 13 Flight, although it would be some time before the new Flight was considered fully operational.

46 S. Holtom, account to the authors, 1 June 2020.

Major Nick Nicholls AAC flew Beaver XP827 from Middle Wallop to Aden to join 15 Flight AAC. (Army Flying Museum)

The all-important flying rate, including Scout conversion training, was hampered to a degree by a lack of dual control fitted aircraft. The decision to remove duals had been taken in 1964, primarily for safety reasons. The risk of a fully equipped passenger, laden with webbing and a weapon, snagging the cyclic or jamming a pedal was just too great. The need to have the capacity to lift an additional passenger was also deemed a higher priority than the luxury of dual controls. At the time, Lieutenant Colonel Graham-Bell, CO of 3 Wing AAC had another, ulterior motive for the removal of duals from the Squadron's Scouts. He had an idea that would gain the Squadron another pilot, as Major Claude Surgeon AAC, a QHI, was due to arrive in Aden. Graham-Bell was waiting:

> It is the practice to send out once a year to all squadrons, a team of experts from Middle Wallop to ensure that standards of flying and maintenance are kept up to the high level that is so necessary in flying, and to make sure that 'tribal customs' are not creeping into different theatres. These experts, or 'Standards Officers' are always referred to as 'The Trappers' and their arrival is anticipated with misgivings, as pilots brush up their techniques prior to a flying test and an hour's oral examination on all aspects of flying.... I ordered the removal of dual controls from the Scouts, on the operational grounds that passengers cannot be flown

> in aircraft with dual controls. Of course, without dual controls the Standards Officer cannot check the pilot under examination and therefore cannot test him.
>
> Claude was not amused when I explained the situation to him and his sense of humour completely failed him when I further explained that not only would he not be able to test us but that the shortage of helicopter pilots in the area meant that he would have to stay in Aden for a month and help out as a section pilot. It was rather a shock for him to learn that the visit of a few days with plenty of entertainment and 'runs ashore' into Aden, had suddenly turned into a month's hard work and with no chance of completing the task for which he had specifically come to Aden.[47]

The biggest constraint on conversion, however, was the Scout's chronically short engine life. The Nimbus engine had a projected overhaul interval of 300 hours. Experience in Radfan had shown that the average life of each of the four engines changed during the Radforce operations was just 127 hours. This was very early days for the Scout and the apparent short life of the engines, although it had improved over time, was still a cause for concern. The austere conditions in Radfan were not thought to be the cause of the chronic shortfall and much REME time was spent identifying technical problems with the Scout power plant. A visit from the Bristol Siddeley test pilot and a team of development engineers eventually identified a design defect in the free turbine generator which was rectified in due course.

In spite of these difficulties, 8 Flight achieved operational status and came online for tasking alongside 13 Flight in February 1965. Major Nunn recalled the nature and intensity of operations:

> During this period, the Flight commenced the type of operations which were to continue without pause for the remainder of my period in command. The majority of the effort was in the Radfan-Dhala area with periodic sorties and short detachments throughout the Protectorate and as far as Beihan. For much of the time a detachment was maintained at Habilayn airstrip (Rad-Force Base) as a joint effort with 13 Flight.
>
> Tasks included reconnaissance of all types, carriage of troops and supplies, artillery observation, casualty collection and evacuation and liaison sorties. These tasks were flown for all units operating up country, including Federal Regular Army and National Guard units. 24 Inf Bde was not operating as a formation at this time, but its units were spread out between up country tours and IS duty in Little Aden. The artillery in particular were deployed widely by batteries and troops, all of which received support from the Flight.[48]

For most of these tasks, the normal working payload of the Scout limited sorties to a duration of 50 minutes of flying between refuels. This led to an innovative new tactic that also went some way towards preserving fuel and wear and tear on the Scout's hardworking engine. Lieutenant David Ralls RCT explained:

> The slow pace of ground movement in the wadi below often meant long intervals between five-minute hops. A necessary expedient to save fuel was to find a high flat spot, land and shut down. The pilot then became a one-man OP until the next leap frogging movement was required. It was a fine judgement to go out on a limb in this way, but the alternative would have been endless dead flying time to refuel and using jealously guarded engine hours.[49]

47 F. Graham-Bell, *Helicopter on the Heights* (unpublished manuscript, 1968), p.104.
48 Papers of F. Graham-Bell, Private Collection, '8 Flight AAC Report, Aden October 1964 – February 1966', p.1.
49 D. J. Ralls, account to the authors, 12 August 2020.

A simple solution to a complex problem – Lieutenant David Ralls in the dismounted OP role, saving vital flying hours. (D. Ralls)

Later in the AAC's history, during the Cold War, the dismounting of Gazelle aircrew from a recce aircraft in dead ground would become fairly common practice. The deployment of a lone crewman with binoculars would allow the unarmed Gazelle to remain in cover from the mobile air defence systems that were expected to accompany advancing echelons of Soviet armour on the North German Plain.

Headquarters 3 Wing AAC

The arrival of 8 Flight AAC from Kenya and the subsequent deployment of troops and platoons operating within the Integrated Air Scheme had expanded the aviation Orbat in Aden beyond that which could be commanded by a squadron headquarters staff. The solution was the formation of 3 Wing AAC at Falaise Airfield and the appointment of an AAC Lieutenant Colonel as Commander Army Aviation Middle East Land Forces. The remit for the post stretched beyond Aden to the Persian Gulf and East Africa. The incumbent of the post was also the CO of 3 Wing AAC. The fledgling Wing HQ had overarching responsibility for the maintenance of aviation standards and the provision of engineering support. Up until September 1965 the appointment of CO was held by Lieutenant Colonel Francis Graham-Bell AAC who was subsequently relieved by Lieutenant Colonel John Hornby RA. Both men were subsequently made Officers of the Order of the British Empire in recognition of their efforts while in command. The Wing HQ retained the 653 Squadron AAC crest and commanded and administered an eclectic mix of cap badges and aircraft types. Each of the AAC Flights was commanded by a major and, although a seemingly top-heavy rank structure for helicopter Flights of four or more aircraft, it reflected the degree of autonomy during Aden operations and the associated levels of responsibility. Lieutenant Robin Grist of the Gloucestershire Regiment outlined the orbat as it was when he arrived in Aden in May 1966:

> When I arrived, 653 Squadron no longer existed and I joined 3 Wing AAC commanded by Lieutenant Colonel John Hornby, Commander Aviation MELF. 3 Wing consisted of 8 Flight with six Scouts commanded by Major Greville Edgecombe AAC, 13 Flight commanded by Major Chris Crouch RTR and 15 Flight with seven Beavers commanded by Major Malcolm Fleming. 8 Flight was part of 24 Brigade, 13 Flight part of Aden Brigade and 15 Flight part of MELF. However, each Scout Flight took it in turns to support Area West, which included Radfan and bordered Yemen, with its headquarters at Dhala. The Scouts were based at Habilayn with one of the British battalions. 8 and 13 Flight rotated at about monthly intervals.[50]

Beaver Operations

Regardless of whatever excitement was happening outside the perimeter fence at Falaise Airfield, the pattern of life, especially the posting in and out of individuals, carried on. Major Malcolm Fleming of the Royal Highland Fusiliers recalled arriving to take over command of 15 Flight AAC from Major Arnold Palmer, East Anglian Regiment.

50 R. Grist, account the authors, 6 June 2020.

> I flew to Aden by charter aircraft and was quartered at RAF Khormaksar before moving to proper quarters in Little Aden about ten miles away, where 3 Wing AAC based. Falaise Airfield, where the Beavers were based together with the Scout and Sioux helicopters, had a triangle of three small tarmac runways (giving six directions for take-off and landing), hangars and offices as well as an Aircraft Workshop all on the edge of the desert, close to 24 Brigade Camp and purpose-built families quarters, NAAFI shop and school, all recently built. The Flight consisted of six pilots, several REME Technicians led by a Staff Sergeant Artificer, a General Duties Sergeant, a few Drivers and Signalman all from the Royal Artillery. Total 30 all ranks, not forgetting seven Beavers on charge, all lodged in a hangar which had a crew room and offices inside. The Wing HQ was located just across the complex, as was the Aircraft Workshop where second line servicing was carried out, with the Armoury next door.[51]

The 15 Flight Beavers had undergone some minor modifications from the standard specification common to Beavers in other AAC Flights worldwide. A short stub exhaust was fitted, replacing the longer silenced version that normally fed the cabin heating system, redundant in Aden. The Decca Navigation System was not fitted. The Sun Air HF radio with a trailing antenna was fitted, filling in the gaps where there was no VHF or UHF communication available. As with the Scout, the Beaver was also fitted with Larkspur B47 and B48 VHF radios to communicate on tactical nets with ground units. Rugged as it was, the Beaver was not impervious to the harsh environment of Aden. The austere conditions on the desert airstrips added to the maintenance bill and the REME workshops kept a close eye on a number of specific areas. Propeller tips and leading edges required regular inspection, as did the undercarriage tyres. The tail plane leading edges were especially prone to damage and fibre glass shields were fitted to the affected areas for protection. Because of the airstrip dust and frequent sandstorms, filters and tyres were changed more frequently than in other theatres. These local modifications to the Beaver were effective, further enhancing the type's reputation for reliability.

The utility of the Beaver had an obvious appeal, as its small size in comparison to other fixed wing types operating in Aden also suited the needs of the SAS. The Beaver could land at rough airstrips and carry out parachute resupply tasks as and when required by SAS patrols operating out in the Radfan or even further afield.

The resident SAS squadron often used the 15 Flight Beavers as part of their training programme. SAS troops were regularly parachuted into the desert by day and by night, and once on the ground they were expected to find their way back to Aden unaided. Regular cooperation led to a consistent mission profile for the Beaver pilots. They would climb to 10,000 feet with both rear doors removed and four fully loaded SAS troopers on the floor. The SAS troopers would then freefall from that height. On one occasion two Beavers were used for a night drop and, as it was difficult for pilots to keep formation in the dark, the SAS dropped on a single letter code over the radio. The SAS were clearly working up a capability that they could add to their range of options in the Radfan, as Major Fleming explained:

> A plan was formulated with RAF Khormaksar radar, that the Beavers would carry out a daylight recce on which they would fly over the proposed DZ and beyond so as not to make the opposition suspicious. With early contact to Khormaksar radar to put the Beaver on course to the DZ, the pilot would count down until overhead the DZ and tell the radar operator to mark his screen accordingly. On the night, the radar operator could tell the pilot

51 Quoted in D. Kellock, *A Wonderful Aircraft* (Hull: Steve Taylor Design, 2016), p.78.

> when he was exactly on the DZ whichever angle he approached from. The SAS commander approved the plan, but unfortunately the operation was cancelled.[52]

The allocation of tasking between rotary and fixed wing assets was defined by the availability of airstrips. The Scout and Sioux flights routinely operated in the areas that the Beaver could not, and *vice versa*. The Beaver worked in the Radfan, servicing deployed forces from the numerous rough airstrips or, in extremis, by parachute resupply. The Beaver also had the range to fly out of Aden, often carrying passengers or freight between Habilayn (50 miles to the north), Little Aden and Al Milah (45 miles to the north). Flights up to Dhala (now Ad Dali), 65 miles north of Aden and close to the border with Yemen, were a regular fixture for 15 Flight. The Beaver was operating comfortably in the space between the intimate support provided to field units by the two Scout flights and the heavier lift provided by the RAF's Twin Pioneers and the four-engined Beverley. As the campaign progressed the Beaver became the first choice for the resupply of the many small, isolated detachments scattered around the fringes of the Aden theatre of operations.

The members of the RAF detachment at the radio station on Perim Island, 75 miles west of Aden, were always pleased to see the Beaver landing with supplies and mail. Other destinations included the RAF air staging post at Mukalla, 280 miles east-north-east of Aden. The 15 Flight pilots got to know almost every corner of Aden and the Protectorate. Logbooks listed landings at numerous FRA outposts, in places such as Al Ghurfar, Mukeriras, Beihan, and Husn El Abr. The nature of flying in and out of these rough, dusty locations was captured on canvas by the artist David Shephard; his painting of a Beaver in the dust depicts a 653 Squadron Beaver operating out of the strip at Wadi Ain close to the border with Yemen.

Occasionally, 15 Flight was tasked to fly Beavers as far west as Addis Ababa in Ethiopia, Djibouti or as far east as Bahrain. In addition to ranging far and wide overland, 15 Flight pilots also worked hard to maintain their deck landing currency. Whenever an RN aircraft carrier appeared offshore its crew would be treated to the unusual sight of AAC Beavers carrying out deck landings. All of the AAC flights took the opportunity to add deck landings to their pilot logbooks. There was a valid reason for these sorties out over the Red Sea, as the carrier's Air Group usually contributed jets and helicopters to operations in Aden. The planning of these combined operations required close liaison and briefing on board the carrier. The ship's company welcomed the appearance of a Beaver or a Scout as they usually carried the ship's mail from the British Forces' Post Office in Aden. A summary of 15 Flight activity over a period of 12 months through 1965 and 1966 listed 2,825 operational flying hours which involved 6,023 landings. During those sorties the aircraft carried 8,402 passengers and 250,675lbs (over 110 tons) of freight.[53]

Aviation Operations Up Country

By April 1965, the pattern of operations in Aden was settling into a regular rhythm and, other than the occasional unusual SAS tasking, the AAC flights were becoming well versed in their respective roles. The capabilities of the Scout and the Beaver were much better understood by user units and the aircraft were being more efficiently utilised. Night flying up country remained a challenge for all concerned, as navigation and landing aids were primitive and the combination of Aden's mountainous terrain, high altitude and extreme heat could be treacherous. Despite the hazards, AAC crews regularly rose to the challenge, such as the typical Aden scenario that unfolded on 12 April 1965.

52 D. Kellock, *A Wonderful Aircraft*, p.79.
53 Anon, '15 Flight A.A.C.', *Journal of Army Aviation*, 1966, p.79.

In the early hours of the morning, the tented camp at Al Milah came under fire from a nearby ridge. Dissidents had occupied the high ground overlooking the British encampment and were firing down into the tents occupied mainly by Royal Engineers. The dissidents were using rocket launchers, LMGs and rifles, and the weight of fire was significant and went on for some hours. The Al Milah camp had not been attacked before and was therefore considered to be relatively safe. It was for this reason that 300 (Parachute) Squadron RE, a Territorial Army unit, had been based there. The Territorials had been deployed to Aden on their annual camp with the task of supporting other RE units involved in improving the local roads. The weekend soldiers were now in the thick of a protracted firefight. The sappers fought back with .30-inch calibre machine guns, LMGs and their SLRs. Squadron Sergeant-Major John Lonergan and Sergeant Cyril Atfield, Royal Army Pay Corps, attached to 24 Field Squadron RE, were both killed. Two officers and three NCOs were wounded and Major Clive Samuel RAMC, the medical officer, risked his life to attend to and rescue two badly wounded men caught out in the open. He was assisted by Private Terry Baker RAMC. The problem now, was getting the seriously wounded back to hospital.

Lieutenant Tom Ogston was the QHI of 653 Squadron and had arrived in Aden in December 1964. On the night of 12 April 1965 Ogston was tasked to provide casevac for the TA sappers at Al Milah. He had had some experience of night casevac the previous February, having recovered a wounded Federal National Guard soldier from a mountain location, landing a Scout using only crossed vehicle headlights as a landing aid. This task, however, was to be more challenging, as described by Major Nick Nicholls:

> Lt Ogston was the pilot of a Scout helicopter which went to evacuate two British Territorial Army who had been critically injured in an attack on their camp at Al Milah. It was imperative that they be moved to hospital without delay.
>
> After a thirty-mile journey in the dark above a layer of cloud Ogston located the camp and descended through a hole in the cloud to land by the light of hand torches.
>
> The accomplishment of such a journey was an incredible feat of navigation and airmanship made possible only by the pilot's skill and complete disregard for his own safety.[54]

As a result of this citation Lieutenant Tom Ogston was awarded a QCVSA. Later the same month another incident took place that gives some insight into the pattern of operations and the prevalent threat at the time. 8 Flight's commander, Major Nunn AAC, recalled the incident nearly a decade later, by which time he had been promoted to Lieutenant Colonel:

> The Flight Commander (on Habilayn Casevac detachment), while attempting to land a small Parachute Regiment patrol at Ad Dimnah to recover equipment, ran into small arms fire from dissidents who had occupied the village. Fire was returned by the aircraft passengers and the aircraft landed safely back at Habilayn with all electrics shot away and an embarrassed Flight Commander, who had lectured his pilots only the previous day on the subject of avoiding unnecessary battle damage. Ad Dimah was subsequently reoccupied by our troops and many resupply sorties were flown in their support.[55]

By the autumn of 1965, the forward operating base at Habilayn was well established and had an air of permanence. The base centred on a rough airstrip that had been graded out of the desert for

54 TNA, WO 373/137/14: Army Form W3121 Recommendation for Honours or Awards, Lt T Ogston AAC, 29 April 1965.

55 Papers of F. Graham-Bell, Private Collection, '8 Flight AAC Report, Aden October 1964 – February 1966', p.2.

Radforce. The omnipresent dust was kept down to a manageable amount by periodically soaking the airstrip's surface with heavy waste oil. The glutinous mix acted as a binding agent on the arid soil and dust. The strip was exposed to the additional threat of infiltration and mining. Dissidents continually attempted to disrupt the lifeline provided by the airstrip. Later, on 21 June 1967, they did eventually manage to evade detection and successfully plant a British Mark 7 mine on the airstrip. The mine was initiated by a taxiing RAF Beverley transport aircraft and the blast wrecked the starboard undercarriage assembly of the heavy-lift aircraft, writing it off.

There were other threats to Habilayn. The vulnerability of helicopters to dissident mortar fire had been recognised from the outset, and significant efforts had been made to protect aircraft when dormant within the bounds of the base by parking them in large pits dug out by the RE. The Wessex, Sioux and Scout could be ground handled into these protective bays at night. With each bay surrounded by extensive sandbagged walls, light maintenance could be carried out in relative safety. Aircrew accommodation was far from luxurious and Scout pilots slept in trenches with a bivouac (bivvy) sheet over the top. The messes and briefing facilities were heavily protected by numerous layers of sandbags. The helicopter flights were located on the eastern side of the airstrip where the Sioux of the resident battalion's Air Platoon were located close to the AAC Flight (8 or 13 Flight on rotation) and the RAF's duty Wessex detachment. Command and helicopter tasking emanated from the office of the BASO. There was also an SAS communications team located in the AAC Operations Room, linking SAS patrols in the field back to the BASO.

An equal amount of effort had been expended on the ground defences of Habilayn. The base was surrounded by a generous number of weapon pits, with each carefully sited position home to a Wombat 120mm recoilless rifle or a GPMG. Those positions were further augmented by the 76mm main armament and machine guns of resident Saladin armoured cars. As ever, one potential threat to the aircraft at Habilayn was the disruption of their fuel supply. The Scouts and the RAF Wessex were refuelled by Bedford bowsers, while the Sioux was fuelled directly by pump from 44-gallon drums of AVGAS.

By this point in the campaign, life as a line pilot in Aden had a routine of its own. Staff Sergeant Fred Small was a RAMC SNCO serving as a pilot with 8 Flight:

> The aircraft when I first arrived were not fitted with intake sand filters. This gave reasonable performance, but the engines only lasted about 300 hours. After the sand filters were fitted the engine life improved but performance suffered, and we all moaned a bit. The loss of performance led to some near misses when landing at altitude – raising the lever did not seem to slow the aircraft down as expected. Communications with ground forces was by the B47 or B48 radios. These had to be tuned manually and frequencies changed in flight. Not easy at the best of times. At night it was a bit of a trial, and it was not unknown to land with the antennae still extended. This attracted much abuse from the REME.
>
> We flew in short sleeve KD shirts, smart cords, and desert boots, with a water bottle, mini flares and a pistol attached to our belts. The aircraft carried survival packs, water, flares and a first aid pack up. Up country, we also carried a Self-Loading Rifle, there was no proper weapon stowage, the butt was usually jammed into the map pocket. On ops we usually flew with the doors off, and rear seat removed to allow rapid deployment, I do not remember any straps for passengers, but nobody ever fell out. The maps that we flew on were rudimentary, just black and white contour maps. We coloured in red for tops, blue for valleys. Eventually better maps appeared, but by then, we knew our patch well.[56]

56 F. Small, Account to the authors, 4 May 2020.

8 Flight AAC Scouts at Thumier FOB loading SAS Troops. (Army Flying Museum)

Typical tasking for the Scout flights now included supporting RAF Wessex employed on night casevac tasking. Successful trials of launching parachute flares from a Scout while airborne had given the aircraft a useful new capability. The Scout could now be used to illuminate a landing site in advance of the much larger Wessex, significantly improving visibility for the Wessex crew on their approach and departure to a casevac extraction point. Captain Holtom describes the configuration of the Scout when roled for flare dropping:

> The Scout was loaded with about 25 slow-falling parachute flares, stacked like logs on the floor in the back and launched down a wooden chute bolted between floor and skid. The dispatcher had a harness and could move around in the back of the Scout, the flares had a ripcord and snap-hook which was clipped to one of the rings on the Scout floor – fortunately the ripcord was not long enough to reach the tail rotor when it pulled out of the flare on release from the aircraft.[57]

On the night of 19 November Captain Holtom was on duty at Habilayn with Corporal Chadwick, a ground crewman on attachment from the RAC as his flare despatcher. Holtom was at Habilayn on a two week roulemont from 8 Flight. It would be an eventful night for both men as Captain Holtom later recalled.

> I was duty standby Scout pilot that evening, we were called out by the BASO to illuminate the Heli Landing Site at Al Milah which had just been heavily mortared. Reports indicated

57 S. Holtom, account to the authors, 1 June 2020.

that there were three serious casualties that needed casevac to hospital in Aden. Al Milah was about six miles South of Habilayn, it was a Sapper camp occupied by a RE Squadron with FRA infantry protection. The Engineers were building the Dhala Road and digging wells for the local tribesmen.

We were quickly airborne in Scout XR634 with our load of flares. The Wessex was still being hauled out of its pit by the RAF tractor. Al Milah was five mins flying time to the South, it was situated in the middle of a valley. The valley was four miles wide and at an elevation of 1,000 feet, either side of the valley, were jagged peaks going up to about 4,500 feet. I set up in a slow (40 knot) racetrack pattern at 1500ft AGL. I was circling to the East of the valley to avoid dropping the flares onto the Wessex. I was below the high ground of course, but at the time it was a clear starlit night and the peaks stood out clearly. I had good comms with the incoming Wessex and asked him to call me when he wanted the HLS on the dirt playing field to be illuminated. He soon called and Cpl Chadwick started launching the flares down the chute at about 20–30 second intervals. The valley was soon brightly lit, and the nearby mountains could also be seen. I saw the Wessex on the ground below with the lights of Land Rover Ambulances around it.

At that moment, without any warning I suddenly entered cloud. The immediate reaction was sheer panic, at that time I had 973 hours of which only twenty was night flying and, I had no instrument rating. I had converted onto the Scout in Aden (from Alouette in Kenya) about a year before and had not flown it under the hood. Because skies were normally so clear in Aden, night flying was relatively easy. But not on this night!

My memory is vague of the antics I got into, I looked at the artificial horizon and thought it had toppled because it was down at 5 o'clock. I must have almost stopped and pulled more torque in the panic; I suddenly rose vertically into clear air and immediately saw that I was very close to the mountains. All this time Chadwick was throwing out the flares and my world in the clouds was a yellow pulsating chaos. I pulled myself together above the clouds and flew back towards the middle of the valley trying to work out what to do next. Chadwick said all the flares had gone. I didn't think I'd have enough fuel to fly back to Aden where I knew there would be no clouds.

Suddenly, there was a hole below me, and I could see the Wessex taking off in the light of the remaining flares. He also called 'airborne' – later over a few beers he said he wondered where I was! I decided to descend through the hole and land at Al Milah which was safer than going back to Habilayn in those conditions. I was at around 1,000 feet when the last flare went out and I came out of the clouds, I could see the lights of the camp and some vehicle headlights, I tried to remember where the playing field was that the Wessex had just left. I set myself up on finals heading for the black patch near the camp with no lights on it.

I switched on the landing light and it didn't work, panic was looming again but I slowed to thirty knots and made the rate of descent as gradual as I could. I did have some outside reference from the camp lights, so I knew which way up was, unlike ten minutes before!

Suddenly without warning I hit the ground, luckily almost in the middle of the dirt playing field. It was a mixture of sump oil and sand, like Habilayn airstrip so there was no flying dust. The Scout skidded to a halt and afterwards I saw I had just missed a large steel goalpost which would certainly have wrecked the rotor had I hit it. I was completely shattered and can only remember people coming to the aircraft and helping me out, my legs like jelly. I went to the Command Post and got Major Jimmy Nunn, my OC, on the phone saying would it be OK if I returned to Habilayn with Cpl Chadwick when it was light. We got given camp beds and crashed out, can't remember where, but we were back at the aircraft at first light.

There appeared to be no damage to the skids, and it had heavy landing checks back at Habilayn with no problems. A week later 628 got peppered with Kalashnikov rounds when

Jimmy was flying it and lost all the electrics, we all laughed and said even he noticed that most of his instruments stopped working! It needed a new loom. It was all part of the adventure of flying in Aden. But I will never forget that night over Al Milah.[58]

The Integrated Flight Scheme in South Arabia

The withdrawal of the Auster from Aden, coupled with the success of the Beaver and the Scout had firmly established the standing of Army aviation in the eyes of those deployed in theatre. There was, however, always a limit to the number of pilots and aircraft available for unit tasking. This was especially apparent when larger scale operations and/or SAS operations absorbed all the flying hours available. Individual infantry battalions, armoured car and artillery regiments had also needed an increasing level of intimate aviation support. During the early days in Aden there was an almost daily shortfall in available aircraft to meet these requirements.

The Integrated Air Scheme, as envisioned under the Partial Air Integration Plan approved on 24 October 1963 was, in principle, ideally suited to operations at this tactical level. The concept of partial integration had proved its worth in Borneo and it was to be refined in Aden. The first air troops arrived in Aden in February and March 1965; this was towards the end of phase one of the Partial Air Integration Plan. The integrated units were equipped with the Sioux. Affectionately nicknamed the 'clockwork mouse', the Sioux was less capable than the Scout, but had several positive characteristics that were ideal for the air troop role.

The Sioux was generally flown by a single pilot, although a crewman could be seated alongside in the three-seat, bubble-canopied cockpit. Alternatively, the pilot could fly solo and, fuel and weight permitting, carry two passengers alongside him (there was no rear seating). The Sioux's bubble-like cockpit was ideal for observation and recce tasks and the pilot had an unrivalled field of view. There was, however, a corresponding lack of shielding from the sun. Excellent cockpit visibility was complimented by a well thought out mix of radios. Reliable communications were essential when working in the recce and observation role, as was intended for the air troops and air platoons. The Sioux pilot was well equipped; he could switch between the Sioux's primary tactical radio, the VHF/FM ARC44, to the PTR170 UHF set for military air traffic and, crucially, FAC missions. The crystal controlled ARC44 VHF/FM tactical radio was often invaluable for radio relay and passing vital information between isolated patrols and base. On one occasion Lieutenant Micky Barnes of the Welsh Guards made a brief climb early in the morning to 12,500 feet to pass a message from Wadi Ayn to Habilayn, a distance of almost 100 miles; the aircraft was noticeably sluggish to respond to the controls at that altitude. Habilayn was using C42 radios with a locally made antenna which enhanced its range. Finally, the Sioux also carried an STR37 Radio for civil air traffic chat.

The Sioux's lift was generated by a two-bladed rotor head, powered by a turbocharged, six-cylinder petrol engine. The turbocharger was essential at the altitudes the aircraft was required to operate at in Aden. Although generally regarded as underpowered, the Sioux's Lycoming engine proved to be a reliable power plant with constant output. Originally designed in 1947, the low-tech design made it straightforward to maintain in the field. That low level of mechanical sophistication further enhanced the suitability of the Sioux for air troop operations. In summary, the Sioux was a low-cost, low-maintenance, and above all, simple to fly helicopter. The Sioux equipped air troops soon made their mark. Manning the new aviation units involved the selection of suitable personnel from the parent unit. Lieutenant Mike Richards QDG flew the Sioux in Aden:

58 S. Holtom, account to the authors, 1 June 2020.

Integrated Air Troops were a valuable asset in Aden, particularly effective when partnered with unit Recce elements like this 4 RTR Ferret armoured car. (The Tank Museum)

> The aircraft was a model of simplicity but its unsophisticated instrumentation and controls (a trusted aid was the balance indicator cord stuck to the bubble!) presented significant challenges to the pilot in the harsh conditions. There was very little power to get out of trouble and flying techniques were developed on the spot to compensate. The Sioux was a great aircraft and well chosen by Army aviation for the role.[59]

Lance Corporal Ron James served in the 10th Royal Hussars (10 RH) and his story is typical of those non-AAC soldiers who manned the air troops and air platoons in Aden. After a spell on Centurion tanks in Germany he moved with his regiment back to Tidworth and the delights of Salisbury Plain. Shortly after the Hussars arrived back in the UK, they converted from tanks onto the Ferret scout car and the heavier Saladin armoured car. A deployment to Aden was imminent, when Lance Corporal James was asked by his squadron commander whether he had thought about becoming a pilot. The Hussars were forming their own air troop consisting of three Sioux helicopters and a number of Hussar officers and SNCOs were already under training. Unfortunately, Lance Corporal James discovered that he was colour blind, and consequently failed the aircrew medical. Shortly after the medical Ron was promoted to corporal and made a troop corporal in B Squadron, just in time to deploy to Aden as an armoured car commander in October 1964:

59 M. Richards, account to the authors, 12 August 2020.

> Life on operations in Aden followed a regular pattern, B Squadron was initially based in a tented camp upcountry in Thumier (later called Habilayn). The camp also functioned as a Forward Operating Base (FOB) for AAC aircraft, I maintained an interest in the regular comings and goings of the Beavers, Scouts, and the Air Troop's Sioux aircraft. The Armoured Cars regularly worked with AAC aircraft as they escorted Royal Engineer and infantry units who were employed clearing routes of Improvised Explosive Devices and numerous Mines planted by dissident groups. Ambushes and firefights with the 'Dizzies' (dissidents) were frequent, and the .30 Calibre machine guns of the armoured cars were a real asset in these situations. The FOB itself was far from being a sanctuary, local villagers would attempt to infiltrate the base under cover of darkness. Most mornings the villagers would return and attempt to recover their dead from the perimeter wire.[60]

As well as general patrolling and route clearance tasks, the Hussars were also involved in supporting specific operations against dissident strongholds. These operations required more firepower and the Saladin's 76mm main armament proved itself to be useful for this kind of task. In March 1965 Corporal James was moved across to the Regiment's Air Troop:

> I was posted in as Troop Corporal responsible for Mechanical Transport and Signals. We were based in Little Aden at Falaise Airfield which was the main AAC Base. It was decided that there would only be three Observers trained, but as Troop Corporal I was included as an additional Observer. Our training was overseen by 653 Light Aircraft Squadron AAC and we had to complete twenty-five hours flying front seat in Sioux or as a crewman on other types. We also had to master the technicalities of aircraft refuelling, aircraft husbandry, signals and the Emergency and Rescue Procedures in the event of a downed aircraft. The main thing was to get our twenty-five hours in. I got flying experience with our own Air Troop Sioux but also on Scout with 653 Squadron and on a Fleet Air Arm Westland Wasp which was on secondment from HMS *Zulu*. The pilots were great because they appreciated that we assisted with refuelling, marshalling, and cleaning. We also took our turn in the Air Traffic Control tower which was great because the Hawker Hunters from RAF Khormaksar would come and do circuits and bumps on the runway and we had to control the airspace. On completion of our training we were awarded our Observer Brevet which we wore on the upper left sleeve of our Khaki Drill jackets. We did lots of flying but we were only allowed three paid Observers on the Air Troop so although I flew, I didn't get the flying pay of 3/6 (17.5p) per day.[61]

The introduction of the Sioux equipped flights at unit level was an immediate success, as it was in Borneo. The typical air platoon fielded three Sioux helicopters, although the QDG deployed their own air squadron to Aden with six, later eight, aircraft. The administrative hub for the various Sioux detachments was co-located with HQ 3 Wing AAC at Falaise Airfield in Little Aden.

The reliability and technical simplicity of the Sioux made the provision of engineering support in the field relatively straightforward in comparison to the Scout. However, after a very promising debut in Southern Arabia, the Sioux was grounded with a debilitating technical fault. The nature of the problem was explained by a frustrated member of the Air Troop of 10 RH:

> We received our first aircraft – three beautifully new Sioux at the beginning of June. At first all went well after theatre conversions; the Troop went to Beihan to carry out a shakedown

60 R. James, account to the authors, 2 March 2020.
61 R. James, account to the authors, 2 March 2020.

> exercise and to practice movement drills. There we ran into trouble and the Sioux developed an odd habit of engine cuts. At first, we thought that perhaps we were closing the throttle too quickly in the rare air (Beihan is 3,500 feet ASL and has ambient temperatures of 40 deg C). We continued to experience engine cuts and the aircraft were grounded for investigation. It was found that fuel vaporisation was the cause and that the fuel was boiling well before the carburettor.[62]

After extensive research and numerous visits from UK based technical teams, the Sioux's fuel vaporisation problem was resolved. Two and a half months later the air troops were back in the air over Aden.

Although inevitably there were occasional incidents of regimental tribalism, the working relationship across the integrated flights and the AAC was consistently good, and friction was rare. The character of these small aviation units reflected that of their parent unit. The Sioux allocated to 1 Regt RHA were, for example, formed into what was designated as AOP Troop, 1 RHA. The ethos of this Integrated Air Troop was very much linked to that of the RA-manned wartime Auster AOP squadrons. Originally formed as part of 19 Lt Regt RA in March 1965, the pilots and REME personnel remained in situ when the Regiment was relieved by 1 RHA. The transition between the two units and subsequent operations was described by members of 1 RHA:

> Our original work up phase went very smoothly, with all ground crew receiving about six weeks training. Towards the end of this period our aircraft were received, and work of a more practical nature could be undertaken, including pilots' theatre conversions. Troop training progressed along approximately the right lines with several single aircraft detachments upcountry on a training/operational basis. The culmination of this training was a complete detachment at Habilayn in the Radfan area for just over a week, when it became obvious to everyone how valuable the Sioux were going to be when based with the upcountry batteries. Approximately thirty DFs [defensive fire targets] were registered or re-registered that week which really underlined the title of the Troop – a most gratifying fact – and this trend was continued. We were then handed over as the first operational Air Troop in this Command.
>
> We then had a glorious four weeks of integrated operational flying until, with a crunch, it all came to a grinding halt with the fuel vaporization problem, which grounded us for two and a half months. That wasn't time wasted, although it got very frustrating, as we got our fly out role jacked up and also tackled Khormaksar over airborne FAC training. This latter part has now been completed and all three pilots are now recognised as airborne FACs. That was not without its moments either as the Troop Commander, orbiting in a dual Hunter, watched another strike pair put down cannon fire only two hundred yards away from the controlling Sioux!
>
> The Regiment has just over two Batteries upcountry spread in two-gun sections over a distance of two hundred miles. We maintain an average of 1.5 aircraft up with these Batteries. Area West, based on the Dhala Road always has one aircraft and Area East based on Beihan some 170 miles away, has one aircraft on a monthly roster between us and 4/7 RDG [4th/7th Royal Dragoon Guards].[63]

Command and control of the Sioux sat firmly with the various parent units. There was, however, close cooperation with the AAC flights for more complex operations. The air platoons and troops

62 Anon, '15 Flight A.A.C.', *Journal of Army Aviation*, 1966, p.79.

63 Anon, 'Air O.P. Troop 1st Regiment Royal Horse Artillery', *Journal of Army Aviation*, 1966, p.79.

were at their busiest when their battalion deployed up country and the Sioux generally deployed with their unit headquarters into Habilayn. This level of intimate aviation support allowed commanders to extend their unit's physical reach and their influence. Having an integral aviation capability significantly enhanced the range of the battalion's observation and reconnaissance capability. Discreet recce by air platoon or troop Sioux, augmented with air recce photographs provided by RAF photo reconnaissance Hunter aircraft, allowed headquarters to build a more complete picture of the terrain over which they were planning to operate.

Individual aircraft were routinely tasked to support companies or platoons deployed to out stations at Dhala, Al Milah or Wadi Ayn. Tasking from these locations was coordinated by the battalion CP and often included aviation support to RE road construction or hearts and minds visits to remote tribal villages carrying the battalion MO. The air platoon could also be employed to carry British Political Officers and the battalion IO to meet tribal elders and to visit isolated FRA garrisons in relative safety. When not deployed up-country, the air platoons were available for tasking to other theatre units that had no aviation support, and the Crater region soon became familiar territory to many Sioux pilots. The Sioux detachments were also a key component in many of the larger aviation operations that were mounted on a regular basis. The provision of security and overhead cover to road moves along the Dhala Road was an important, asset-heavy task coordinated by the BASO cell. Lieutenant Micky Barnes was a Sioux pilot who flew on many of these escort operations, and his description gives insight into the joint nature of aviation operations in Aden:

> A fairly frequent event was support for any convoys travelling up from Aden to Habilayn, especially when travelling to Dhala…4–8 man picquets would be positioned on strategic hills along the route and leap-frogged forward as necessary using Scout or Wessex. A Sioux would be airborne to provide immediate AOP or FAC support. Towards the end of our time upcountry Hunters would be on airborne alert with constant updates from the airborne FAC, including sometimes a dummy pass over a previous, or likely trouble spot just as the convoy was nearing that place.[64]

At this stage one could argue that the Flight Integration Scheme was having the desired effect of relieving pressure on AAC manning and preserving Army aviation itself. Lieutenant Robin Grist's path into Army flying was similar to many would-be aviators in the mid–1960s. While serving as an infantry platoon commander in the Gloucestershire Regiment he took the plunge and applied to his CO to be released for flying training. Like many COs, he realised the potential of the Air Integration Plan and was keen to integrate an air platoon into his battalion. The major selling points of the Integration Plan were twofold: the parent unit did not lose promising manpower; and at the end of the process, a new capability was gained. Grist recalled his CO's enthusiasm for the idea of a platoon of Sioux helicopters manned by his own men. It was also considered equally important that the platoon be commanded by one of his own cap-badged officers. The application was signed, and Lieutenant Grist subsequently reported for flying training:

> All AAC Flying Training took place at Middle Wallop, twelve weeks in Chipmunk, twelve weeks in the Hiller and twelve weeks in the Sioux. It was an extremely time efficient system lasting nine months in total. I was posted to command the Air Platoon of the Northumberland Fusiliers in Aden but during the Sioux phase of my training it was decided that it made more sense for the current 2IC of 8 Flight, who had experience flying in Aden, to go to the Fusiliers and that I should convert to Scouts and join 8 Flight.[65]

64 M. Barnes, account to the authors, 4 May 2020.
65 R. Grist, account to the authors, 6 June 2020.

It is evident that some thought was already being given to balancing the levels of experience, both in the cockpit and in command. The swapping of pilots between the Northumberland Fusiliers Air Platoon and 8 Flight AAC was logical and mutually beneficial. Clearly operational requirements were already reshaping the boundaries of the original concept for the Integrated Flight Scheme. On the surface, the relationship between the AAC flights and the unit air troops and platoons worked well in Aden. It was also however, evidence of the view, held by many at the time, that integrated air troops and platoons were a temporary phase in the development of Army aviation. The majority view was that the aircraft and the pilots were ultimately destined to be absorbed into the AAC as was proved to be the case when recentralisation began later in the 1960s.

Flying to the Limit – SAS Operations

Almost from its introduction into service in Aden in 1964, the Scout had established itself as the preferred method of travel for SAS troops working out in the Radfan and the wider Protectorate. This close working relationship with the SAS often presented the Scout pilots with the opportunity to test their flying skills and tactical awareness to the limit. Although the general tempo of operations in Aden was not as frenetic as it had been in the days of Radforce, the SAS role in Aden had expanded, and with it came a corresponding demand for aviation support. Lieutenant David Ralls had joined 13 Flight as a freshly trained Scout pilot in August 1966. His candid recollections of his first SAS insertion task, led by Major Chris Crouch, are testament to the challenges of single pilot operations in Aden:

> A regular task was the insertion at dusk of SAS patrols, often on high saddles from which they would move at night to hide out and observe in the days which followed. All four Scouts, led by Chris, would stream to the operational area to carry out diversionary dummy landings before eventually dropping their heavily laden occupants at the designated spot.
>
> It seemed unwise to place the least experienced and most apprehensive pilot at the back of the stream but so it was. The others were slick with their quick start drills and, on radio silence, already hurtling off at 100 knots. Catching up was the first challenge as daylight and the limited fuel on board ebbed away. Eventually the final drop-offs were completed. A very shaky take off followed by a nervous climb between mountains into semi-darkness should have brought a sense of relief. The reality was stark terror. There was no sign of the other aircraft which were by then well on their way back to Habilayn – without lights and still on radio silence. A 360-degree turn served only to heighten a growing sense of disorientation. This was single pilot stuff in the raw.
>
> Scout pilots of that era will recall that there was no compass on the instrument panel. Use of the horizontal WW2 bomber type main P12 compass could be tricky. In any case, to an addled brain it didn't appear to be giving the right information. Reference to the standby compass served only to increase the workload without bringing enlightenment. The horrible truth dawned that in addition to being 'uncertain of position' (and not just temporarily) any sense of the direction in which base was to be found had gone. A brief attempt to read the jumble of contours, pocket torch in mouth, was curtailed by the map flying out of the doorless cabin for good.
>
> The only good thing was a light-ish night which, above the surrounding mountains, gave a discernible horizon. Less good was the sudden blaze of flashing red panel lights which brought the news one hoped never to receive. These were the dreaded 'winky-wanky' lights, not previously experienced, which announced that precisely twelve minutes' fuel remained:

not 'roughly twelve minutes' and absolutely not a second more. The brain said, 'Start the clock and cancel the lights, as taught.' But they would be back. With a vengeance.

A degree of panic had set in which verged on paralysis.

It was time to attempt the final ergonomic challenge (and 'final' seemed pretty likely) of the Scout at that time. In the absence of any response from Habilayn to the UHF line of sight radio, and in the knowledge that that there was no air traffic ground station within seventy miles, an attempt should be made to use the B 47 tactical radio. The sensible thing would have been to try to talk to the SAS on the frequency selected and hope they had an HF (high frequency) link back to base. Ingrained training brought the wrong answer: we could not compromise a covert operation. In any case word would not reach the Flight in time to do any good.

The ergonomic challenge was to attempt to retune the B 47. This meant the pilot looking down, peering below his left armpit, turning a small handle to wind the ribbon dial to the new frequency and looking up occasionally to see whether the aircraft was still the right way up. Amongst many uncertainties there could be no doubt of one thing: Any attempt to retune the B 47 would complete the sense of partial disorientation and decide the matter once and for all.

The red lights erupted again; was that eight minutes' fuel left? There came a rare moment of clarity in that sorry saga of personal confusion. There was something which could be done. It would make sense to use the remaining fuel to climb above the higher ridgeline, hope that some discernible lower ground showed up in the starlight, try to calm down and prepare for the engine-off of a lifetime (perhaps). The direction was no longer important, so which way was the light wind into which the earlier landings had been made? The answer was from the East though after dark it had probably died anyway. Never mind, climbing at sixty knots and heading East was a decision of sorts.

And then joy of joys, not too far away – perhaps five miles – was a stream of parachute illuminating mortar flares erupting in the sky. Oh, dear and wonderful friends who had thought of doing that. As it turned out one of the returned Scouts was being hastily refuelled for Chris to get airborne again with all lights blazing to point the way home. In the event the flares did the trick.[66]

Operation Flash-Bulb Charlie

One notable SAS support task had taken place earlier that year on the morning of 2 February 1966. At 0727 hours three Scouts from 13 Flight AAC, led by their Flight Commander, Major Christopher Crouch, were tasked to support D Squadron SAS on Operation Flash-Bulb Charlie in the Haushabi country to the north of Al Milah. Sergeant Michael Jones, R Sigs, was the duty pilot at Habilayn and he remained in the ops room to coordinate 13 Flight's activities with the RAF BASO, 3 Wing AAC and the SAS Ops team. The SAS were monitoring the activity of a group of 12 uniformed and armed men that were moving tactically and believed to be dissidents. Major Crouch and his crews were briefed on the group's activities by Major Glyn Williams, the Commander D Squadron SAS. Confirmation that the suspect group were dissidents, and not patrolling tribal guards, was required before the SAS could open fire on them. A plan was quickly formulated and the Scouts (one armed) rapidly embarked D Squadron's 16 Troop and lifted out in

66 D. J, Ralls, 'Arabian Frights', in R. Mallock, *The Likes of Leicester* (Self-Published, 2013), pp.35–41.

The first Attack Helicopter – The armed Scout carried two fixed forward firing GPMGs, and a pintle mounted door gun. (Army Flying Museum)

line astern. The three remaining troops of D Squadron were already on the ground, waiting to act once visual confirmation of weapons had been received.

Major Crouch took the lead, accompanied by the D Squadron Ops Officer who was map reading and working the SAS command frequency on the B47 radio. The second (armed) Scout was flown by Captain Russell Wright RA, the third by Major Jake Jackson AAC. The flight set off at low level towards the last known position of the suspect group; each aircraft carried four SAS men. The plan was essentially an advance to contact. D Squadron reasoned that tribal guards would not engage the approaching formation, dissidents would immediately open fire.

Flying at 10 feet, the Scouts rounded a steep *jebel* and Major Crouch spotted the 12 men just 200 yards away. Warned by the distinctive sound of the approaching Scout rotors, the dissidents scattered for cover. Within seconds the aircrew confirmed that the dissidents were well armed and hostile, engaging the heliborne force with a combination of LMG and rifle fire. The lead Scout was hit by the incoming fire, a round entering the cockpit and passing beneath Crouch's right foot. He immediately banked away from the enemy fire position and the three aircraft then dropped into the cover of dead ground. Reacting quickly to the situation, Major Crouch and the SAS commander decided to continue to use the cover, and flew on towards a drop-off position, 175 yards behind the enemy force. From here, the SAS could establish a cut-off position, blocking the escape of the now scattered dissidents. The other two Scouts unloaded their passengers before Wright lifted out of cover to strafe the dissident position, covering the other two unarmed aircraft as they lifted out of the dead ground. Wright's Scout was hit in the tail and Jackson's aircraft took a round through one of its main rotor blades. The after action review reported that Wright had succeeded in killing a dissident area commander and wounding his deputy.

What followed was a protracted running battle. The 13 Flight Scouts were used to redeploy SAS troops from fire position to fire position, maintaining contact with the retreating dissidents. The battle lasted through the morning and into the afternoon, and Major Crouch remained in the air almost continuously, directing his Flight and keeping Williams aware of the location of the enemy. Meanwhile Jackson lifted the SAS headquarters onto a *jebel* that gave them an overview of the battle area. Wright continued to engage enemy fire positions and guided the troop-carrying Scouts along safer routes to insert the SAS patrols into new fire positions. The Scout crews were able to manoeuvre the remaining three troops of D Squadron, maintaining contact with the dissidents, allowing them to leap-frog each other thereby maintaining constant pressure. Jackson, acting on his own initiative, picked up an 81mm mortar and crew and positioned them close to the SAS CP, adding some useful firepower to the mix.

During the ongoing firefight Crouch's Scout was holed by small arms fire a second time. Crouch had slipped back to Al Milah to refuel and collect an A41 radio, and Staff Sergeant Alec Baird AAC, another 13 Flight pilot. He intended to read Staff Sergeant Baird into the battle and then handover his aircraft to him. Major Crouch would then locate himself at the D Squadron CP and run his own aviation command net on the newly scrounged A41 set. He was, however, forced to drop Staff Sergeant Baird off at a friendly position and respond to an urgent call to lift SAS troops to a new position. During this move he heard that an SAS SNCO, Sergeant Smith, was badly wounded and required evacuation by air. At some risk, Major Crouch flew his aircraft onto one of the SAS forward positions to carry out the casevac from an exposed position while under fire. It was at that point in the afternoon that his Scout was hit by enemy fire for the third time. Undeterred, Crouch flew Sergeant Smith out of the firefight and into the Casualty Collecting Point at Habilayn, from where he was subsequently evacuated to the RAF Hospital at Khormaksar. The firefight raged on until late afternoon, the Scouts refuelling at nearby Al Milah and returning to re-join the battle throughout the day.

At 1000 and 1030 hours respectively the SAS were reinforced by two standby Wessex carrying an airlift of troops from 1st Battalion, King's Own Yorkshire Light Infantry (1 KOYLI). The two RAF aircraft continued to lift in troops, ammunition, and water throughout the day. Further reinforcements arrived in the form of a 15 Flight Beaver flown by Captain David Swan AAC. Swan took on the important role of airborne relay station overhead the battle and maintained a good communication link back to Habilayn. Major Jimmy Nunn, OC 8 Flight arrived in a fourth Scout and joined in with securing the perimeter, escorting the Wessex, and completing resupply lifts as required. Late in the morning the intensity of the battle subsided although dissident snipers did maintain accurate fire on the SAS squadron CP at irregular intervals.

By 1400 hours all firing had ceased. SAS observers could see the body of a dead dissident in the wadi and suspected there to be at least one additional wounded dissident in the same area. All of the Scouts landed in cover behind the SAS *jebel* CP and the pilots gathered there for a joint briefing with the D Squadron command team. The *jebel's* location offered a commanding view of the battlefield and the area was methodically scanned for 55 minutes. The dissident snipers proved to be elusive, still managing to generate harassing fire. The SAS wished to gain positive identification of the dead dissident and to kill any remaining dissidents. An overland approach would be exposed to fire as it crossed what was an expanse of open ground. The decision was taken to adopt a bold approach and use a Scout to drop a four-man snatch patrol behind the bluff that appeared to mask the dissident snipers. It was Major Jackson's turn to fly:

> When he was ready to lift off, concentrated mortar and LMG fire was directed onto the dissident and above it. Jackson flew in and out unharmed. Under covering fire from the Jebel, the SAS carried out a quick firefight, and returned over the hill with two bodies. (One of them

later identified as the local area commander – the other, his second in command). Jackson returned and collected the four SAS and the two bodies – again without any bullet strikes.[67]

The POR from D Squadron describes Jackson's role in the operation in some detail:

> The helicopter flew the patrol to a position in dead ground 150 yards short of the enemy dead, under mortar and LMG covering fire. The pilot, who gave a magnificent performance of tactical flying, immediately flew away and 'stood off'. The patrol quickly contacted the one wounded – but active enemy and killed him. The patrol then came under fire from possibly up to two more dissidents a further 300 yards to the South. Mortar and LMG fire were brought down in the general area, as the enemy could not be pinpointed. Under this covering fire the pilot returned and quickly extracted the patrol and two dead enemy, who were dragged to the helicopter, together with a weapon and some equipment...[68]

By 1530 hours all enemy fire appeared to have ceased and observation detected no movement. In order to clear the area before dusk, it was decided to repeat the successful snatch operation on the next feature, this time deploying an eight-man SAS team from 17 Troop with two Scouts flown by Major Crouch, and Captain Wright with the armed Scout. Supported by mortar and LMG covering fire, the two Scouts flew forward to drop the SAS behind a rocky outcrop to the south of the enemy positions. The 13 Flight POR described the flight:

> Crouch and Wright flew tactically and were hidden from the feature until the very last moment – when they flew line abreast – the armed Scout firing its GPMGs – to the lea of the objective. Despite the excellent covering fire, a party of about 3 dissidents – previously undiscovered – opened fire on the helicopters from a position halfway up the feature. Crouch's aircraft received a bullet through the fuel tank, and upon receipt of an urgent call from Comd SAS [Commander SAS] – broke off and returned to the secluded wadi. Wright broke off in the opposite direction, having received a burst of Bren fire. This holed both scavenge oil pipes and the tail cone. He was not aware that he had come under fire.[69]

The 17 Troop men on both aircraft returned fire at the dissidents but reported no hits. Both aircraft aborted the lift and disembarked their troops. Major Crouch returned to Habilayn to pick up the Commander of Area West and two REME technicians, WO2 McGarry and Corporal Weltch. He then returned to Al Milah to re-join the operation. At the refuel point, the ground crew and REME were working continuously to keep all four Scouts involved in the battle. The Flight's REME AQMS, WO2 Martin McGarry played a key role in keeping 13 Flight in the air. McGarry was later awarded a Queen's Commendation in recognition of his efforts, an extract from the original application indicates the importance of his actions on the day:

> During the initial lift of interception patrols, and whilst a casualty was being evacuated, an armed Scout and a second Scout were hit. Later in the action a third Scout was also hit, and an aircraft hit earlier was holed twice more. AQMS McGarry kept the aircraft of his Flight flying

67 AFM 182: 13 Flight AAC, Report of Army Aviation Participation in Operation Flash-Bulb Charlie, dated 17 February 1966, p.3.

68 AFM 182: 13 Flight AAC, Flash-Bulb Operation Charlie, D Sqn SAS Report on Contact, dated 17 Feb 1966, p.5.

69 AFM 182: 13 Flight AAC, Report of Army Aviation Participation in Operation Flash-Bulb Charlie, dated 17 Feb 1966, p3.

> by sheer genius. As aircraft came in to refuel carrying out ad-hoc and makeshift repairs since it was vital to the success of the operation that the full number of aircraft be kept serviceable to enable the encircling technique to be put into operation.
>
> One aircraft was hit in both oil pipes and was forced to make a precautionary landing close to the engagement area. Without hesitation, AQMS McGarry immediately volunteered to go out to the aircraft which he effectively repaired using bandages and masking tape.
>
> A second aircraft with holed fuel tanks was repaired by temporarily plugging using lead pencils.[70]

McGarry recorded his own candid view of his work that day:

> Scout XP600 was hit by a burst of automatic small arms fire, several bullets passing through the self-sealing fuel tanks. The pilot headed for base and came in with his aircraft performing like a watering can. "Fix it as best you can and tank it up." Was the pilot's instruction.
>
> The holes looked just the size of a pencil, and so it proved; a trial plug from one I had in my pocket completely sealing a hole. A quick whip round produced as many pencils as bullet holes and in about five minutes the Scout was refuelled and airborne.
>
> The pilot was quite reluctant to exchange his 'lash up' for a permanent repair – he reckoned that the HB2 leaded fuel gave a better performance![71]

Back at the wadi the battle continued, with just a few hours of daylight remaining there was a possibility that the three surviving dissidents would be able to slip away under cover of darkness. Any further ground operation to cut them off risked potential casualties, so the SAS elected to call in an RAF Hunter strike. Eleven minutes after an initiating call from the Squadron Commander on his SARBE radio the Hunters were overhead. The airstrike using cannon and rockets was judged successful. Two 13 Flight Scouts continued to assist with resupply lifts as the SAS withdrew and 1 KOYLI moved into night ambush positions supported by a single 105mm Howitzer from Chestnut Troop, 1 RHA. By nightfall D Squadron and all 3 Wing AAC aircraft had withdrawn from the wadi area leaving D Company of 1 KOYLI in position. The KOYLI troops reported and engaged dissident movements overnight, estimating the dissidents to be about 20 in strength.

At first light on 3 February 13 Flight Scouts returned to the area to recover any dissident bodies but none were found. Major Jackson did however land on and recover abandoned documents and some items of dissident clothing and equipment. The documents were considered to be highly sensitive and to have intelligence value at the time. The operation was assessed to have been successful and the POR submitted by D Squadron, under 'Close Helicopter Support', highlighted 13 Flight's contribution:

> a) The rapid and successful redeployment of the Sqn by Scout, which resulted in the encirclement of the enemy was only brought about by the coolness, determination, and initiative of 13 Flight AAC.
> b) It was considered necessary to redeploy elements of the first Troop into the vicinity of the enemy to engage them by fire and thus to pin them down; at the same time to place smaller elements on the higher ground. This calculated risk must be taken if the enemy

70 TNA, WO 373/137/37: Army Form W3121, Recommendation for Honours or Awards, WO2 AQMS M. McGarry, REME, 19 February 1966.

71 Kennett & Tatman, *Craftsmen of the Army: Story of the Royal Electrical and Mechanical Engineers*,p.352.

Lance Corporal Jim Turner RTR was among the first ground crew locally trained to become an Air Door Gunner in 13 Flight AAC. Jim was later Squadron Sergeant Major of 653 Squadron and Regimental Sergeant Major of 3 Regiment AAC. (J. Turner.)

are to be contained by fire. If the troops are only redeployed on the high ground out of effective SA [small arms] range (400 yards) the enemy will split and disappear.

c) It is the first time that an operation of this kind involving Scouts so intimately has been carried out. Although considerable thought and training was carried out between D Sqn and 13 Flt before this operation many lessons were learnt. With more extensive combined training and cooperation this type of operation can be carried out with greater success and with minimal danger to Scout Helicopters.[72]

The after action review of Operation Flash-Bulb Charlie concluded that the SAS had accounted for two-thirds of the 12-man gang. Four had been killed, a further four were wounded, and two senior leaders were captured among the wounded. The action was regarded as a huge success and the role of Major Crouch and 13 Flight were regarded as pivotal to the success of D Squadron on the ground. Recognition was forthcoming and Major Crouch was awarded the DFC for his handling of the Flight through the day and his own personal bravery. Major Williams was awarded am MC for his leadership and courage during the battle and Trooper Brian Dodd, who was with 16 Troop of D Squadron, was awarded the MM for his part in the fighting.[73]

Helicopter tactics were constantly evolving in Aden and Aviation Force Packaging, as practised during Operation Flash-Bulb Charlie, was becoming an established concept:

> Operations have varied considerably. We have generally found that three Scouts backed up by a Wessex produces the most viable force. One Scout is armed with fixed forward firing GPMGs and one internal swivel mounted GPMG, and carries the Flight Commander, the Ground Force Commander and the Air Gunner in the back. The other two carry four armed men each. The Wessex, close behind, carries up to a further fourteen men. This force can swoop down on a target spotted by hidden Ops, or on random targets, and give a good account of itself until reinforcements arrive.[74]

Overdue Action

Even when not directly engaged in operations such as Flash-Bulb Charlie, the risks and dangers were ever-present for aviation aircraft and their crews and passengers. On 6 December 1966 Captain John Fleming RHA was flying a Sioux from 1 RHA AOP Troop carrying two passengers, Captain John Sharpe RHA, himself a former Auster AOP Pilot, and Gunner John Cain. The three men sat in the cockpit with Gunner Cain carrying his SLR and nursing a weighty resupply of grenades and ammunition strapped onto the Stokes litter. The task for the day was to fly Captain Sharpe around the gun positions of Chestnut Troop, 1 RHA. As the sortie progressed the Sioux approached Musaymir, close to the border with Yemen, when the aircraft suddenly plunged earthwards and spiralled out of control. The stricken Sioux crashed into a wadi bed. Immediately after hitting the ground the fuel tanks above the cockpit caught fire, the burning fuel and exploding ammunition killing any of the three passengers that may have survived the crash. The ferocity of the fire destroyed most of the airframe and made it almost impossible for the subsequent crash investigation to determine beyond doubt the reason for the crash. When Captain Fleming had

72 AFM 182: 13 Flt AAC, Flash-Bulb Operation Charlie, D Sqn SAS Report on Contact, dated 17 February 1966, p.5.

73 *London Gazette*, Dated Friday 15 July 1966.

74 Anon, '8 Flight Army Air Corps', *Journal of Army Aviation*, 1967, p.78.

failed to report in by radio a second Sioux, flown by Sergeant Forde QDG, was launched to search for what was then considered to be a missing aircraft. In spite of his best efforts, Sergeant Forde did not locate Capt Fleming's aircraft and was forced to return to base. Service families lived close to the airfield and, after several hours of fruitless searching, when Sergeant Forde landed, he was met on the dispersal by John Fleming's wife who was obviously anxious for news of her missing husband.[75]

The search operation was coordinated by Major Crouch and Lieutenant Grist was one of the pilots involved:

> What followed showed how incredibly difficult it was to find a crashed aircraft. Nothing was found on the 6th, and we reassembled at dawn on the 7th and were each given a 2x3km area to search at low-level. I think there must have been about fifteen helicopters searching, perhaps more, and yet it took all day, but by 5pm the crash was found. I had already searched six areas but so happened to be at Habilayn refuelling so was tasked with flying the 2iC of 1 RHA and the doctor out to the wreck which had caught fire on impact. A Company of Irish Guards set out to march to the site and guard it. It took ages to get the bodies into body bags and load them into the aircraft by which time it was pitch dark and we were in quite a deep bowl and I was exhausted, shaken and nervous of getting airborne without hitting one of the surrounding hills. Fortunately, there was an RAF Shackleton overhead and the pilot offered to drop a huge flare when I was ready to take off. It made it easy and then Richard Grevatte-Ball, who had been on the same flying course as myself but was a Beaver pilot, realising I was in trouble made me follow his lights home otherwise one disaster might have been two.[76]

Initial investigation of the wreckage revealed that Gunner Cain's SLR was missing and that one of the bodies, as well as being burnt, had been mutilated as was the custom of tribesmen in the region. This led to speculation that the Sioux had been shot down by ground fire, rather than crashing due to mechanical failure or pilot error. The subsequent accident inquiry could not determine the reason for the crash. But the fact that tribesmen had arrived so quickly on the scene, stolen the SLR and mutilated a body in the manner that they would traditionally inflict on a fallen enemy, indicated to the inquiry panel that Sioux XT125 had, in fact, been brought down by ground fire. Lieutenant Brian Preston AAC, who was flying a 13 Flight Scout, remembered the day clearly:

> We had been involved in a large-scale operation all day. Our Scout helicopters armed with three Machine Guns were providing 'top cover', lifting troops to the high ground, ferrying stores to the forward locations and prepared to evacuate casualties if necessary. Sioux helicopters were also being well-used directing the fire of the guns and acting in the classic recce and observation role. During the afternoon, a Sioux from 1 RHA with the pilot, John Fleming, and his two passengers were reported missing over the mountainous area to the West. A large-scale search using all available helicopters in the area was immediately started and it was not until the next day that the crashed helicopter was found. Sadly, it had been shot down and there were no survivors.
>
> Three bodies then had to be transported back to the main airport at Khormaksar. Every available pilot had flown all through the day and we were all tired, I personally clocked ten hours searching that day. However, my flying hours were not the highest and so I was declared 'available' and was duly briefed to fly the three dead bodies back down to civilisation. I was

75 S. Holtom, account to the authors, 1 June 2020
76 R. Grist, account the authors, 4 July 2020.

invited to take an ailing Scout from 8 Flight, since its engine hours were nearly expired, and it could be usefully deposited with Base Workshops for a Major Service.

….We dismounted the three Machine Guns and replaced them with a litter on each skid… We loaded a body into each 'litter' and lay the third, wrapped in a blanket, across the rear bench seat secured by the seat belts. This rear seat was virtually a canvas stretcher anyway, with a strip of padded material fastened to the bulkhead with press-studs at about shoulder height to form a backrest. All doors would remain off since, although night was approaching, the heat and humidity never abated and the refreshing breeze whilst flying was always very welcome after the sweltering conditions on the ground…. I climbed up to a reasonable altitude to avoid the jebels and random shots from trigger-happy natives. I set course and anticipated the first welcome glimpse of the permanent flame burning from the BP Oil refinery chimney in Little Aden, always a comforting sight, and confirmation that map, compass, and pilot were in agreement.

I levelled off at about 3,000 feet, flying a strange aircraft with its engine hours nearly expired, wondering idly whether it would last the trip, but otherwise mentally drained. Scything through the night sky: the soft spice-tainted breeze wafting in from all sides inducing a pleasant slightly soporific feeling; the instruments emitting a friendly glow; the good old Nimbus turbine humming contentedly, at one with the environment. Alone with my thoughts, and three bodies being carried homewards through the starry night sky…. Then I felt a gentle tap on my shoulder…. What the…!

Hairs stood on end. Heart raced. Hands and arms tightened on the controls. Backside clenched the seat. Wide eyes stared into the darkness. No one could have invented a quicker way to interrupt a reverie! With trepidation I forced myself to look round. There flapping gently in the breeze, was one end of the unfastened rear seat backrest…[77]

Some weeks later a patrol from 45 RM Cdo were in the vicinity of the crash site hurrying to assist a convoy of RE vehicles that had been ambushed by tribesmen when, en route, they themselves came under fire. The marines identified the ambush position as being a sangar built into a hillside cave and they mounted an aggressive counter-attack against the position. The assault was led by the Troop Commander, Lieutenant Knott RM, and it was Knott who closed on the cave and posted a grenade which silenced the enemy fire. Inside the cave the marines found a single dead tribesman and a search of the body revealed that their ambusher's rifle included working parts that matched the serial number of Gunner Cain's missing SLR. The evidence appeared to corroborate the inquiry findings that Captain Fleming's Sioux had been shot down.[78]

ABFAC

If ever an illustration was required of the value of the Air Integration Plan and the Sioux, a prime example took place over the Christmas of 1966. Twenty-two-year-old Lieutenant Micky Barnes was a Sioux pilot serving with Air Platoon, Irish Guards at Habilayn. Air Platoon tasking carried on over Christmas with the traditional battalion tasking, interspersed by Santa sorties. The latter involved the flying the Battalion's Father Christmas (one of the Company Commanders), the Chaplain, CO, and other visitors around the outlying piquet posts that protected Habilayn. Christmas was considered to have been relatively quiet, although there were some reports from

77 B. Preston, 'One Starry Night', in Mallock, *The Likes of Leicester*, pp.97–99.
78 TNA, AVIA 101/680: Accident Report, Sioux XT 125 Helicopter, 6 December 1966.

the FRA garrison at Awabil Fort (20 miles north of Habilayn) of occasional incoming mortar fire and some small arms fire. The fort was situated 12 miles northeast of Dhala and the attacks on the FRA created some concern, with contingency plans including new orders for the Air Platoon. Late afternoon on 28 December Barnes was ordered to pre-position from Habilayn to Dhala as an ABFAC to support the fort. The call that the fort was under attack came as sunset was nearing and Lieutenant Barnes took off with Captain Sean O'Dwyer, the Irish Guards mortar platoon commander, who was also a trained FAC, on board. During the 10 minute transit to the fort Lieutenant Barnes received a situation report over the radio including the grids of three active dissident fire positions on a *jebel* overlooking the fort. As he approached the fort, Barnes established contact with a pair of RAF Hunters overhead. Identifying the dissident positions would, however, prove to be a challenge, as Lieutenant Barnes noted that the positions were on a mountain ridge which ran back five miles uninterrupted to the border with Yemen. The sides of the ridge were so steep that it was almost impossible for the FRA below to direct their mortars onto the dissident positions sited about a third of the way down the side of the ridge. Any attempt to employ mortars ran the very real risk of dropping bombs onto the outskirts of a nearby village.

The lay of the land also presented Lieutenant Barnes and the circling Hunter pilots with another problem. When viewed from Awabil, the enemy were on the mountains due west and the sun, therefore, shone straight into the eyes of any approaching pilot. Barnes worked his way towards a suitable position to observe the target:

> The difficulty in getting initially myself and then the Hunters to see the first target where the only possible attack direction was flying into the setting sun. Another practical difficulty was the different gridding between the 1:100,000 and 1:50,000 maps of the area. Later, I was questioned as to whether the target was our side of the border or in the Yemen. Transferring the grid reference from the 1:50,000 map, which was not an authority on the border to the 1:100,000 gave the location as our side of the border but in fact it was inside Yemen.[79]

Seeing the Hunters overhead, the dissidents elected to withdraw towards the safety of Yemen. Suddenly Lieutenant Barnes saw a large cloud of smoke appear on top of one of the *jebels*. The Hunter pilots also saw the smoke. Suspecting that this was the result of timely FRA target marking, or dissident mortar fire, Barnes called the Hunters and cleared them to make a strafing run using the smoke as a reference point. Because of the terrain and the location of the target, the incoming Hunters were forced to make their run in from the east. A conventional attack would have resulted in the Hunter pilots being blinded by the setting sun. They quickly adapted their tactics and attacked the targets from a very steep dive, pulling out just a few hundred feet above the *jebel*. A follow-up attack was made later using rockets on the same position.

The action was not over, another suspected mortar position was identified further north along the ridge. With the sun sinking steadily, Barnes flew his Sioux into the wadi and used the main *jebel* to shade his view of the new target. Lieutenant Barnes guided the Hunters into the Wadi Hajil, which was by then in the shade of Jebel Awabil, before they were able to see and strike the target. This proved to be a tall order and the first run resulted in cannon bursts on a small hill in front of the Sioux. At this point, Captain O'Dwyer voiced his concerns about empty cases from the Hunters hitting the Sioux. To avoid this, Barnes was holding the Sioux at very little airspeed, out of ground effect in a wadi with no ability to move sideways, at an altitude of 6,000 feet. A situation he later described in a letter home as 'not funny'.

79 M. Barnes, account to the authors, 4 May 2020.

The Hunters came in for a second run. This time they entered the shadow so that they could identify the target before they engaged. Barnes reported movement on the ridgeline, dissidents heading north towards Yemen, and the Hunters attacked again. With no effective communication with the FRA in the fort, Lieutenant Barnes decided to assess the effectiveness of the Hunter strikes himself and to ensure that there were no further active positions. He recced the target areas and likely positions without incident. With the light fading, he followed the wadi that led towards Yemen, the most likely withdrawal route for any surviving dissidents.

> I was just about to turn back and call it a day when we were met by some very bright flashes estimated at between 600–1,000 metres away, there were also some small flashes coming from the same general area. I pulled the aircraft round so quickly that I over-rode the hydraulic assistance on the cyclic control and dived the aircraft into what I hoped was dead ground. I was already quite low.[80]

Still under fire, Lieutenant Barnes transmitted a contact report and ordered Captain O'Dwyer to drop smoke grenades to indicate their position to the still loitering Hunters overhead. With Lieutenant Barnes taking evasive action to avoid incoming fire, Captain O'Dwyer could not place the smoke accurately. The lengthening shadow of the *jebel* also made it impossible for the Hunters to identify the smoke. Once clear of fire, O'Dwyer quickly plotted the coordinates of the fire positions on a 1:50.000 map and passed the grids to the waiting Hunters. Strike Wing Headquarters responded to the grids with a request for clarification that the targets were on the Aden side of the border. Barnes recalled:

> My 1;50,000 map placed the target about 200 metres inside the Yemen, but at the foot of the map there were a few words stating that this map could not be used as an authority on the position of the border. Transferring the grid to my 1;100,000 map placed the target very clearly on this side of the border, and there was no little note at the foot of the map.[81]

Meanwhile a 13 Flight Scout had arrived overhead. Major Crouch, OC 13 Flight, was flying the Scout with the BASO in the left-hand seat. Major Crouch called Lieutenant Barnes and told him to break off the action as light was fading rapidly and the Sioux had no blind flying instrumentation. The action carried on around Barnes and O'Dwyer and two more Hunters arrived overhead to join the first pair in the attack. Barnes was just setting himself up to lay down a line of smoke to aid another Hunter run when they called him to inform him that they could see that the Sioux was under fire. Barnes was oblivious to the fire due to the noise all around him. At this point Major Crouch again directed Lieutenant Barnes to break off and return to Dhala before nightfall. Reluctantly, Barnes turned the Sioux away from the wadi and began to head back to Dhala. Given the nature of the day's events it was not that straightforward. The fight was not over, as Barnes explained:

> On my return, which for tactical and border considerations was not a straight line, I saw the last target silhouetted and it had the audacity to fire occasional bursts at the Hunters which appeared to be two thousand feet above it. I could not possibly let this happen. I gave the Hunters instructions with a new direction of attack and then from nearly 1,000 feet above a plateau dropped the 'most superb' (comment received later from the Hunter pilots) line of

80 M. Barnes, account to the authors, 4 May 2020.

81 M. Barnes, account to the authors, 4 May 2020.

Lieutenant Micky Barnes Welsh Guards (with beret) pictured with members of 10 (Air) Platoon Irish Guards. On the extreme right is the OC Air Platoon, Captain Simon Gordon-Duff SG. (M. Barnes)

> smoke. By this stage I had run out of all smoke grenades apart from yellow ones, which are not normally appropriate to use in the desert. But with the last rays of sun these turned out to be perfect. The Hunters picked up the line of smoke and hit the target first time with cannon. I confirmed that they were on the right target, told them they were clear to attack with rockets and warned them that accurate fire was coming up at them from the position. Their reply to the fact that they were under fire was 'So are you'.[82]

Finally, fire from the dissident positions ceased and all aircraft returned to their respective bases. The local Political Officer reported back those casualties on the first two enemy positions numbered five dead, three of them Egyptian soldiers, and four wounded, one of whom was also an Egyptian soldier. Investigation later revealed that the third position was in fact 400m beyond the Yemeni border and there was therefore no immediate intelligence available on casualties. Lieutenant Barnes later reflected:

> The Hunters also came under heavy fire from three different positions, and I have a vivid memory of seeing them coming under AA fire – the tracer straddling them. They were still

82 M. Barnes, account to the authors, 4 May 2020.

> able to nullify the dissident attack and destroy an anti-aircraft gun position in the process. During the whole engagement I do not remember hearing a crack of a bullet. But the Hunter pilots who I saw a few days later.... assured me that they heard the crack of bullets coming over the radio – this was despite me wearing a throat microphone which I did not think would pick up the crack of hostile fire. They also told me that they saw numerous flashes of presumably rifle fire coming up at me from various positions.[83]

Reports did suggest that it had been the site of a Russian 12.7mm quad mounted anti-aircraft gun. Empty cases of that calibre and weapon remains were recovered from the site by a subsequent FRA patrol, including one burnt-out round recovered with the tracer incendiary bullet still intact which was presented to Barnes by the commander of the FRA fort as a memento. Whatever the nature of the weapon system, Lieutenant Barnes was subsequently recommended for the DFC for his actions but was awarded a Queen's Commendation for Brave Conduct.[84]

Night Casevac

As the final summer in Aden progressed the temperature climbed and the frequency of dissident attacks escalated. The demand for the speedy and safe evacuation of casualties increased at a commensurate rate. Although all Army aviation aircraft types could be employed in this vital role, the external litters on the Sioux were viewed as being a last resort, for use only if no larger aircraft were available. Service flying regulations often restricted the use of RAF Wessex helicopters from being more gainfully employed, and that left the two Scout flights as the most regular option for casevac tasking.

Major Greville Edgecombe, OC of 8 Flight AAC, nicknamed 'Low level Greville', was a much-respected aviator among his peers in theatre. He was also recognised as a brave and capable leader, both in the air and on the ground. As 8 Flight Commander, Major Edgecombe was involved in numerous incidents, and he was awarded the AFC for one of many casevac sorties. The citation captures the timeline of events and the part he played when flying with Captain Robin Grist:

> At approximately 0200 hours on 19th February 1967, following a dissident attack on DHALA, a request was made to HQ Area West in HABILAYN to arrange casualty evacuation for two Priority One cases – One a Mulazzim [junior officer] of the Federal Regular Army, the other a marine of 45 Commando RM. Both men had been very severely wounded, and the Medical Staff considered that their chances of surviving the night were very slight unless they were evacuated as soon as possible.
>
> Major Edgecombe, OC 8 Flight AAC was asked if he would attempt the evacuation, though a night evacuation from either the British or Federal Army camps at Dhala had never been achieved before. At HABILAYN there was complete cloud cover but a weather report from DHALA revealed that, though some cloud could be seen moving across the plain, there was no cloud cover at DHALA itself. Undaunted by the adverse weather and the fact that there was no moon, Major Edgecombe, with his customary selfless approach to his duties, decided to make the attempt using a dual control Scout helicopter with Captain Grist as second pilot. The Scout helicopter is equipped with radio and a full instrument panel, but it has no avionic navigation equipment; the pilot therefore has to rely entirely on a bearing and distance calculation. Between

83 M. Barnes, account to the authors, 4 May 2020.

84 TNA, WO 373/137/40: Army Form W 3121, Citation, Distinguished Flying Cross, Lt M Barnes WG, dated 28 December 1966.

Looking relaxed with a borrowed, and easy to stow M16 rifle, Major Greville Edgecombe, OC 8 Flight AAC in Aden. (Army Flying Museum)

HABILAYN and DHALA the country is very mountainous, and the ground rises from 2,000 feet to 4,500 feet with several intervening peaks of 5,000 to 5,500 feet. For any small aircraft night flying in the area is always extremely hazardous even in good weather conditions.

After taking off from HABILAYN, Major Edgecombe flew his aircraft through 1,000 – 1,500 feet of cloud until he was able to obtain some horizon even though only from the stars and cloud below him. He had asked for assistance from the radar at RAF Khormaksar, but they were unable to see him on their screen. After about 15 minutes of flying a descent was made towards the British camp at DHALA where the troops had positioned four Gooseneck Flares and an Angle of Approach Light, but the final approach had to be made between two dark hills into a small pad surrounded by sandbags. This was difficult. A landing in the Federal Regular Army camp would have been impossible and it was therefore necessary for the aircraft to wait on the pad while the Mulazzim was brought to the British camp. It was a long journey. Dissident fire had ceased some time before but movement around the camp continued to be reported and the British troops kept up harassing fire in order to discourage any further movement or an attack on the aircraft.

The journey to Aden was some fifty miles. Major Edgecombe flew his aircraft above the mountains and the clouds and flew on this course for the next thirty miles. When clear of the mountains he then had to make a descent through the clouds which by now had thickened to some 2,500 feet, and eventually landed his casualties on the helicopter pad outside the main entrance to Khormaksar Beach Hospital.... .Refuelled with intention of recovering 3 light casualties at first light.... other aircraft had completed the task.[85]

85 TNA, WO 373/138/18: Army Form W 3121, Citation, Air Force Cross, Maj Greville John Brooks Edgecombe AAC, 14 March 1967.

The Temperature Rises

The spring of 1967 was an increasingly tense time. Uncertainty mounted on the streets as to the viability of Aden within the Federation of South Arabia. In May, the British Commonwealth Secretary, George Thompson MP, announced that British forces would leave Aden in January 1968. The announcement immediately heightened tension between the NLF and FLOSY. The situation deteriorated steadily. As a direct result, the British authorities became concerned for the safety of British civilians and dependents living in and around Aden itself. The port and its environs were the responsibility of HQ Aden Brigade. The Brigade was determined to maintain control of Aden and viewed the reassurance of the British population as essential. Dissident attacks on the British community were assessed to be most likely at night. An enhanced programme of anti-dissident patrolling was introduced. This involved aircraft from 3 Wing AAC with tragic consequences. Lieutenant David Ralls recalls the loss of Scout XT635 and the four men on board at Al Jabai on the night of 5 May 1967:

> Our sister flight, 8 Flight, was up-country covering the Habilayn/Thumier Radfan commitment and we were at our base airstrip at Falaise Camp in Little Aden. One commitment was night patrolling in support of 39 Infantry Brigade of their urban area of responsibility. Lt David Morgan RCT had just joined 13 Flight having completed the very comprehensive QHI-led theatre familiarisation course. He was accompanied on the task by Cpl Christopher Gibb REME who was acting as Crewman. Cpl Gibb was trained to handle the flares and had flown on this type of task before. There were also two passengers on board the Scout, Marine John Dunn of 45 Commando and Trooper Gilbert Isles of 22 SAS.
>
> Part of the task was to drop four-inch parachute illuminating flares in areas close to expat housing. Presumably, HQ Aden Brigade thought this would reassure British and other civilians working in the local economy that the security situation was to a degree contained. Debatable though that was, and dangerous though the sortie undoubtedly was, 3 Wing had accepted the task as a more or less nightly event. As with much else individual pilots were used to working their way into the trickier stuff cautiously and in the most circumspect way that circumstances allowed.
>
> The inherent dangers of the flare dropping procedure were: the risk of bungling the arming and dropping of flares; the considerable amount of magnesium being carried; the potential loss of the pilot's night vision when a flare ignited as it fell away from the aircraft; the distraction of a single pilot having to manage two and sometimes three radios when supporting ground troops in controlled airspace; the overall pilot workload compounded by apprehension; and the unforgiving nature of the Scout's ergonomics when the pressure was on. There was a high risk of pilot disorientation immediately after flares ignited for which the pilot had to brace himself by giving absolute concentration to keeping straight and level, clear of obstacles and sometimes on instruments, until full night vision returned. To these dangers could be added the risk of hostile fire, actual or perceived, when it was not unusual for the urban area to be criss-crossed with tracer and visible firefights on the ground regardless of whether the aircraft itself was targeted.[86]

Whatever the cause, XT635 flew into the mountainside. All four on board were killed. The aircraft was thought to have crashed due to pilot disorientation, compounded by the high cockpit workload and, possibly, hostile fire in the vicinity.

86 D. J. Ralls, account to the authors, 17 September 2020.

Aviation Counter-Attack – A Day in May 1967

By May 1967 Lieutenant David Ralls had settled into the rhythm of life as a line pilot in Aden. He was destined to be at the centre of events on a particularly challenging day on 30 May 1967. He later wrote a letter home to his family, entitled 'A Day in May', recounting his experiences on what had been a notable day for Army aviation in Southern Arabia:

> I was up country at Habilayn, as one of four Scout line pilots of 13 Flight Army Air Corps which was part of 45 Commando Group. I slept in a 160-pounder ridge tent shared with my friends and fellow pilots Brian Preston (13 Flight) and Mike Perry (1 RHA Air Troop). An hour before dawn I rolled out of my camp bed without disturbing the others and, at the open-air facility about twenty yards away, got on with the morning ablutions. The day was off to a good start because the most significant of these was of a good to firm consistency (it was not always so) and I felt perky and well. We had not been attacked during the night and had slept quietly. The heat of the day had not yet kicked in. The sky was clear, and first light was only a ridgeline away. Breakfast would be later – much later as it turned out. Now a reasonably experienced pilot, with eight months of continuous counter-insurgency operations, I looked forward to whatever the day might bring without too much apprehension.
>
> The working dress for pilots was our cherished aviator's light blue beret, pocketed when the flying helmet went on, a khaki twill shirt, locally tailored cotton cord trousers, desert boots and a web belt. Holster patterns varied to taste. I had a Second World War tank crew open holster with 6 cartridge loops, containing a Smith and Wesson 0.38 revolver and worn on the right hip well clear of the Scout flying controls. Others had bought private purchase 9mm Luger or Walther pistols. On the left hip was an infantry 44 pattern webbing pouch which contained two Self-Loading Rifle magazines taped together. These supplemented two more magazines, also taped back to back for quick reloading, already loaded into my rifle which would shortly be taken from the Flight command post to clip in the holders by the aircraft instrument panel. This meant that there were 80 rounds of rifle ammunition quickly to hand if the aircraft came down in hostile territory – actually the only kind of territory available.
>
> The kettle in the command tent came to the boil. Lance Corporal Twyman, seconded from the [11th Hussars], was to be my companion for the day, as he often was, and had a mug ready. He was a highly expert and courageous Air Gunner who operated the most important of the three belt-fed General Purpose Machine guns fitted to the duty armed Scout. This GPMG was mounted on a swivelling pintle arrangement on the starboard side of the aircraft where the air gunner shared the same sideways view as the pilot. The pintle gun could be reloaded in flight but the other two, forward firing and operated from the pilot's control column, could not. Ammunition for these two was limited to 200 rounds each in a preloaded box. LCpl Twyman's meticulous loading of the belts gave a high proportion of tracer to allow a visible stream when 'hosepiping' towards the target. He was equally meticulous in cleaning the guns of which he was rightly proud.
>
> He knew exactly the degree to which they should be lubricated, in the conditions of the day, to minimise clingy sand adhesion whilst still keeping the action working. I cannot recall a stoppage, which is remarkable. Paul Twyman was a total professional whose coordination, accurate shooting and imperturbability were exceptional.
>
> We could not know that we were about to have the most exciting day of our soldiering lives. The armed Scout, which today was XR629, would be the first to be airborne at no notice as the initial response to any unexpected incident. In addition, it would provide top cover as required to planned operations by 45 Commando Royal Marines, a fighting unit we greatly admired and to which we were very close. As the day dawned, we were at the aircraft carrying

> out our pre-flight checks. For the pilot, an aiming mark was chinagraphed onto the windscreen measured by a couple of hand-spans from the centre of the bottom edge according to individual preference. The measurement was decided in airborne marksmanship practice to deliver converging shots from the two guns at 300 yards, not unlike zeroing a double express big game rifle but even noisier. Switches were made as the checks were completed. Nobody else would touch the controls. All that remained was to strap in, turn on the master switch and press the button to commence the automated wind-up of the mighty Rolls Royce engine. With a few short cuts the Scout could be airborne very quickly.
>
> We scrambled at 0835 hours. A Royal Engineers convoy had been ambushed in the Wadi Matlah about twelve miles to the South. The situation was unclear because the Sappers were not carrying VHF or UHF radio – a prerequisite for working with other ground troops or aircraft. Their call for help had come by HF radio and we had not been alerted to their being on the move. All messages had to be relayed through our Flight command post. As I flew to the scene it became clear that the convoy had taken casualties and was pinned down by a dissident force of unknown size. I decided to land near the convoy in what appeared to be dead ground. A Sapper ran across and was able to wave a rough direction for the source of enemy fire. The Scout was shot at as I headed out at very low level and, according to the Flight log, at 0850 hrs identified the main enemy position. The aircraft was positioned for a side-on machine gun attack by LCpl Twyman which kept the opposition quiet for a bit.[87]

Reports of the battle around the convoy at Wadi Matlah triggered an impressive response. Just 50 miles away a pair of RAF Hunters scrambled from RAF Khormaksar as Mission 269. A Sioux from the 45 Cdo RM Air Troop flown by Lieutenant Ian Uzzell RM was also manoeuvring to join in the counter-attack on the ambushers. The Sioux also carried Second Lieutenant John Meardon RM, a trained FAC. The 45 Cdo Recce Troop was also preparing to embark onto a 78 Sqn RAF Wessex led by Lieutenant Terry Knott RM. At the same time, Captain Ronnie Matthews, 2IC of 13 Flight was approaching in a utility Scout to extract casualties from the ambush site:

> By 0855 hrs, fifteen minutes after we had scrambled, there was quite a lot going on. To my genuine astonishment I seemed to be on top of things, helped no doubt by adrenalin and reassured by the company of a fine and imperturbable Air Gunner. It was normal for the first Scout on the scene to control the action for as long as seemed sensible. It dawned on me that I was now commanding a serious and potentially successful counter-attack. A couple of enemy rounds had found their mark. Something had thudded into the rear of the aircraft and a couple of sharp cracks marked the passage of bullets across the open cockpit close to my nose. The white pilots' helmet was fine for reflecting the sun but offered a rather good aiming mark as we made the first starboard-on attack run. My poor description of the enemy positions to Ian and John in the Sioux, and no time to fix a grid reference, left them without the information they needed to direct Mission 269. The Hunters had arrived and were keen to get on with things before running low on fuel. LCpl Twyman prepared a smoke grenade marker which we dropped about 400 yards from the main enemy position. I gave the Hunters their attack instructions for the initial run-in with cannon fire. The corrections for two subsequent attack runs were using information from the convoy by HF radio to 13 Flight HQ, thence to my B47 radio, then passed on to the lead Hunter on my UHF radio. This was not ideal, but it worked.

87 D. J. Ralls, account to the authors, 17 September 2020.

With the dissidents now under heavy attack, Ronnie flew his Scout low up the valley from the South to attempt the first casevac but was waved off by the Sappers who had spotted a Blindicide (anti-tank) sangar position which was now pinning them down. My Scout was shot at again from a position which Twyman and I could not identify. John Meardon in the Sioux spotted the source, a group of men taking cover in a cave on high ground, and with great skill directed a strike by Mission 270 which destroyed the cave and its occupants. The Sioux engine had by now been hit and was overheating so Ian had to break off for Habilayn.

The decisive phase of the battle was the dropping by Wessex 'Juliet' of Terry Knott and his Recce Troop to assault the larger element of dissidents' head on. I led them in using the forward firing machine guns to keep heads down. We marked a landing point for the Wessex with a smoke grenade and peeled off to give the Marines a clear run. And run they did, despite having jumped from about 12 feet from the Wessex in a high hover because of the terrain. This was classic Commando stuff, highly aggressive and professional, with all the opposition dealt with in minutes. Mission 270 remained to give top cover, having been warned by me not to engage the area previously struck and now occupied by Recce Troop.

It was now 0945 hrs and the situation at last allowed casualty evacuation by Ronnie and a second Scout to be attempted. The convoy had lost two killed and there were nine wounded. The two other Scouts lifted seven casualties. I was low on fuel but there was time to pick up the last two. We were all back at Habilayn by 1015 hrs – ninety minutes after the initial scramble.[88]

Ralls and Twyman set about refuelling and rearming XR 629 only to find that the hit that they had sustained during the firefight had damaged the engine starter solenoid, the Scout would not start. The two men transferred to XT 641 and rejoined the operation around the ambush location. It was during the subsequent three hours of flying that the marines recovered Gunner Cain's SLR, stolen by the dissidents from Captain John Fleming's crashed Sioux six months earlier. Finally, after a full day, Lieutenant Ralls and Lance Corporal Twyman returned to base and managed to secure breakfast. Both men were recommended for awards by the CO of 45 Cdo RM. David Ralls received the DFC for his part in the action. Acting Lance Corporal Twyman was recommended for the MM, but he received a Mentioned in Despatches. The citation, possibly the first for an AAC air gunner reads:

On 30th May 1967, Lance Corporal Twyman was the senior Air Gunner of 13 Flight AAC at Habilayn in the South Arabian Federation. His duty was to man a machine gun in the back of an armed Scout helicopter.

This gun is a most important part of the armed Scout since it is on a flexible mounting. The remaining armament is fixed to fire forward as aimed by the pilot. While the pilot is manoeuvring his aircraft close to the enemy, his safety depends on the speed and accuracy of fire by the rear gunner.

Early that morning a convoy of Royal Engineers vehicles was ambushed by a dissident force of about thirty men in the Wadi Matlah to the South of Habilayn. An armed Scout was called for and Lance Corporal Twyman went as the gunner. The aircraft was over the ambush position five minutes later, but it was unable to get in touch with the ambushed troops because their radio had been put out of action.

The pilot tried to land to obtain an up-to-date report, but the aircraft was immediately shot at. In order to relieve the ambushed troops, it was essential to pinpoint the enemy so

88 D. J. Ralls, account to the authors, 17 September 2020.

> that accurate fire could be brought to bear on it. It was some minutes before Lance Corporal Twyman, or the pilot could identify the enemy position. Then, in order to indicate the target to fighter aircraft which were over the area, the pilot flew over it and Lance Corporal Twyman engaged it with the machine gun, during this run the Scout was hit by enemy small arms fire close to the gunner's position. Nevertheless, the Scout circled and Lance Corporal Twyman dropped a smoke grenade to mark the target.
>
> Guided by this smoke, the fighters carried out three attacks. When these were over, the Scout again flew over the area to assess the effect of the air strikes. As it did so the helicopter again came under heavy fire from an unseen position. Lance Corporal Twyman continued to engage the enemy until the arrival of a Wessex carrying marines of 45 Commando RM.
>
> Still under fire, the Scout led the Wessex into a landing site to the North of the target.
>
> For the remainder of the action, which lasted until late evening, Lance Corporal Twyman continued to give gun support during the landing of follow-up troops and assisted in the evacuation of the ambush casualties and enemy dead.
>
> A subsequent search of the area revealed three enemy dead and evidence that a number of others had been wounded.
>
> Dedicated to his job, Lance Corporal Twyman has given outstanding service in perfecting techniques for the use of machine guns in Scout helicopters. To date he has logged over 200 hours flying while manning the waist gun and has two suspected kills to his credit. Throughout the operation on 30th May Lance Corporal Twyman displayed courage, coolness, and professional competence of the highest order in extremely hazardous flying conditions and in the face of very accurate enemy fire from strong positions.[89]

There was a postscript to Lieutenant Ralls 'Day in May' letter:

> It was fortunate, and not always the case, to have begun the day with a good night's sleep and no 'gippy tummy'. By nightfall we had flown pretty much all the advertised Army Aviation roles of the time: reconnaissance; armed action; command and control; limited movement of men and materiel; and casualty evacuation. Our antics were regarded as one of the first occasions when the AAC led a determined counter-attack involving other arms and, as it happens, all three Services which had a decisive outcome. The relationship with 45 Commando was for me formative, with friendships which extended to Northern Ireland on later tours and beyond.
>
> I was by then reasonably experienced, with eight months of continuous counter-insurgency operations and a good number of stimulating flying hours in the logbook. A few months later, by then an acting captain as 2IC of the 'new' 653 Squadron in Sharjah, I heard that I had been awarded a DFC. Paul Twyman had a Mention in Despatches though recommended, rightly, for a higher award.
>
> Our battered Scouts had been sent for base overhaul. Those longest in the Middle East were found to weigh, variously, between 60 and 100 pounds more than the basic weight shown in the aircraft record books. This difference was the volume of sand which had found its way into their innermost recesses. It turned out that we had flown over maximum all-up weight more often than we thought.[90]

89 TNA, WO 373/138/66: Army Form W 3121, Citation, Military Medal, A/LCpl Twyman, 1 July 1967.
90 D. J. Ralls, account to the authors, 17 September 2020.

Ironically, XR629 soldiered on after Aden and many years later was deployed to the Falkland Islands with 3 Commando Brigade Air Squadron on Operation Corporate. Almost 15 years to the day after its heroics in Aden, on 28 May 1982 near Goose Green the venerable XR629 was shot down by an Argentinian Pucara ground attack aircraft. The pilot, Lieutenant Richard Nunn RM, was killed and his air gunner, Sergeant Bill Belcher RM, was badly wounded, ultimately losing his leg. Nunn was posthumously awarded the DFC, the second pilot to receive the award as a result of actions while flying XR629.

See Map 11

The Crater Battle – 20 June 1967

Events in Aden were continually influenced by external players, with President Nasser of Egypt being the most active of those wishing to drive the British out of Southern Arabia. International events took a dramatic turn on 5 June 1967 when the Israeli Government launched an offensive against Egypt, Syria, and Jordan. What became known as the Six Day War ended in victory for Israel on 10 June. As a result of the fighting, the Egyptian Air Force was destroyed and Nasser's reputation in the Arab world was severely damaged. Nasser used Cairo Radio to launch a vociferous propaganda offensive claiming that the Israeli war effort had been funded by the British. This anti-British propaganda had the same effect as petrol on a fire in Aden. The already tense atmosphere on the volatile streets of Aden deteriorated further and even large numbers of the usually loyal FRA became fervently anti-British. Perhaps the final catalyst for what was to follow was a speech given by the British Foreign Secretary, George Brown, on 19 June. The contents of the long speech included the announcement of a new constitution for an independent South Arabian Government and critically, the withdrawal of all British forces by 9 January 1968. This immediately set a timetable for what would effectively be civil war. The two dissident organisations, the NLF and FLOSY, had a deadline by which they must destroy their rival, topple the Federal Government, and occupy the power vacuum left by the now imminent British withdrawal. The repercussions of the speech were almost immediate.

At dawn on 20 June 1967 the NLF issued a statement denouncing attempts by the British to hand over control of security operations in Aden to an Arab army and called on 'all Arab officers and men to resist this dirty imperialist plot.'[91] It was going to be a long and bloody day, even by Aden standards, when the QDG Air Squadron were tasked to support the 1st Battalion, The Royal Northumberland Fusiliers (1 RNF) in and around the Crater. 1 RNF were nearing the end of their tour in Aden and were in the process of handing over to the incoming 1 A&SH. Captain Mike Richards QDG was flying a Sioux that day:

> I started early with a great breakfast at 0500 which I remembered with gratitude as I did not eat again until supper. My first sortie was to install a 1 RNF picket on the side of Inscription Hill, across Crater from Temple Cliffs. There was a lot of early activity which was unusual and several small disturbances which I was asked to monitor. After shifting the picket along the ridge and a further couple of recces I went to refuel. An extensive task followed in support of 1 PARA in Sheikh Othman which was getting lively with SA, grenade and Blindicide attacks. Sergeant Martin Forde took over the Crater cover.[92]

91 LHCMA, Papers of Major General Dunbar, File 2/3, NLF Communiqué, 20 June 1967.
92 M. Richards, account to the authors, 12 August 2020.

Forde was tasked to fly a two-man observation team from the RNF to an OP overlooking the Crater District on Temple Cliffs. 1 RNF were based at Waterloo Lines, a barracks located beside RAF Khormaksar. The OP position was seven minutes flying time away from the RNF Battalion HQ HLS at Waterloo Lines. After an uneventful flight, the two-man team were dropped at the OP position shortly after 0700 hours. Forde was tasked to return and extract the OP party at 1200 hours the same day.

While Richards and Forde went about their business, the NLF statement was having an almost instantaneous effect in the charged atmosphere of Aden. At 0800 hours a mutiny erupted at Lake Lines, the SAA training depot near Khormaksar. Amid the confusion some SAA recruits under training fired across into a nearby British barracks. This incident was the initial flashpoint for a series of escalating events that would eventually mark the day out as one of the most significant in the whole saga of the British in Aden. Although the SAA officers managed to re-establish control of the troops who had fired on the British barracks, it added confusion to the already heated atmosphere. On hearing about the incident at Lake Lines, an unknown official in the Ministry of the Interior alerted the SAP at nearby Champion Lines. The message stated that the SAP should standby for potential deployment on riot control duties. This warning order and the news from Lake Lines were both confused and misinterpreted by the SAP. The story that circulated at Champion Lines was that it was British troops who were attacking the SAA barracks.

From this point onwards the situation deteriorated rapidly. By 0950 hours rioting SAP recruits had barricaded themselves into Champion Lines, seized weapons and captured a truckload of ammunition. The mutineers positioned a Bren gun in the minaret of the barracks mosque and fired into the British troops in the adjacent Radfan Camp. The Bren also fired onto the nearby airfield. At 1015 hours the mutineers murdered a member of the Aden Police Force and his driver, they also killed a passing British civil servant. The situation escalated further when the defence platoon from 1 LANCS, who were manning the sangar positions around Radfan Camp, returned fire on the mutineers. Orders were received from Brigade Headquarters that the LANCS were not to return fire. Second Lieutenant Angus Young LANCS, the 21-year-old platoon commander, was killed by enemy fire while attempting to move between sangars to relay the order.

At 1030 hours events became even more chaotic as an overloaded three-ton truck packed with 19 British soldiers appeared on the road. The driver and passengers were returning from an early morning on the ranges, unaware of the mutiny or the threat from the minaret. The mutineers opened fire on the helpless soldiers from 60 Squadron RCT at just 100 yards range. The RCT men abandoned the truck, retrieved some ammunition, and engaged the mutineers. The firefight lasted for over three hours; the survivors of the ambush were trapped in an incredibly exposed position. During the course of the fighting, the RCT had seven men killed and a further six wounded. Staff Sergeant Butler RCT, who had rallied the 60 Squadron men under fire, died of his wounds that night. The situation had escalated rapidly, and all available aviation assets were called in to assist the embattled ground units in the Crater. Mike Richards was involved:

> I was getting into the swing of sniper spotting when Brigade diverted me urgently to Radfan Lines for casevac work. There I went straight into the middle of the multiple fire fights around the barrack areas. Individual units and QDG armoured cars were extracting the casualties to Radfan Lines from where I flew six double litter loads into Khormaksar Beach Hospital, there were a lot of serious gunshot wounds, mostly upper body and head, from 1 PARA, KORBR, RCT and LANCS.[93]

93 M. Richards, account to the authors, 12 August 2020.

Sergeant Forde returned to Temple Cliffs at 1200 hours as planned and took off with Fusilier John Duffy RNF in the cockpit, carrying the OP Team's A41 Radio and Lance Corporal Jim Keightley RNF prone in the litter on the side of the Sioux. During the flight back to Waterloo Lines, Forde reported crowds barricading the entrances to the Crater district with buses. He also reported two burning Land Rovers outside the Police barracks, with Arab rioters throwing objects into the flames. On his return, Keightley also reported unusual activity outside the armed police barracks. Shortly after landing and making a verbal report, Sergeant Forde was re-tasked to immediately re-position the OP party at Temple Cliffs. The two Fusiliers dutifully climbed back into Forde's aircraft and the Sioux took off shortly afterwards and headed back to Temple Cliffs. At about 1250 hours the Sioux was flying at 700 feet above the Queen Arwa Road. As Forde made his approach to Temple Cliffs the aircraft staggered under a prolonged burst of automatic fire from the ground. Forde was wounded, Keightley was a helpless spectator in the litter on the side of the Sioux:

> I looked up to the pilot when there was a big bang by my head. A bullet had gone through the canopy of the helicopter and shot the pilot through the knee. We were just about to land when somebody shot the rear rotor blade and, of course, at that point it went into a spin and we crashed into a gully. Both my legs were hanging off below the knee.[94]

Although badly wounded, Sergeant Forde maintained control of the damaged Sioux and managing to keep the aircraft level. During the descent he managed to transmit a mayday which was acknowledged as he flicked the twin alternator and battery master switches off in an attempt to prevent a fire on landing. As the aircraft neared the ground one of his skids caught the edge of a gully below. The aircraft grounded heavily and slid down the slope to the bottom of the gully. As it came to a halt the volatile AVGAS fuel in the damaged Sioux immediately burst into flames. Fusilier Duffy reacted quickly; despite the flames and incoming sniper fire he pulled the badly wounded Keightley from the wreckage. He then returned to the Sioux and extracted Forde from the cockpit. He also recovered his rifle and the A41 radio, the latter was used to call for rescue on the 1 RNF command net. Luckily for the survivors of the downed Sioux help was at hand. Richards remembered hearing the call:

> As I got airborne with my sixth load, I heard Martin Forde's Mayday. I climbed to observe into Crater and saw a vivid flash on Temple Cliffs. I reported this to ATC and 1 RNF, simultaneously with the picquet I had placed earlier. I was unfortunately not in a position to help as my port litter casualty was bleeding badly from a serious head wound and I was holding the dressing on with my left hand, then the fuel warning light chose that moment to illuminate. I headed back for the hospital and with great relief heard Fusilier Duffy report the situation on his A41.
>
> On turning away, I contacted an RAF Wessex on the ground at Khormaksar, asking for help. He played a blinder, collected a 45 CDO team from near the runway and extracted the three crash survivors while under fire. I saw the Sioux explode before the rescuers got there, Martin would have been dead if Duffy had not got him out quickly. Greville Edgecombe and SSgt Small's Scout arrived ahead of the Royal Marines, his group moved the wounded to the Wessex. All did a superb job and we subsequently intervened to stop the RAF pilot, whose name I have regrettably forgotten, from falling foul of his system. Champagne had a hand in this two days later when we met.[95]

94 Quoted in N. van der Bijl, *British Military Operations in Aden & Radfan*, p.174.

95 M. Richards, account to the authors, 12 August 2020.

Staff Sergeant Fred Small, 8 Flight AAC (right) was on duty throughout the infamous battle for the Aden Crater. (Army Flying Museum)

The Scout mentioned was one of two 8 Flight Scouts operating from Waterloo Lines that day. The Flight Commander, Major Edgecombe, was also on the ground acting as liaison to 1 RNF at their battalion headquarters. Staff Sergeant Fred Small was one of the Scout pilots on duty:

> Shortly after our arrival at Waterloo Lines we were told that a Sioux had been shot down over the Crater. I set off to the reported crash site with Major Edgecombe and Lance Corporals Wilkinson and Brown on board. I flew direct to the Crater, followed the Pass, located the Sioux, and landed just uphill of it. Major Edgecombe and the two Lance Corporals jumped out of the aircraft carrying a stretcher and ran across to the wrecked Sioux. They were under rifle fire from the Armed Police Barracks as they ran from under my rotor disc. I remained in the Scout with my rotors running ready to extract casualties. At the same time a Wessex had arrived carrying Royal Marine Commandos who promptly debussed and set about returning fire to the Police Barracks and recovering the Fusiliers. Major Edgecombe and the two JNCOs stretchered Martin Forde to the waiting Wessex for casevac. My three passengers then re-embarked on the Scout and I lifted out of the Temple Cliffs heading back to Waterloo Lines.
>
> I was then tasked to take Colonel [Colin] Mitchell [CO 1 A&SH] and two others to Crater so that they could observe the scene of the ambush. We overflew Crater at about 2,000 feet (to avoid small arms fire). On arrival we observed the ambush scene. There was a bus parked across the road with the Land Rovers of the handover parties adjacent. There were bodies but no signs of life. That sortie was 1hr 15 mins – much longer than the norm. We returned to Waterloo Lines where my passengers disembarked.[96]

96 F. Small, account to the authors, 4 May 2020.

It was a dramatic incident in what was to be one of the most eventful and costly days of the Aden campaign. In the immediate aftermath, Lance Corporal Keightley was evacuated to medical aid; he survived surgery but lost a leg in the process. Fusilier Duffy was later awarded the DCM for his actions throughout the incident. The crash and Duffy rescue were later immortalised in a painting by Terence Cuneo entitled 'Temple Cliffs, Aden, Yemen 20 June 1967', the original hangs in the Fusiliers Museum of Northumberland in Alnwick. The Temple Cliffs shoot down and rescue was also covered in comic book form by *The Hotspur* the following year.[97]

The reality of the day was, however, a far grimmer affair. Mike Richard's logbook testifies to the type of flying undertaken in the aftermath of the mutiny:

> Between 21 and 29 June I flew twelve sorties for 1 RNF, initially monitoring the movement of corpses from the ambush sites with a view to recovery operations, then resupplying their standing picquets. By 23 June, the Argylls were in the saddle and the planning of a suitable response began, leading to the now famous re-entry operation. I flew fourteen sorties for the Argylls, six of them with Colonel Mitchell. I also flew a couple of Royal Marine recces as a beach landing option was briefly considered and two further casevac sorties. Aircraft attracted ground fire on most sorties over these dates.
>
> At one stage our radio communications were being jammed and I flew a couple of detector sorties with Royal Signals personnel. Finally, we had an RAF photo specialist attached who I flew to get planning coverage. Throughout this Mitchell kept me involved in the planning progress which was quite an insight. Our tasking came from 1 RNF, then 1 A&SH, and sometimes from 24 Brigade direct.[98]

20 June 1967 remains one of the darkest days the in post Second World War history of the British Army. A few days later, on 25 June, the 22 military, and one civilian dead from the mutiny were buried in Aden with full military honours. Another two soldiers killed in the days after the mutiny were also buried. It was the greatest loss of life suffered by the British Army in a single day since the Korean War. The actions of Fusilier Duffy at Temple Cliffs had narrowly prevented 3 Wing AAC and the QDG having to mourn the loss of Sergeant Forde. Feelings understandably ran high among the garrison and at home in the UK. Extensive recce and intelligence activity suggested that a brigade sized operation would be required to reoccupy the Crater. The NLF was believed to be continuing to coordinate the dissident defenders and the mutineers. Any operation on this scale against prepared defences was sure to be costly in both military and civilian casualties. Nevertheless, preparations continued, and reinforcements were airlifted into Aden from the UK. Tasking around the Crater continued as Mike Richards described.

> While decisions on a response to the Crater set back were in hand, fighting got vigorous along the escarpment and at Main Pass, the western entrance to the town. Several attempts to recover the ambush bodies were repulsed on 21st June and two QDG armoured cars were badly shot up, the crews injured, and the troop leader temporarily blinded. Several of my sorties were devoted to identifying firing positions and directing fire. QDG were using their .30 Machine Guns to good effect which was good for morale. Our firepower contribution came from my observer, Trooper Haile's SLR and bombing runs with CS and smoke grenades, neither very effective due to the air currents in the area. We also spent time tracing the movement of the bodies of three Fusiliers from a party that went to assess the ambush and

97 *The Hotspur*, issue number 235, 9 March 1968.
98 M. Richards, account to the authors, 12 August 2020.

> never returned. Many unpleasant rumours circulated but finally they were returned by a loyal police officer having been dragged around the town during the night.[99]

The reoccupation of the Crater was planned and expertly executed by Lieutenant Colonel Mitchell and 1 A&SH. His plan was carried out on 3 July and the Crater was back under British control by the evening of the next day. Operation Stirling Castle was a welcome success that did much to lift British morale in Aden and at home. The Argylls, supported by 45 Cdo RM and the armoured cars of the QDG, retook the Crater without incurring casualties. The operation remains a controversial episode to this day, as does its architect, Mitchell, forever known afterwards as 'Mad Mitch'. A detailed narrative of events lies outside the remit of this history, however, as it was primarily an infantry-based operation with armour support.[100] Having worked closely with him, Mike Richards described Mitchell as, 'a good soldier and an inspiring leader who was greatly respected by his tough battalion. I learned a lot from working for him.'[101]

The mutiny was a clear indication of just how volatile the situation in Aden was becoming. There was no prospect of the situation stabilising, let alone improving. British political initiatives to encourage and support a federal solution proved futile. It became increasingly clear that the NLF and FLOSY were determined to destroy each other and seize control from the Federal Government. British plans to evacuate the non-essential personnel were accelerated and contingency planning for an early withdrawal got underway.

Operation Big Lift – 23 July 1967

The struggle between the two warring factions did not distract them from attacking British troops and installations. As the summer progressed the dissident fighters adopted more sophisticated tactics and were increasingly better armed. The Belgian Blindicide anti-tank weapon was now popular with the NLF guerrillas. A much-improved version of the Second World War US Army Bazooka, the Blindicide was an ideal weapon for use against British Land Rovers and fixed positions, such as observation towers. The hand-held launcher fired an 83mm rocket out to an effective range of 400 metres. One of the early countermeasures employed was to fit 'Blindicide Screens' to OP towers to deflect the rocket, or to detonate it before the projectile reached the walls or windows of the manned position. The fitting of these screens became routine business for the RE both in Aden itself and up country. There were, however, locations where the NLF threat to any team of sappers was so great that the construction of screens was impractical. These sites were increasingly targeted by NLF Blindicide attacks. One such position had been identified by Major Joe Starling, the second in command of 1 Para. The observation tower known as OP4, at the police station in the centre of Sheikh Othman, a sprawling township a few miles north of the of Aden, was subject to almost constant attack:

> …the difficulty at OP4 was to fit such a screen over the fighting floor of the tower. To attempt to carry out the construction work involved – much of it being exposed to enemy fire – would have invited an unacceptable number of casualties. The obvious solution was to construct a screen on a frame and lower the whole thing over the tower by helicopter…[102]

99 M. Richards, account to the authors, 12 August 2020.

100 For more detail on full operations at that stage of the campaign see van der Bijl, *British Military Operations in Aden & Radfan,* A. Edwards, *Mad Mitch's Tribal Law: Aden and the End of Empire* (Edinburgh: Mainstream Publishing, 2015) and C. Mitchell, *Having Been a Soldier* (London: Hamish Hamilton, 1969).

101 M. Richards, account to the authors, 12 August 2020.

102 Quoted in Walker, *Aden Insurgency*, p.213.

So began Operation Big Lift. The task of manufacturing the bespoke Blindicide screen was allocated to 60 Field Squadron RE. The task of delivering what became known as 'The Thing' into one of the most volatile hotspots in Aden fell to 13 Flight. Lieutenant Brian Preston AAC recalled a rehearsal on 22 July 1967, where the first question was whether…

> … someone could find a pilot stupid enough to hang the thing underneath a helicopter and hover with it over the tower in the dark to allow it to be secured before the 'baddies' woke up in the morning. The day before the planned event, a crack team was formed: two Scout Pilots (yours truly and Lieutenant Ronnie Matthews AAC), together with two off-duty Corporals (Westgate and Finch). We flew to a sandy field next to the Sappers' tented camp to inspect the screen and identify its aerodynamic qualities…. It was a vast hexagonal structure made from scaffolding, with heavy duty pierced steel lattice, maybe 10–15 feet high, welded around its sides. The diameter was far greater than the Scout's 32-foot, 3-inch rotor diameter and the weight was technically assessed as 'reasonably heavy'!
>
> The plan was for the helicopter to hook onto the four steel hawsers attached to the top of the monstrosity and attempt to lift it. Some brave unsuspecting souls would then duck under the swaying rig to grab hold of the dangling ropes to guide it into position…. Health and Safety had not been invented.[103]

After some discussion, Corporals Westgate and Finch were delegated the task of handling the ropes while Matthews took on the role of ground liaison via a hand-held radio. Preston had the dubious honour of being the pilot to 'hook up to 'the beast' and fly with it'. There was a rehearsal in the sapper field and Preston recalled his view from the cockpit:

> I started the aircraft.
>
> "How do you read?"
>
> Clatter, clatter, hissss, "With diffi(clatter)culty!"
>
> The radio was far too close to the helicopter and was receiving as much engine and rotor noise as Ronnie was receiving sand from the downwash.
>
> He moved position.
>
> "How now?"
>
> Whirr, clatter, "a bit (clatter) better!"
>
> I hovered over the Sapper's work of art and took the strain…
>
> "Wumper, wumper, hisss" said the radio.
>
> "Say again?" I said, as the enormous scaffold ring appeared, swaying gently either side of the hovering helicopter, the men underneath grappling with flailing ropes.
>
> "They (hiss) have mot pit" (clatter)
>
> "What do you mean mot pit?"
>
> "Ot it, ut it!" (Crackle)
>
> It was hard work playing guessing games while trying to tame hundredweights of gyrating scrap iron under one's belly.
>
> "Did you say cut it?"
>
> Hissss, "ot', 'ut', 'it" (Wumper) click.
>
> So, I released the load…
>
> A writhing mass of steel careered into the sand like a giant rocket-propelled pastry cutter, nearly imprisoning the startled handlers.

103 B. Preston, 'Dress Rehearsal', in Mallock, *The Likes of Leicester*, pp.81–3.

> The radio went silent.
>
> Westgate and Finch went silent.
>
> The Scout landed and went silent.
>
> "I was trying to say; they've got it", said a sand-caked Ronnie.
>
> That was the first and only rehearsal...in daylight and without even a mock-up watch tower.[104]

The next day at first light the lone 13 Flight Scout appeared over Sheikh Othman with Preston again at the controls. Major Starling and 1 Para troops in the town below looked on. Starling describes the operation:

> ...promptly at first light, the Scout helicopter appeared, flew over the surprised inhabitants of Sheikh Othman and hovered over the Police Station with the 'thing'. Those on the ground passed the word to the pilot to drop, and the 'thing' came to rest within a few inches of its designated position.
>
> The 'thing' proved successful and although many more Blindicides were fired at the OP, no further casualties were suffered. Operation Big Lift was an excellent example of what can be achieved by inventive minds, skilful flying, good timing and plain old-fashioned guts.[105]

Subsequent reports speculated on the number of casualties that the new screen had prevented. There was, however, more immediate feedback for Preston from CO 1 Para on the ground in Sheikh Othman. Shortly after Operation Big Lift was completed 13 Flight received the following signal from 1 Para:

> FROM COMMANDING OFFICER PD PLEASE CONVEY CONGRATULATIONS AND THANKS TO 13 FLT FOR VERY SKILFUL FLYING IN POSITIONING BLINDICIDE SCREEN ON TOWER OF SHEIKH OTHMAN POLICE STATION PD[106]

Drawdown

While operations continued through the summer of 1967, there was a steady flow of personnel and equipment out of Aden. HQ 3 Wing AAC moved out of Falaise Airfield in Little Aden and established itself at RAF Khormaksar. The AAC flights followed as the facilities at Little Aden were handed over to the Federal Government. Inevitably, the reduction of the British presence on the ground created a higher demand for aviation support as the remaining force elements were being asked to do more with less manpower and mobility. The withdrawal of those units that fielded an integrated air troop or platoon also reduced the number of helicopters available. Towards the end of the year there were no longer any Sioux operating in Aden, only the Beavers and the Scouts remained. Lieutenant David Ralls' memories of the time gives some insight into the undercurrent of tension that prevailed:

104 Quoted in Walker, *Aden Insurgency*, p.213.

105 Quoted in Walker, *Aden Insurgency*, p.213.

106 B. Preston, 'Dress Rehearsal', in Mallock, *The Likes of Leicester*, p.81.

> By the summer of 1967, the armoured car and artillery detachments had been withdrawn from the Protectorate to the secure perimeter of the colony of Aden far away. A top-level decision was made, perhaps in London, to demonstrate trust in local forces by leaving Scout detachments in place. The pilot, REME technician and ground crewman who made up the detachment were increasingly vulnerable. We slept, sometimes restlessly, in a shed, yards from our aircraft, a Smith and Wesson revolver under the pillow and a Self-Loading Rifle inside the sleeping valise.[107]

The concern was justified as the autumn was marred by the loss of another 13 Flight Scout and its crew. The missing aircraft was Scout XT641 from 13 Flight flown by Sergeant John Baulcomb AAC. The aircraft was reported missing on 3 September 1967 when it failed to return from a task to Mayfa'ah, 200 miles northeast of Aden. Sergeant Baulcomb was flying Major Peter Gooch of the Somerset and Cornwall Light Infantry, seconded to the SAA, on a mission to visit local tribal leaders in an attempt to persuade or deter them from joining any rebellion. When the aircraft failed to return and was classified as overdue, the two men were listed as Missing in Action. David Ralls remembers overdue action being initiated by 13 Flight's commander:

> Overdue action for XT641 was initiated by the Ataq Detachment AAC radio operator during the afternoon of 3 September using HF radio to 13 Flight HQ at RAF Khormaksar. Fixed wing aircraft went to the area but found nothing. Major Mike Sharpe AAC issued a warning order to the Flight to be ready for a night transit to the area to arrive at first light on 4 September. We had three remaining aircraft. Those of us who had carried out the same detached duty immediately before Sgt Baulcomb knew the ground well. We suggested that the Flight should deploy in the field to one of our pre-recced night crash-out locations. These were for contingency purposes in case the three-man detachment (pilot, technician, and radio operator) had to get clear of Ataq quickly if attacked by the increasingly unpredictable Federal Regular Army.
>
> I flew with Mike to take him to a preferred location on high ground between Ataq and Mayfa'ah where a fighting OP could be positioned overlooking a wadi from which we would operate during the day. This was the only occasion when Army aviation in South Arabia deployed in the field away from a secure operating base. We took 4-inch recce flares expecting to remain deployed for a night and day search, as well as the armed Scout fit. Mike was a fine leader who, without fuss directed a lively operation, keeping spirits up, directing our low-level search sorties, and ensuring that we had effective and coordinated top cover. We searched all day at very low level supported by our Beaver Flight and Hunter attack aircraft. There was much shooting but none of our Scouts was hit.[108]

An extensive search operation was launched. A wide array of aircraft were employed in the search, including RAF Shackleton and photo-recce Hunter jets from 1417 Flight RAF that scoured the flight path and the surrounding area for wreckage. Initial flights found no wreckage but reported ground fire from the tribesmen in the suspected crash area. One of the Hunter sorties did, however, identify a dark area in the desert that photo-interpretation suggested marked the site of a large fire. It was decided to bring in Beavers from 15 Flight to carry out a low level search of the area. Two Beavers flew up the next morning and, with RAF Hunters circling overhead as top cover, began their search. Lieutenant Richard Grevatte-Ball RCT was flying one of the Beavers that day:

107 D. J. Ralls, account to the authors, 17 September 2020.
108 D. J. Ralls, account to the authors, 17 September 2020.

> I was involved in this sad operation, where together with one other Beaver, I piloted a Beaver from Khormaksar up to Ataq on 4 September 1967. Both aircraft then conducted a low-level search down the usual flying route from Ataq to Mayfa'ah. As we neared Mayfa'ah we were soon suspicious as we could see groups of Arabs firing at us from the ground with their rifles. This was confirmed when a bullet hit the other Beaver, entering through the door window panel, narrowly missing a middle seat observer (who luckily had just turned his head towards the centre of the passenger area to 'relax his neck') and passing out through the cockpit roof just behind the co-pilot's seat – without causing serious damage to anything vital.[109]

The search was inconclusive and later in the day all aircraft were recalled. The complete absence of a wreck or the bodies of the unfortunate crew suggested that the Scout had possibly been captured and hidden, or more likely destroyed. Search operations continued for another two days but neither the Scout nor its crew were found. Intelligence gathered later suggested that Sergeant Baulcomb and Major Gooch had in fact been captured while on the ground at Mayfa'ah. The two were murdered by tribesmen and their aircraft set on fire. The two bodies were then buried, along with the burned-out wreck of the Scout. The only trace of the crime being the scorch marks correctly identified by the RAF Photo Interpreters.

Whatever conclusions were drawn from the search, the dissident tribesmen in the area were considered too numerous and too hostile to mount a ground search for the burial site.

Intelligence sources suggested that the wreck and the crew had been buried in a nearby wadi using a mechanical digger. The suspected wadi was a well known site for the extraction of building sand, consequently it was covered with numerous almost identical mounds of sand. The wadi story was apparently confirmed when a piece of the Scout wreckage was handed over as evidence by a local. However, any recovery operation would be a protracted, and high risk, undertaking that would require a sizeable ground force, supported by a continual air presence. Much as everybody involved wanted to recover the bodies of the two missing men, the required forces were no longer present in Aden. The timing of the incident was also unfortunate as peace talks were underway in Switzerland. The British Government lacked any appetite for fresh casualties, and the political will to launch fresh military operations in Aden had long since evaporated. Nothing was to be allowed to hinder a British withdrawal from the troubled colony, as recalled by David Ralls.

> Several local people came to our camp with rumours of the aircraft having been destroyed on the ground with its occupants still on board, after dropping a local dignitary in the old fortified town of Mayfa'ah. There was apparently some corroboration being sifted by the intelligence community in Aden. It was concluded that, sadly, there was no chance of Sgt Baulcomb or his passenger, Major Gooch who was attached to the FRA, being found alive. By nightfall it was decided that the Flight should withdraw before becoming too vulnerable in an increasingly hostile environment. Mike Sharpe and I left last and flew together. We were tired and subdued, and very conscious of the distressing fact that we had lost two out of six line pilots in not many months. We arrived safely at RAF Khormaksar around midnight on 5 September, having flown what turned out to be the last Scout up-county sortie in South Arabia.[110]

The story of the incident was sanitised and presented to the public with a softened narrative tone. Sergeant John Baulcomb's obituary appeared in the AAC Journal of 1968. 'It was whilst serving his second detachment in Ataq, Southern Arabian Army HQ (Area East) that he was sent to

109 Quoted in R. Deacon, *Tales from The Frontline: Middle East Hunters* (Barnsley: Pen & Sword, 2020), p.330.
110 D. J. Ralls, account to the authors, 17 September 2020.

Maifah (Mayfa'ah) village with the local Emir, and he and a British Major were killed by the bullets of frightened tribesmen.'[111]

The lack of bodies made official confirmation of the deaths problematic, and Sergeant Baulcomb's wife Alice clung to the hope that her husband had been captured and was being held as a prisoner. She waited hopefully in their married quarter at Middle Wallop for his return. Inevitably the deaths of the two men were eventually confirmed and Alice Baulcomb left Middle Wallop and moved to Southampton. The welfare support provided at the time was considered generous, but contrasts starkly with what has been expected and provided during more recent operations. In addition to qualification for a War Widows Pension, Alice Baulcomb received £60 from an AAC whip-round, and she was also provided with an Army three-ton truck and driver to assist with her move to her new home. To date, neither the Scout nor Sergeant Baulcomb and Major Gooch's bodies have been found.

Beaver Bombers

As the last months in Aden passed slowly by, the physical footprint of British forces in the colony reduced at a steady rate; the need to do more with less increased at a corresponding rate. The dwindling number of troops and aircraft had to be as agile as ever, which brought new life to an old idea: the concept of the Beaver in the close support role as a light bomber. Back in 1964 after Operation Nutcracker, Lieutenant Colonel Graham-Bell had lamented:

> My one real regret is that we were never allowed to use the Beaver in an offensive role. We had bomb racks fitted and could easily have done some effective low level bombing, but my request to use the Beaver was rejected on the grounds that this was an RAF task not an Army task![112]

The concept of the Beaver bomber was resuscitated at the twilight of the Aden campaign with the Beavers being fitted with four Light Series Bomb Carriers under each wing for flare dropping. The carriers were also compatible with RAF 25-pound fragmentation bombs. The main practical problem with the idea of employing the 15 Flight Beavers to drop bombs was the lack of a bombsight on the aircraft. The designers of the Beaver had also not envisaged their aircraft being employed as a dive bomber; it was simply not designed to make steep, Stuka-like bombing dives. Lieutenant Allan Blaxland AAC was flying Beavers with 15 Flight:

> The aircraft was fitted with two pairs of 'bomb racks'. Each pair was positioned under the wing, just outboard of the top of the strut. Max load for each rack was 250lb, total 1,000lb. Release was either electrically via a pickle switch on the control column, or manually via a handle. The problem with these racks was that, should a hang-up occur, you were left with up to 500lb hanging from one rack – a 100% overload on the rack and a serious problem for aircraft balance. Most people watched carefully and if a hang up occurred, the manual release handle was pulled immediately.
>
> The racks were hardly ever used – it was such a hassle preparing the load and then fitting it all to the wing it was really not worth the vast amount of extra effort with serious drag airborne. The aircraft had a terrific performance anyway, and so the shortest strip could nearly

111 Anon 'Obituary of J. Baulcomb', *Journal of Army Aviation*, 1968, p.23.
112 Quoted in Kellock, *A Wonderful Aircraft*, p.75

> always be used, obviating the use of parachutes. I think I only used the racks once in Kenya, to supply drop to the KAR [King's African Rifles] who were stuck forward and back by floods, with no strip available.[113]

Nevertheless, 15 Flight persevered, and the targeting problem was overcome with the judicious use of a chinagraph cross on the cockpit Perspex window. Initial trials and some training did take place using inert practice bombs up-country. The drawdown and withdrawal in Aden were, however, accomplished without the need for the Beaver to add low level bombing to its long list of capabilities.

As the British withdrawal from Aden and handover to the Federal Government loomed ever closer, 15 Flight were given an important task. Concern mounted for the safety of the British officers seconded to the FRA. The once loyal Arab soldiers that made up the FRA battalions were now aware that the British would be gone from Aden by the end of 1967. Aden faced an uncertain future, and those that had, to date, worn a federal uniform must choose a side in order to safeguard themselves and their families. The new Aden would be dominated by either the NLF or FLOSY, and the two factions were already fighting each other for control of post withdrawal Aden. It was already apparent that neither group would show mercy to any Arab maintaining loyalty to the old regime. The imminent handover, therefore, placed all of the FRA's British officers in a potentially hazardous situation. In circumstances that would be familiar to some AAC personnel years later in Iraq and Afghanistan, many of them lived among their Arab soldiers. Most were located in isolated outposts, miles away from the protection of the nearest British unit. Any move to initiate an early withdrawal of the same officers would be construed politically as a lack of faith in the new government and could lead to mutiny or the collapse of the FRA. Some form of contingency for the emergency extraction of individual officers had to be established.

An extraction plan using 15 Flight was developed. A single Beaver was maintained at immediate readiness to fly out and extract FRA officers from any of the FRA outposts. The plan was simple; the Beaver would land on at the resupply strip close to the FRA garrison in question. This was not without risk, given the murder of Sergeant Baulcomb, along with Major Gooch, when his Scout was on the ground weeks earlier. The solution was to ensure that each Beaver sortie was covered by at least one RAF Hunter circling overhead as a show of force. The Beaver pilot was issued with a hand-held SARBE UHF radio so that, while on the ground, through the handset he could maintain direct communications with his guardian angel overhead. The extraction worked well, and all of the FRA officers were extracted safely. The presence of the Hunters provided an ample deterrent, the Beaver the ideal method of extraction. Many of the extracted officers were visibly upset after leaving their men in haste and under such circumstances.

The Wing was also steadily drawing down its strength. Falaise Airfield was vacated on 7 September 1967, with the remaining aircraft and personnel moving to RAF Khormaksar. This was part of a wider withdrawal plan seeing British units withdraw from the Protectorates and falling back into the bases in the environs of Aden itself. The much-reduced operational footprint allowed for a corresponding reduction in aircraft numbers and, as a result, the Scouts of 8 Flight AAC were shipped back to the UK. Flying operations carried on with the six remaining 13 Flight Scouts working alongside 15 Flight Beavers to meet enduring tasking requests. An atmosphere of tension and uncertainty prevailed, as Lieutenant Grevatte-Ball recalled.

> The Army Air Corps element was much rapidly reduced; 8 Flight and its Scout helicopters had already returned to the UK, leaving the Scouts of 13 Flight to prepare for a move to Sharjah

113 A. Blaxland, account to the authors, 4 July 2020.

> in the Arabian Gulf .... 15 Flight was endeavouring to coax four of their Beaver aircraft into RAF Belfast transports for the move back to the UK, leaving three Beavers, together with the Beverleys of 84 Squadron, as the only aircraft to continue to fly up country in support of the rapidly mutinying South Arabian Army. These trips were now flown with the nagging doubt as to which side the Arab troops would be on; the Government, NLF or FLOSY. The final up-country sortie flown by the AAC was flown by a Beaver on the evening of 12 October.[114]

The last weeks in Aden had also seen 15 Flight involved in the training of the SAAF, the fledgling air arm was due to receive its own US Specification Beavers prior to British withdrawal. Some of the AAC Beavers had been loaned to the SAAF pilots for training until the new SAAF Beavers arrived during the last week of September.

The impending British withdrawal did nothing to reduce dissident activities against the last British bases in Aden. Dissident attacks on base perimeters were emboldened and particularly prevalent at night, which generated a requirement for flare dropping in and around Little Aden and Aden. The Beaver had a flare dropping capability, and the provision of illumination for patrols was regular business during the drawdown period. The Beavers flew with the rear seat folded back and the centre seat removed. The door opposite the retained seat was removed and a crewman was carried in the back to ensure that the flare's static line was securely anchored to one of the rigging points on the cabin floor. The Beaver was authorised to drop flares from its wing carriers, this, however, limited the number of flares that could be carried. To avoid removing the door a procedure was developed locally that involved dropping the four-inch flares through a hole in the cabin floor. The hole was created by the removal of a blanking plate for the antennae of the redundant Decca Navigation System. Major Fleming commented on flare dropping sorties, 'The flares burnt for about three to four minutes and were very effective in illuminating the ground. Sorties usually lasted one or two hours, but one was recorded to last four hours. Without the door at night these sorties could be cold affairs!'[115]

The AAC presence in Aden finally came to an end on 13 October 1967. A single 13 Flight Scout and the three remaining Beavers made a combined final flypast before the Scout headed out to sea to land on HMS *Fearless*. The departure of the last Scout signalled a reorganisation of flight numbering. As part of the restructuring of the AAC component of MELF post Aden, the Beavers were retitled 13 Flight AAC. The 15 Flight association with the Beaver was maintained, however, as the Flight reformed as a Beaver Flight at Topcliffe, North Yorkshire in 1968. The new 13 Flight was to remain in the Middle East, operating the Beaver from Sharjah, which was, at the time, one of the Trucial States protected by Britain under a longstanding treaty. The transit to Sharjah in the Trucial States, the final chapter of the AAC involvement in the Aden Insurgency, was not without incident.

The Flight of three Beavers took off from Aden on 13 October, their flight plan following a pre-recced route that was broken into three legs. The initial leg involved a transit from RAF Khormaksar to Salalah in Oman; the second leg went to Oman's capital of Muscat; the final transit went from Muscat to Sharjah, a total distance of nearly 1,500 miles. Each of the aircraft had been fitted with a ferry tank to extend its endurance and each carried a REME Air Mechanic as a passenger. The three pilots making the sortie were Lieutenant Grevatte-Ball, Lieutenant Nigel Ironside AAC, and Sergeant Jim Cant REME.

The first leg was going smoothly until Ironside's aircraft (XP778) had a propeller oil seal failure. The oil rapidly sprayed over the windscreen and washed into the cockpit through the open cabin

114 Quoted in Deacon, *Tales from The Frontline*, p.335.
115 Quoted in Kellock, *A Wonderful Aircraft*, p.83.

door window. Ironside was covered in hot oil and aware that he had only a few minutes in the air before his engine seized; the formation was forced to divert to the airfield at Riyan Mukalla in Yemen. Between them the three aircraft carried the mechanics, the tools and the spares required to replace the seal once on the ground, a task that would normally take about 30 minutes. The three aircraft landed and disgorged their REME mechanics to assess the state of XP778. The problem was that the diversion airfield, until recently an RAF outpost, was now in the hands of disaffected local Arabs of uncertain loyalty. A single English-speaking local guarded the group as they set about replacing the seal, as another group of more hostile Arabs took pot shots at the repair team. Lieutenant Grevatte-Ball gave his Flight Commander's perspective of the incident:

> ...I instructed the affected pilot to land and told the third Beaver to fly back to Aden to radio our misfortune. I then landed behind the prop oil failure aircraft. It took about half an hour to replace the seal .... we were soon surrounded by rifle waving Arabs. It suddenly dawned on me that a DC-3 standing nearby was the Air Djibouti aircraft that they had recently hijacked. Daylight time was fast running out, the Arabs were getting more restless – "you cannot go" their leader said "until our boss arrives". It was getting ugly. Suddenly, very high in the sky I spotted a Hunter that had been scrambled from Khormaksar to come to our aid. Quickly, into my UHF radio, I explained our somewhat dire predicament. "OK" he replied, "I've only got fuel enough for one low pass. I'll fire my cannon and if I hit your Beaver as well – my apologies."[116]

The threat of the Hunter was enough, there was no need to make a cannon strafing run with all the risks it entailed. There was a sudden change of heart among the hostile Arabs. The two Beavers were given permission to leave, and all concerned scrambled aboard their respective aircraft and made a hasty departure. The reunited formation made it to Salalah early that evening. A subsequent daylight inspection of XP788 revealed a bullet hole going through the port tail plane and elevator torque tube. Unfit to fly, the damaged Beaver was later picked up by an RAF C130 Hercules and flown to 13 Flight's new home at RAF Sharjah, where the Flight remained for three years. The other two serviceable Beavers landed at Sharjah on 22 October. With the force elements that had made up 3 Wing AAC now dispersed around the Middle East or in transit back to the UK, there was no further need for a Wing HQ. The Wing (in effect 653 Squadron) was disbanded on 19 October 1967. So ended the AAC presence in Aden, which had been eventful until the last. Those involved in operations in Aden had made remarkable progress in the development of Army aviation.

The importance of the campaign was summarised by David Ralls, who continued to enjoy a successful career after Aden; he eventually became a Brigadier and was appointed Commander Aviation BAOR:

> Aden in the mid–1960s was the proving ground for the remarkable Westland Scout – the first rugged multi role Army helicopter. The Scout heralded the full evolution of Army Aviation as a future combat arm, integrated in the all-arms battle. Had spurs not been earned in South Arabia, and subsequently in Cold War Germany with Scout SS-11, there would have been no Lynx and certainly no Army Apache.[117]

---

116 Quoted in Deacon, *Tales from the Front Line*, p.336.
117 D. J. Ralls, account to the authors, 17 September 2020.

Drama until the last AAC sortie; an unplanned diversion into Riyan Airport required top cover from an RAF Hunter. (Army Flying Museum)

Captain Robin Grist, later destined to command 6th Airmobile Brigade and to become Director Army Air Corps, also reflected on the lessons learned in Aden:

> In Aden I had learnt the importance of a rapid casevac, a lesson developed spectacularly by the RAF Chinooks in Afghanistan; the importance of sand filters to protect gas turbine engines in sand and was dismayed to discover that this requirement had been dropped from the Lynx on the grounds that 'they would not be required in Western Europe'. Getting them for the First Gulf War was a challenge. I had also learnt the value of fitting the most powerful engine available to enable a helicopter to fly effectively in hot and high conditions.[118]

Those lessons would remain relevant to AAC aircraft and operations for a further 50 years.

118 R. Grist, account to the authors, 4 July 2020.

Chapter 8

British Guiana and Belize

British Guiana

See Map 12

On the north-east coast of South America, Guiana had been a colony since the British took possession of three adjacent Dutch territories in 1796 during the French Revolutionary Wars. At the London Convention of 1814 the colonies were formally ceded to Britain and in 1831 Berbice, Demerara and Essequibo were unified as British Guiana. The colony's economy was based on sugar cane plantation agriculture which depended initially on slave labour and, after the abolition of slavery, on poorly paid indentured workers brought over from India. From the beginning of the twentieth century unrest based on labour disputes became an increasing feature of life in British Guiana. Political reforms after the Second World War led to the emergence of labour based political parties and after elections in 1953 the left-wing People's Progressive Party (PPP) took control of the House of Assembly. The British Government, fearing communist influence in the colony, declared a state of emergency and sent troops to British Guiana, including a single Sycamore HC11 helicopter from 1906 (Helicopter) Flight RAF, embarked on the aircraft carrier HMS *Implacable*.[1]

Following further elections in 1961, in which the PPP remained the dominant party, British Guiana became self-governing in terms of domestic matters, although the British military presence remained. The British Government remained responsible for the defence of the colony and in 1962 neighbouring Venezuela announced that it was rejecting the 1899 boundary agreement and renewed its claim to all of Guiana west of the Essequibo River. In the same year widespread riots, strikes and disturbances broke out as a result of complex racial, social and economic conflicts. The British Government increased the military presence which, in 1964, included an AAC contingent.[2]

On 10 June 1964 24 Flight AAC, commanded by Captain Gilbert Macintosh AAC, deployed to British Guiana. Three Alouette IIs left RAF Gutersloh in RAF Armstrong Whitworth Argosies, while four pilots and 23 ground crew flew together with the Flight's stores in an RAF Bristol Britannia. After an overnight stop in Gander, the Flight and its aircraft arrived at Atkinson Field, approximately 30 miles south of the Guianese capital, Georgetown. Helicopters were vital to operations, particularly in the undeveloped interior of British Guiana. The country had a population of approximately 620,000, around 90 percent of which lived along the narrow coastal strip at the north-east of the country. Over 160,000 lived in the capital, Georgetown. Half the population were of East Indian origin, a third Afro-Caribbean and the remainder included Portuguese and Chinese immigrants and indigenous Amerindians. The racial and ethnic mix aggravated the

1 Warner, *First in the Field*, p.85.

2 For further information on British Guiana see T. J. Spinner, *A Political and Social History of Guyana, 1945–1983*, (Boulder: Westview Press, 1984).

Lieutenant M. Collins of 24 Flight AAC supervises the refuel of his Alouette at Atkinson Field in 1964. (Army Flying Museum)

long-term grievances over labour conditions, pay and representation. Those that wished to disrupt the local and British governments through sabotage, rioting, arson and murder had ample opportunity to move and hide with relative impunity in the heavily jungled interior. The main routes from the coast into the interior were via the rivers Essequibo, Mazaruni, Demerara and Berbice but they were not always navigable and took a long time to travel upstream.[3]

As 24 Flight arrived in the country the unrest reached its peak when a wave of murders and arson all but destroyed the small town of Wismar (now Linden), 50 miles inland up the Demerara River. Within two days of arrival the Flight conducted its first operational sortie when it was sent to help disperse a crowd in Wismar, which was successfully achieved by a very low 'beat-up' by an Alouette over the crowd. Shortly after a number of political activists were arrested in Georgetown and the Flight was tasked to fly them to the Mazaruni prison situated in a remote location on the huge Essequibo River. The flight took 30 minutes, saving an 11-hour journey by boat. The flying was challenging and dangerous. Landing sites were often very confined and unlit at night. The surface was often swampy and pilots were frequently not sure how far the skids of the Alouette were going to settle when they landed. In addition, flying over the vast tracts of jungle gave very few options in the case of an emergency, but fortunately the 'reliable Artouste engine gave little cause for worry'.[4]

3 AFM 123: '24 Flight AAC, Commanders Diary Narrative', February 1965.
4 Anon, '24 Flight A.A.C. British Guiana', *Army Air Corps Association Journal,* 1965, p.15.

24 Flight carried out frequent troop-lift tasks during the first two weeks in the country until the arrival of the County-class destroyer HMS *Devonshire* along with two Wessex helicopters. Two experienced AAC pilots, Major John Tippen AAC and Major 'Warby' Warburton AAC, were seconded to the RN for the operation and, after a brief Wessex conversion at Culdrose, flew out to Georgetown to join *Devonshire*'s Flight.[5] They were initially allocated XM923 '89', a rather tired Wessex that had been cruising round the Caribbean for an extended period and embarked on wide range of operational tasking.

> We flew 89 (when we could start it) to Georgetown, Buxton, Mahaicony, Blairmont, New Amsterdam, Skeldon, McKenzie, Wineparu, Mazaruni, Bartica, Mahdia. We flew to the missions, St. Cuthberts, Santa, Santa Rosa. We flew up and down the Demerara, the Essequibo, the Potaro, the Mahaicony and the Berbice. We flew political prisoners, murderers, beds, blankets, bullets and blacks.[6]

Some of the weight of routine liaison flying was taken up by WO2 Alan Rolling AAC who was not a member of 24 Flight but was in British Guiana flying a locally commandeered Cessna. At the same time 24 Flight was also maintaining its flying tempo, taking part in an increasingly wide range of tasking. 1310 Flight RAF had arrived with Whirlwind helicopters and a high degree of coordination was developed between the three services' aviation elements:

> The biggest operation the Flight was involved in during their stay in the country was Operation 'White Horse', which took place in August. Three Alouettes, two Wessex and a Whirlwind 10 swooped at dawn on a 15 mile stretch of Mahaicony Creek, a notorious terrorist area. We carried men of the Queen's Own Buffs and the Devon and Dorset Regt, who in addition to their airborne forces, also moved in on the area in lorries, boats and some even on horseback. Some 200 persons were rounded up, and nearly 40 detained as suspected murderers and arsonists.
>
> The Flight also evolved a system of commanding riot squads from the air, in case of serious rioting in Georgetown, and practised dropping tear-gas from the aircraft, although it never became necessary to use this method in anger, even though there were some serious bomb outrages in Georgetown in June and July.
>
> Casualty evacuation by helicopter became increasingly popular. The first few the Flight carried out were for military personnel, but later we began to receive requests from civilians. Many of these took place at night and although often involving comparatively short distances the lack of navigation aids and lights on the ground sometimes made it quite difficult. On one occasion one of our ground crew, who was spending some leave at an Amerindian village walked 25 miles through the bush at night to awaken the Duty Pilot for the evacuation of a sick woman to Georgetown.[7]

On 7 December 1964 general elections were held in British Guiana. 24 Flight assisted the local government by ferrying ballot boxes and election officials to the more isolated polling stations. The Alouettes were also busy flying the press and various government and military officials as well as maintaining reconnaissance activity to give early warning of any associated violence. The elections passed off peacefully, and while the PPP returned the highest number of votes, other parties were able to form a coalition with a working majority. It was 24 Flight's final act in British

5 AFM 123: 'News and photos from 24 Flight in British Guiana', undated.
6 Anon (Alpha Zero Seven), 'Far Away Places', *Army Air Corps Association Journal,* 1965, p.15.
7 Anon, '24 Flight A.A.C. British Guiana', *Army Air Corps Association Journal,* 1965, p.17.

Sergeant Gerald Sedgewick of 27 Flight AAC at the controls of an Alouette on a sortie over British Guiana in 1965. (Army Flying Museum)

Guiana as they were replaced by 27 Flight, commanded by Major Freddy Legg AAC, who took over on 7 January 1965.[8] Six months later 27 Flight were replaced in turn by 25 Flight, initially commanded by Major Hugo Leech RASC and then Captain Robert Homeshaw RE.

By the time 25 Flight arrived in British Guiana in June 1965 the situation had improved considerably. Although the Afro-Caribbean and East Indian populations were still distrustful of one another, the racial and political violence had abated. It was replaced by a crime wave including examples of robbery with violence at cutlass point in Georgetown. Flying tasks were largely liaison sorties flying government and military officials into the interior along with troop-lift sorties during cordon and search operations for arms and explosives. Despite the reduction in operational tempo, however, flying into the interior still presented unique challenges as discovered by pilot Sergeant Orne:

8 Anon, '24 Flight A.A.C. British Guiana', *Army Air Corps Association Journal,* 1965, p.17.

> After landing on an apparently solid area of ground in the middle of a swampy field he noticed that the skids were slowly sinking through the thin crust of hard earth and had to carry out the fastest Alouette start ever made, the aircraft reluctantly leaving the ground to the accompaniment of loud sucking noises...
>
> Casevacs often involve transporting pregnant women from mission hospitals or villages in the interior to the main hospitals in Georgetown for major surgery in the dead of night. This has caused a certain amount of concern on the part of pilots in that, with the internal stretcher layout, the pilot tends to be a lot closer to the scene of the impending arrival than the male nurse sitting behind him. In fact we're thinking of using sterile flying gloves in future.[9]

With the general reduction in violence and the PPP no longer in power the British Government was able to return to a timetable for Guiana to achieve independence and a withdrawal from the country as soon as practicable. In November 1965 a conference on independence held in London quickly reached agreement on an independent constitution and Guyana achieved full independence at midnight on 26 May 1966. The British Government began the drawdown of military forces in the former colony and 25 Flight flew its final sortie in the country on 6 July 1966. The following day the Commander British Forces Guyana, Colonel G. J. A. Dewar, issued a Special Order of the Day:

> 25 Flight Army Air Corps have now completed a most successful operational tour in this country. I am sure all members of the Flight can always reflect that they are the only Army Air Corps Flight to serve in the South American Continent. At all times 25 Flight has maintained a high record of air readiness. This excellent state of affairs has only been made possible by the whole hearted enthusiasm and cheerful willingness to work long hours by the ground staff. The British Battalions have benefitted greatly in the past two years by the presence of the Flight and the resourceful skill of the pilots. The Flight leaves with more than 2,000 flying hours in Guyana to its credit and in the knowledge that it has done a first class job at all times. All those, of all ranks, who at any time have served in 25 Flight in British Guiana/Guyana have just cause to be proud of doing so. Good luck to you all in the future and thank you from all of us who have been your passengers.[10]

Despite the AAC's organised commitment in Guyana coming to an end the Corps was represented in-country for a little longer. HQ British Forces Guyana retained the use of the Cessna light aircraft for liaison duties and the HQ's establishment retained a 'Light Aircraft Pilot' provided by the AAC. The short-lived post was filled by Captain Norman Gould AAC who operated the hired, civilian Cessna 185B Skywagon VP-GCC. The aircraft was based at Ogle airstrip five miles from Georgetown and was entirely maintained by its civilian owner. It proved to be a remarkably reliable aircraft and in the four months it operated on behalf of HQ British Forces it flew over 200 hours with only one minor radio fault. Sorties were flown between the score of authorised airstrips scattered over the northern part of the country. The aircraft was returned to its owner and the Light Aircraft Pilot post was disestablished in late October 1966, marking the end of the AAC's brief sojourn in British Guiana and Guyana.[11]

---

9 Anon, '25 Flight A.A.C. British Guiana Garrison', *Journal of Army Aviation,* 1966, p.85.

10 AFM 123: 'Flight Commander's Log for period ending 6th July 1966', undated.

11 AFM 123: 'Light Aircraft Pilot's Log – HQ British Forces Guyana for the period 1 July 66 – 31 July 66', undated.

A 25 Flight AAC Alouette taking off from a village HLS in British Guiana in 1965. (Army Flying Museum)

A group photo of 25 Flight AAC in British Guiana in 1965. (Army Flying Museum)

Belize

See Map 13

Belize is a small country (8,866 square miles), a little larger than Wales, on the eastern coast of the Central American isthmus. It is bordered by Guatemala to the west and south and Mexico to the north. The interior of the country is largely covered by semi-tropical forest and its coastline is dotted with hundreds of low-lying islands known as reefs or cayes. Having been a haven for pirates in the seventeenth and eighteenth centuries the British appointed a supervisor over the area in 1786. Throughout the eighteenth century the Spanish attacked Belize every time war broke out with Britain, but in 1836, after the emancipation of Central America from Spanish rule, the British claimed the right to administer the region. In 1862, Great Britain formally declared the area a British Crown Colony subordinate to Jamaica and named it British Honduras. The country was granted self-governing status in 1964 with a new constitution although Britain retained responsibility for defence, internal security and foreign affairs. Although the colony was renamed Belize on 1 June 1973 there remained a major barrier to full independence.[12]

Neighbouring Guatemala maintained a longstanding claim to part or all of the territory of Belize stretching back to a British/Guatemalan treaty of 1859. In 1948 Guatemala threatened to invade and British infantry companies were deployed to British Honduras to help protect its territorial integrity. Ten years later, on 21 January 1958, a company of the 1st Battalion Hampshire Regiment reacted to a small force from Guatemala that had crossed the border and raised the Guatemalan flag on British territory. A single platoon was sent out to investigate, located the intruders and exchanged fire with them. The intruders were actually Belizean insurgents aided and encouraged by Guatemala. They were poorly organised and ineffective however, and some 20 members of this self-styled Belize Liberation Army were rounded up and detained. In 1970, following further Guatemalan sabre rattling, the UK Spearhead Battalion was sent to British Honduras and the British troop level in the colony remained at a battalion plus for another five years.

The first Army aviation elements to arrive in British Honduras during the era of the Integrated Flight Scheme, were three Sioux helicopters provided by the Air Squadron of the 14th/20th Hussars. The detachment arrived in the country in the spring of 1972 with three pilots and a small number of ground crew and REME technicians. With the recentralisation of the Integrated Flight Scheme already underway, the commitment was soon taken over by the AAC and responsibility for manning the detachment was handed to 3 Regiment AAC. A detachment of 12 personnel from 663 Squadron, commanded by Captain David de Gonville Bromhead of the Royal Regiment of Wales took over from 14th/20th Hussars in January 1973. The Sioux were based on Belize Airfield (six miles north-west of Belize City) across the runway from Airfield Camp, in a purpose-built hangar designed to provide some protection in the hurricane season. Belize City, the most heavily populated town in the country, housed over 50 percent of the total population. The capital, Belmopan, in the centre of the country was nearly 45 miles south-west of Belize City and had been founded in 1970 as a planned community to protect the administration from the worst effects of the hurricane season. The rest of the country was sparsely populated and largely blanketed by semi-tropical forest. The broad coastal plain climbs to the Maya Mountains, just over 3,000 feet high, and the Mountain Pine Ridge. The underdeveloped infrastructure in the country meant the helicopters were vital to maintain effective communications between scattered military units and in providing life-saving casevac support. The detachment achieved approximately 30 hours flying per aircraft per month.

12 For more information on the history of Belize see A. Twigg, *Understanding Belize: A Historical Guide* (Madeira Park, BC: Harbour Publishing, 2006).

3 Regiment planned to allocate a squadron to the commitment for six months at the time. Personnel from the nominated squadron manned the detachment for three months with a staggered changeover halfway through the tour. This helped to spread the experience of flying in Belize's challenging environment as widely as possible. The difficult operating conditions that led to the aircraft being so important also made them and their aircrew extremely vulnerable. Of particular concern was the issue of aircrew survival in the case of a forced landing. The jungle in the country was considered to present more difficult going than that encountered in Malaya and aircrew isolated on the ground would have to wait an indeterminate length of time before an infantry patrol could reach them for extraction. Survival packs were considered vital along with shotguns so they could shoot animals and birds for food. There was also a request to replace the Sioux with the Scout as soon as possible. The lack of ambient light at night meant there was often no discernible horizon, which had forced the Sioux pilots to abandon several night casevac sorties. A case was submitted, but it was thought it would be some time before Scouts became available given the concurrent demands for the airframes in Northern Ireland and for anti-armour training in Germany.[13] There was also a suggestion that the Beaver could be deployed to take up routine 'milk-run' tasking, the country having an abundance of light aircraft landing strips.[14] Even on routine tasking, operating the Sioux in the hot, humid conditions into and out of jungle clearings was challenging, and required skill and complete concentration from the aircrew:

> With little or no wind most of the time, I find the only safe way of getting out of holes in the jungle is vertically; it takes such a long time to establish translational flight during a towering take off. I've seen so many people nearly come unstuck by launching off into a towering technique and then realizing at the last moment that they're not going to make it. Trying to abort backwards when you're in amongst the treetops fighting for rpm strikes me as a particularly dubious pastime. The situation today is no different from usual...
>
> All secure, passenger strapped in, headset on, okay for sound, wind up, take-off checks including lookout behind and particularly above, radio call for take-off clearance, ease in the power.... Come up gently, hanging right skid down which is to be expected, coming up to full throttle, rate of climb pretty slow, but going up vertically, now at full throttle .... concentrate on maintaining .... as little control movement as possible. Rate of climb very slow, about 40 feet to go to the treetops and it feels as though it wants to turn right – tail is clear so I'll let it feed round and slipstream itself – no sense in fighting it since more left pedal will probably drain rpm which will mean a slight reduction in collective, thereby reducing the small rate of climb that I do have. It's now gone through about 90 degrees and stabilized which means possibly a couple of knots of wind on the nose, although not apparent on the treetops. I'll keep going vertically to 20 or 30 feet above the trees so that I can afford to lose a little height during the transition. That's it, ASI indicating, translational lift becoming apparent and now I can reduce some power and continue the climb out.[15]

655 Squadron took over the detachment in November 1973 just as British Honduras became Belize. Three Sioux (XT172, XT541 and XT810), three pilots and a detachment of ground crew and REME departed McMunn Barracks in Colchester, loaded onto a RAF Short Belfast and flew to Belize via Gander in Canada. At least some of the observations from earlier in the year had been

13 AFM 123: 'Visit to British Honduras by Commandeer 2 Wing Army Air Corps, 30 April – 3 May 1973', 1 June 1973.

14 Anon, '653 Sqn AAC – Belize', *Army Air Corps Newsletter*, No.43, March 1974, p.5.

15 Anon, 'Sioux Performance or Lack of it!', *Army Air Corps Newsletter*, No.48, January 1975, pp.12–13.

acted upon and the aircraft were fitted with Stokes litters for casevac duties. A comprehensive tropical survival kit was strapped to one of the litters, which included a Greener police shotgun and ammunition. Although the aircraft had no navigational equipment installed, an RAF tented radar site, Butcher Radar, on the opposite side of the runway from the AAC detachment, was able to provide surveillance radar approaches and other services. Along with the detachment commander, Captain Chris Buisseret RA, were Sergeant Bob Morris Para and Staff Sergeant Peter Cole RAMC as the three pilots who initially deployed from 655 Squadron. Staff Sergeant Cole later described the opportunities and challenges of flying in Belize:

> Daily aircraft tasking normally involved liaison flights down to the Belize/Guatemala border area near the town of San Ignacio, with the aircraft landing at the 2nd Battalion Royal Green Jackets company base at Holdfast Camp, which was situated a few miles east of the town. Passengers, dispatches and freight would be carried on these flights. After landing at Holdfast the flight then proceeded to the south following the eastern branch of the Belize River for some 15nm [nautical miles] until at the junction with the Rio On, the aircraft would head eastwards following the watercourse up onto the Mountain Pine Ridge area. The next landing was at the HLS adjacent to the Royal Signals re-broadcasting site close to a trig point at 3,600feet. On departure from the rebro site the aircraft routed north to fly down the Sibun Gorge and along the Sibun River to pick up the Western Highway back to Belize Airport. These flights would take between 2.5 and 3.5 hours.
>
> Other tasking included VIP flights, such as carrying H.E. the Governor of Belize down to the new capital at Belmopan (under construction) or flying visiting staff officers and Generals around the various military installations. A number of Casevac flights (both day and night) were completed from Holdfast, the rebro site and Stann Creek [now Dangriga], and regular liaison flights were made to Punta Gorda, Tower Hill [immediately south of Orange Walk], Blancaneaux Lodge [10 miles southeast of San Ignacio] and Millionario [30 miles south of San Ignacio].[16]
>
> Flying the Sioux in Belize presented no operational difficulties, and a full load of fuel and two passengers could be carried on take-off from Belize Municipal Airport to the tasked en route destinations, namely Holdfast Camp at San Ignacio and the rebro station up on the mountain pine ridge. We could normally complete the round trip with our departure fuel (48 Imperial gallons) in a couple of hours. The Sioux was powered by the TVO 435 B1A 270 HP Lycoming 6 cylinder turbo-supercharged piston engine and could maintain take-off power to an altitude of 12,400 feet in ISA [international standard atmosphere] +25°C conditions, and as the highest landing site was circa 3000 feet it presented no problem. Our flight planning did take into account any Weight/Altitude/Temperature limitations that could arise.
>
> Most of the flying was conducted at low level and after departure from Belize Municipal Airport ATC, and outside the range of Butcher Radar we continued to communicate with military units on FM. The radios fitted were the ARC44 VHF, STR37E VHF, and the PTR 170 UHF. I do remember picking up New York taxi cabs when sitting on a helipad at 3000ft, due I assume, to atmospheric ducting, and I also remember speaking to RAF Upavon (call sign Architect) who manned the Air Support Command HF link in Wiltshire. Most of our flying was carried out at a low level and if following the Belize River, it was fascinating to observe the wildlife, particularly the Iguanas sitting in trees alongside the river. Once you had climbed up to the mountain pine ridge area it was a total change in scenery until flying back down the Sibun Gorge where the jungle environment resumed alongside the banks of the

16 P. Cole, account to the authors, 15 October 2020.

> Sibun River. Once back in range of Butcher Radar we would invariably request a Surveillance Radar Approach recovery back to the airport.[17]

BRITFORBEL

In 1974 664 Squadron took over the detachment and a year later Guatemala descended into civil war and moved troops to its eastern border. The UK decided a more robust stance was required to deter any aggression from Belize's neighbour. A second battalion was deployed to reinforce what would become known as British Forces Belize (BRITFORBEL). They were supported by a close reconnaissance troop of Alvis Scimitar and Scorpion armoured vehicles, a field battery of 105mm Light Guns, a squadron of engineers, several Blowpipe surface to air missile detachments, an RAF Regiment Rapier battery and a detachment of RAF Pumas. The main strike threat was provided by six RAF Hawker Siddeley Harrier GR.1As from No.1 Squadron RAF that would become a permanent detachment designated No.1417 Flight RAF.

As the footprint of BRITFORBEL spread the three AAC Sioux remained a critical lifeline to remote outposts in the interior and isolated OPs on the ridges overlooking the Guatemalan border. The associated landing sites were nearly always small and restricted, and often strewn with tree stumps and boulders, the latter being the cause of an accident that wrote off one of 664 Squadron's aircraft. On 2 August 1975 Sioux XT155 departed Salamanca Camp in the interior with the pilot and a REME technician on board. Twenty-five minutes later the aircraft, carrying four replacement radio batteries strapped to the Stokes litter, approached an OP landing site on a ridge measuring just eight by four yards square.

> The weather was good and the pilot carried out a normal steep approach from the West into the light easterly wind. The approach terminated in a low hover above the landing site and from this hover the pilot lowered the aircraft onto the ground. Both skids touched the ground but the aircraft was not firmly settled and it seems likely that at this stage the rear of the starboard skid became wedged against a boulder. The pilot was not happy with the aircraft in this unstable position and he raised the collective lever to lift it into the hover again. The port skid and the nose of the aircraft rose, the pilot applied forward cyclic and he may have overridden the hydraulics; in any event he believed that the control system reverted to manual. The starboard skid came clear of the ground, the aircraft lurched upwards and forwards and the pilot's attempts to stabilize the aircraft resulted in over-controlling and it started rocking about the lateral axis. At this stage the aircraft was completely out of control, it drifted backwards and to the left, the skids hit the ground, the main rotor struck some saplings and it went backwards off the edge of the landing site and down the slope. On the way down the slope it turned to the right and finally came to rest inverted and on fire.
>
> At some stage during the journey down the slope the pilot jumped out, but the technician was left hanging upside down in his safety harness. Despite the fact that his right foot was trapped between the aircraft and a tree, he got free and fortunately neither he nor the pilot suffered more than a slight injury. The aircraft was burnt out.[18]

Although the Integrated Flight Scheme was in its last throes, air squadrons were retained in the Germany-based cavalry and tank regiments in the armoured reconnaissance role that occasionally

17 P. Cole, account to the authors, 21 October 2020.

18 Anon, 'Flight Safety – Accident (2)', *Army Air Corps Newsletter,* No.65, December 1976, p.30.

still fulfilled overseas detachments. In 1976 1 RTR's Air Squadron took up the reins of the Belize detachment supporting 2 SG. As a 'rest' from jungle patrolling, 2 SG organised an expedition to climb Belize's highest point, Mount Victoria; the first attempt by a British mountaineering party. From their base camp the main body took 10 days to traverse the jungle covered foothills, resupplied on the way by a Sioux from 1 RTR. On reaching the ridge below the final peak RE sappers with the expedition built a wooden helipad to assist with resupply and in case casevac was required. As the team began its attempt on the summit of Mount Victoria Lieutenant Howard Gater-Smith RTR was flying Commander BRITFORBEL to Belmopan when he received an urgent radio call from the 2 SG expedition:

> "Quick, get down here fast, the bloody mountain's on fire!" I cast a gaze in a southerly direction and sure enough there in the distance a rising cloud of black smoke could be seen billowing skyward. The transit took about twenty minutes during which time I was briefed on the situation. Apparently, the mountaineers, making good progress had stopped for the proverbial smoke-break. Alas, one of the smokers failed to extinguish his fag-end resulting a short while later in a small bonfire erupting into a raging inferno. The speedy abseil descent caused rope burns to the hands of at least one of their number. Furthermore, the fire began to spread rapidly towards the heli-pad thereby endangering the lives of the thirty or so soldiers gathered nearby.
>
> In the adrenalin pumping excitement and with the density altitude at five and a half thousand feet, my first approach was too high and too fast. The second attempt proved good enough and I wobbled onto the pad. The aircraft which I was piloting had been fitted with a dual set of controls prior to some mutual flying training so I instructed one soldier to sit inside and the other to lie flat on the Stokes litter, both on the right side of the helicopter. As I had kept the rotors at flying revs I was unable to get out and strap him in. I'm sure that in the situation he was none the wiser and probably quite glad to be leaving his precarious situation. I gingerly pulled up the lever and established a hover to prove that enough left cyclic control was available and then committed the aircraft to a towering take-off. It was uneventful but caused a particularly uncomfortable feeling in the gut.
>
> A deep valley situated a couple of miles to the north provided us with a landing spot where the rescued soldiers debussed onto a small shale island in the middle of a river. Somewhat alarmed by the first passenger load I restricted further lifts to one human and two rucksacks until the next refuel stop when the dual controls were removed by a technician. Thus, so continued the rescue operation until all the men had been safely deposited on the island. As I flew in to pick up the last remaining pair, I could not help notice that the fire was perilously close to the pad. With last light approaching fast I informed the group that we would return on the morrow to pick them up and ferry them back to civilisation.[19]

Although possibly not as dramatic as the Mount Victoria rescue, the conditions on foot during 2 SG's routine jungle border patrols meant casevac tasking was a constant requirement:

> The heat and humidity took an awful toll of the infantry, 12 soldiers being casevaced on the first day, and their state of misery was not helped by the sight of Tpr Carney, looking very cool in his observer's kit and dark glasses leaning against a helicopter eating a watermelon. In fact one platoon commander complained that his entire platoon had asked for a transfer to the Royal Tank Regiment.[20]

---

19 H. Gater-Smith, unpublished account, passed to the authors 20 November 2020.
20 Anon, 'Blue Water – White Rum', *Army Air Corps Journal*, 1976, pp.107–108.

But a transfer at that point would have been to no avail. The Integrated Flight Scheme finally came to end shortly after 1 RTR Air Squadron's Belize detachment, and the final four Royal Armoured Corps air squadrons were disestablished and absorbed into AAC squadrons. At the same time the AAC detachment at Airport Camp, up until that point an operational tour, was formally brought on to the British Army's establishment on 1 November 1976.[21] At around the same time the Scout replaced the Sioux with the AAC Belize detachment. When operating in Belize's hostile environment the AAC had to work mutually with the infantry to ensure everyone's safety, as recalled by Captain Paul Edwards who deployed to Salamanca Camp in the jungle with 3rd Battalion, Queen's Regiment:

> Most of it was under canvas with logs placed to form a floor for camp beds. Rats and snakes were often chased out and these flimsy structures had a tendency to flood during heavy rain .... From Salamanca, we would conduct recces in the AAC Scout and deployed patrols by Puma deep into primary rainforest near the border. Guatemalans were known for crossing the border either to intimidate local villages or smuggle items such as chewing gum. Either way the groups were thirty plus and heavily armed. We patrolled as a platoon for weeks at a time with the general mission of hearts and minds, surrounding villages and communicating with the tribal head before pulling back, observing and moving on...
>
> On one patrol we allocated five days to clearing and improving a very large and quite long [helicopter] landing site. Having established our A frames [shelters] and a defensive position, taking advice from the AAC pilots we concentrated on removing or reducing the heights of trees on the approach and overshoot. The task was such that I radioed back for PE [plastic explosive] and det [detonation] cord to speed up the process. The request was denied but within hours a Scout turned up with the cabin filled with chain saws and fuel!... Speed of back up or casevac in close, primary or secondary rainforest made such landing sites essential and we needed to know which was closest at any stage of a patrol. As we were about to leave the above landing site, I had to pull that card in order to casevac one of my team by Scout...
>
> The Scouts provided a vital service in challenging terrain. The weather was hot, humid and very changeable at times. Point navigation would have been a challenge and there was and probably still is, a set procedure for locating a patrol. On the ground we would cut a pole and strap a coloured smoke grenade to it. Then make contact with the Scout on UHF SARBE [search and rescue beacon], pull the pin on the smoke before asking the pilot "what colour smoke do you see?" If he came back with the correct colour e.g. "I see blue smoke", we would come back with "correct", and he would know it was us. An engine failure over the jungle would likely have been catastrophic, whether over primary or secondary rain forest. Some landing sites were no more than makeshift wooden structures probably requiring gentle lowering of the collective or even keeping a little pitch/power on before departing again. It was demanding flying.[22]

Any issue with the Scout's Nimbus engine while in flight would probably result in a difficult landing in a riverbed or onto a track and leave the crew and any passengers isolated and having to fend for themselves until rescue. As a result, the question of jungle survival was a constant consideration.

21 AFM 123: 'Authority for Formation – Army Air Corps Detachment Belize', 27 January 1977.
22 P. Edwards, account to the authors, 4 October 2020. Edwards later transferred to the AAC.

> We must consider our actions following an avoidable, or even our own induced accident, and the possibility, remote as it may sometimes appear, of enemy action...The question of where to go should the engine fail is not academic though. At least the bottom of a valley might conceal a stream; availability of water assumes crucial importance during the dry seasons – yes, Belize has two a year...
>
> The possibility of making a safe forced landing in large areas of Belize is remote. Two extreme circumstances can be envisaged. Either the crew emerge from a broken burning aircraft with what was attached or immediately at hand during escape, or the aircraft survival pack will be recoverable and other useful items may be salvageable. In the first case the only guaranteed survival aids are those carried by the pilot, on his person.[23]

The OC AAC Detachment had a taste of what a forced landing could entail while providing FAC for live Harrier ranges, with the range safety officer as a passenger, on 27 October 1977. Having shut down at a jungle strip to refuel, the Scout would not restart. The OC relayed his problem over the radio via the Harriers to Butcher Radar and a Puma dropped off three REME technicians. They found that the Scout would not fly again without a full engine change and would need recovering to Airport Camp. The issue that now faced the small party was that the Harrier exercise had ended and the Scout's radio could not reach any other station to alert anyone of their situation. With evening approaching it was unlikely that anyone would reach them until the following day, and so the decision was taken to break open the survival pack. The SARBE was activated, and a meal, tea and coffee were made using the last of the available water. The group had built a shelter and prepared for a night in the jungle when they heard a Hercules approaching. It had picked up the SARBE signal and when it arrived overhead the OC related his predicament to the crew. A Puma was diverted from other tasking and arrived just before darkness to undersling the Scout to Airport Camp. An hour later the stranded party lit flares and a 'T' landing aid made up of Benghazi Burners to guide in the same aircraft to transport them home. The OC later complained that on breaking open the survival pack he found that several items had been 'pilfered' by persons unknown and that had his situation been more serious or protracted the missing items could have made life much worse. On checking all other survival packs within the detachment other deficiencies were also found. It was pointed out that several Army pilots had crashed in the Malayan jungles not so many years before and that some were not found by searching aircraft, but with their survival packs (without which they would almost certainly have died) they had made a safe return.[24]

In 1979 a journalist flying with the OC AAC Detachment, Major Jeremy Wheeler AAC, noted that while the Scout achieved 80 to 90 percent serviceability in Belize, the hot and humid conditions placed a great strain on engines and the Scout was often restricted by how much fuel, freight or passengers it could carry. It was hoped that the situation might be improved by the replacement of the Scout with the Gazelle by the end of that year.[25] In fact the arrival of the Gazelle was viewed with a degree of trepidation within the AAC Detachment because, although it was generally accepted that its performance was superior to the Scout, it was less useful and more fragile when it came to moving infantry and equipment around the jungle. Once in Belize, however, the Gazelle was generally viewed favourably by aircrew and technicians alike.

---

23 R. S. Elwell, 'Jungle Survival in Belize', *Army Air Corps Newsletter,* No.95, September-October 1980, p.11.

24 Anon, 'Comrade, is your Conscience Pricking?', *Army Air Corps Newsletter*, No.77, January-February 1978, pp.7–8.

25 Anon, 'British Air Power over Belize', *Air International*, November 1979, p.226.

> Before the introduction of the Gazelles, many people (both inside and outside the Army Air Corps) had severe doubts as to the effectiveness of the Gazelle as a replacement for the Scout in Belize. Now that the Gazelle has been working in the theatre for about 14 weeks most of those reservations have been completely dispelled. The Gazelle obviously has some limitations but they are out-weighed by its advantages of increased speed, endurance and passenger comfort for all but the largest of persons.
>
> For the Gazelle to maintain its developing reputation, a timely supply of spares is absolutely vital. To date things have not been too bad apart from a short spell around Christmas when the supply chain appeared to falter. The efforts of all those involved in establishing the supply of spares is greatly appreciated, but there are still many items which could become critical.
>
> From the piloting point of view, the Gazelle is a delightful aircraft to handle. Its performance in relatively hot and sometimes high conditions cannot be faulted. When properly balanced and tracked, it flies unbelievably smoothly. Its lack of ground clearance is a fact the pilot has to keep firmly in mind when landing due to the rough terrain, and now that the weather has become drier, the protective strips on the main rotor blades are lasting for reasonable periods of time.[26]
>
> The Gazelle is coping very well indeed, even though we have to admit to the odd problem. During the rainy season the sheer quantity of water that fell from the sky was a little disconcerting. Visibility would fall to an unmentionable figure. The next morning the aircraft wouldn't start: a touch of rheumatism. Now it is hot and hazy, and one has to go very high indeed before running out of torque rather than T4 [engine temperature]. Some engines do run hotter here, which is giving rise to voluminous correspondence with Aircraft Branch.[27]

The hot running of engines did cause problems initially, with 20 engines having to be changed a year. The excessive heat was caused by the accumulation of dirt and salt in the engine due to the atmospheric conditions prevalent in Belize. In response, technicians with the Detachment pioneered a method of washing out the combustion chambers and compressors of the Gazelle's Astazou engine using pressurised water. The novel approach was subsequently evaluated by Rolls Royce and the monthly compressor wash or 'comp wash' became a routine event in the working lives of Gazelle pilots for the next 40 years.[28] The atmospheric conditions were not the only hazard that technicians had to deal with. During an aircraft inspection a two-foot python was discovered coiled up in the boot of a Gazelle, which had to be dismantled before the snake could be safely removed and then killed.[29]

On 21 September 1981 Belize achieved full independence and the threat from Guatemala seemed to be receding. With the arrival of the Gazelle the Flight's establishment was increased to four aircraft, one of which was held in reserve. The Flight strength was also increased to 20 all ranks, all on six-month unaccompanied tours, apart from the OC (who was also the Flight QHI and a line pilot) who was on a two-year accompanied tour. There were two further pilots, two aircrew and five ground crew. The remaining 10 members of the Flight were REME technicians and stores men. Although often viewed as a glamorous, 'sunshine' posting not everyone appreciated what Belize had to offer, as Flight Commander Major Sam Drennan AAC bluntly explained in a letter to a fellow officer.

26 Anon, 'AAC Detachment Belize', *Army Air Corps Newsletter,* No.92, March-April 1980, p.26.
27 Anon, 'Belize Flight AAC', *Army Air Corps Newsletter,* No.94, July-August 1980, p.17.
28 G. Smith, 'Jungle Jumpers', *Soldier,* undated, p.42.
29 Anon, 'Flight Safety – an Unwelcome Passenger', *Army Air Corps Newsletter,* No.97, January-February 1981, p.48.

> Belize City I presume has decayed quite considerably since you were here and most of it looks like it will fall down any moment. There are about four eating establishments cleared for use by the British in Belize City and we all avoid dark streets when downtown as the locals are likely to lift watches, wallets and perhaps stick a knife in you for good measure.... Suffice to say that I cannot imagine a worse posting from the personal comforts and social amenities point of view, but the job is quite good and I am left to do what I please.... Take care. We send you greetings from the world's anus.[30]

As Major Drennan had conceded, what Belize lacked in social amenities was compensated for by the operating environment. Although many aspects of the detachment were beginning to become routine, operating the Gazelles in Belize remained challenging and the tasking was varied but critical in supporting what was now known as British Forces Belize's operations on the ground, as reported by the Flight.

> Due to the nature of the terrain and the lack of suitable roads the role of the Flight within the Force is a most vital one. It would not be an overstatement to say that many activities and operations could not even be contemplated without the certainty that an AAC helicopter will appear when required. The environment is hostile to both men and machines but particularly for single-engine helicopters operating at relatively low altitudes over the jungle. With temperatures often at the top end of the aircraft's operating band and relative humidity at or near 100%, controls are less responsive and the power available margin much smaller. The type of pilot who fits into the 'cowboy' category would soon end his career in Belize.
>
> The variety of tasking is extensive and includes recce and observation, FAC, Command and Control, resupply, search and rescue and casevac. The lack of roads and poor quality of the few there are makes command and control a particularly important role. Commanders or their representatives and the many visitors could rarely complete their tasks any other way. Re-supply is also important. There is a limit to the amount a small jungle patrol can carry on a two week expedition. They are always grateful when, having reached an RV, a Gazelle comes fluttering into the clearing with water, rations, mail and anything else they need; well, almost anything else. Emergency re-supply is commonplace. We either have to arrange to meet at the nearest HLS or depending on the cargo, it is either dropped or lowered through the jungle canopy. Water seems to be a favourite here.
>
> Search and rescue here is not the sort of task the reader would associate with the SAR [Search and Rescue] boys of the RAF and Navy. Our version is looking for the patrol that is uncertain of its position. If the patrol's radio is serviceable we use our 'white man's magic' and home in. If not then life becomes more interesting but we've yet to fail. There is an almost unrivalled scope for the airborne FAC and it is unlikely that a fully qualified FAC would go home at the end of his 4 months without having entered several strikes in his log book. Casevac is arguably our most important role. Dehydration and cut limbs being the most common medical problems. Snake bite cases are rare but because of the particularly vicious and venomous nature of some of the snake species in Belize such a case warrants special concern. Normally the quicker we get the doctor there to start the treatment the more chance the patient has of pulling through. The morale of the men on the ground is that much higher when they know the AAC will not let them down.
>
> Most aircrew, with the exception of those in the Far East and Cyprus, will rarely have put into practice some of the more exciting techniques they learned in training. Really confined

30 AFM 259: Letter, 'Major Drennan, OC 25 Flight AAC to Major Skinner, OC 16 Flight AAC', 12 June 1987.

25 Flight AAC's dispersal and hangar at Airport Camp near Belize City. (Army Flying Museum)

areas and pinnacle landings are just two techniques which AAC pilots in Belize cope with almost every day. Few of the landing sites here are not confined areas and border Ops and rebro sites are naturally on the highest point in the area. Over the jungle our rule is to follow tracks and river beds wherever possible even though it means deviating from the heading. A forced landing onto a track or river bed is more likely to be survivable than to plunge through the jungle canopy.[31]

A word on the hazards of jungle flying…Imagine yourself as the Gazelle pilot who made a successful landing in a jungle clearing on a pad 30 feet square, having suffered fuel limiting, rotor decay and sustained engine surge during the final 100 feet of the approach. He had spent the previous 20 minutes flying over jungle covered terrain, correctly and meticulously

31 Anon, 'A Little Goes A Long Way', *Army Air Corps Newsletter,* No.114, November-December 1983, pp.6–7.

> using our SOP [standard operating procedure] of following tracks or river courses. With the exception of a 2 to 3 minute period immediately prior to the incident, when an over jungle transit was needed to reach the HLS, the aircraft could have been landed on a relatively open site, rather than the jungle canopy, had the FCU [fuel control unit] fault occurred earlier. We don't fly over trees unnecessarily in this environment.[32]

25 Flight AAC

Early in 1984 the AAC Detachment Belize was formally renamed 25 Flight AAC, taking its title from the last AAC unit to leave British Guiana 18 years before.[33] The tasking, however, remained unchanged, as did the flying conditions as experienced by a journalist when taken on a sortie by Sergeant Martin Bird Para.

> The weather was typical of the wet season, cloud base at 500ft and intermittent heavy showers. Following one of Belize's only three passable roads with thick jungle on either side of it, one gained a good impression of the inhospitable terrain which worsened as the helicopter flew west towards the hills surrounding the border area, some 60 miles from Belize camp. The Gazelle had to skirt around the hills which met the low cloud in places, until ZA735 came to a hover close to an Army observation post some 400 yards from the border crossing at Menchos [Melchor de Mencos]. Little could be seen apart from a large map on the other side which showed Belize as part of Guatemala and an apparently deserted airfield in the distance.[34]

Guatemala finally recognised Belize's independence and established diplomatic relations in September 1991. As a result, the British Government decided to end its defence commitment to Belize on 31 December 1993. The withdrawal was to be phased and it was agreed that the force would be reduced from 31 December 1993 to about half strength, until 1 October 1994 when it would be withdrawn leaving a just a training unit in the country. British Army Training Support Unit Belize (BATSUB) consisted of approximately 65 personnel, established to provide support to jungle training packages for British Army units. An AAC presence was to remain in Belize as 25 Flight AAC. Although no longer an operational unit, with the withdrawal of the RAF Pumas and particularly arduous training taking place in the jungle, 25 Flight's role remained critical. That was particularly the case in the casevac role as described by Sergeant Mark Rutherford AAC.

> "Casevac! There's a British soldier with heatstroke stuck on a cliff-edge in the middle of the jungle and he can't hold down water." A helicopter was scrambled but it soon became apparent that the casualty was very weak and they had no drip kits. It was understood that a guide was to meet the flying doctor in a partly dried up riverbed. En route to the scene, the casualty's condition worsened and he was to be carried down to the riverbed. The aircraft landed on the river bank with the doctor and closed down. Darkness fell; the patrol still had some five hundred feet to descend. Hindered by their six and a half foot casualty and dense black jungle, coupled with a very steep gradient, they shouted for the aircraft to provide light. Equipped with only a landing light, light was provided. The patrol indicated and updated

32 Anon, 'AAC Detachment, Belize', *Army Air Corps Newsletter*, No.112, July-August 1983, p.27.
33 AFM 123: 'General Staff Redesignation Order', 5 April 1984.
34 D. Oliver, 'Belize Detachment', *Aircraft Illustrated*, February 1986, p.71.

> their position by means of firing miniflares, the light being constantly directed towards the flares as it was not possible to see the patrol. With both the aircraft and the casualty eventually down on the riverbed, the casualty was treated before being flown back to Airport Camp. A five-hour casevac.
>
> That was just one casevac. 25 Flight conducted a total of twenty-eight casevacs in the first three months, one of which was a double aircraft casevac to extract three [Belize] Defence Force soldiers vomiting blood. The others were: difficulty in giving birth, crushed finger, three cases with spinal injuries, internal bleeding and vomiting, machete cut to hand, machete cut to leg, broken elbow, five suspected broken ankles, kidney pains and vomiting, heart attack, two cases of fever, two scorpion stings, four cases of heat exhaustion, heatstroke, unconscious due to allergic reaction to drugs and, finally, severed fingers due to a night diving accident.[35]

As well as supporting BATSUB, 25 Flight AAC also provided support to the Belizean Government which extended to providing casevac sorties for the civil population. With the many isolated communities in the forests and on the cayes, anyone requiring urgent hospitalisation relied on the Flight's helicopters to get them to an urban centre in a timely fashion. Birth complications frequently called on a 25 Flight sortie as described by Sergeant Richie Robinson AAC.

> I lifted off for Punta Gorda [on the coast 100 miles south of Belize City] in my Gazelle with the unit's doctor and midwife and arrived at the town's landing strip an hour later. After five minutes wait the ambulance arrived and the doctor assessed the 'casualty' as 'fit to fly'. He said that there was a possibility of the baby being born en route, although he did not think so! I lifted off from Punta Gorda strip shortly afterwards with four persons on board, the estimated time en route to the Municipal Airport in Belize City being fifty minutes. During the journey the 'casualty' was moaning quite a lot; the midwife and doctor were talking to her, timing contractions as she was in labour.
>
> Approximately eight minutes from our destination, the doctor stated there should not be any big contractions for another eight minutes and we should be on the ground by then. No sooner had I acknowledged this than the baby girl was born. The doctor was worried about the mother as there seemed to be a lot of blood. I was asked on the radio by another Gazelle crew about the nature of the casualty. I replied "We are having a baby!" Crossing Belize International's area, I gave my 'personnel on board' figure as four; this of course had to be changed to five, much to Air Traffic Control's amusement. I had requested the ambulance about forty miles (twenty minutes) out. It arrived two minutes after we had landed. The midwife stayed with the mother and baby, the doctor returning with me to Airport Camp.... The aircraft had collected a serious casualty from a road traffic accident approximately two weeks previously. He died en route so it was with much pleasure I can state that mother and daughter are doing fine.[36]

Major Ian Curry RAMC was completing a flying tour as a Specialist in Aviation Medicine when he spent an attachment with 25 Flight. As a qualified doctor as well as a pilot he had particular insight into the casevac sorties he conducted.

35 M. Rutherford, 'With 25 Flight Army Air Corps in Belize', *Army Air Corps Newsletter,* No.168, July-September 1995, pp.18–19.

36 R. Robinson, 'Casevac – A Birth in Flight', *Army Air Corps Newsletter,* No.172, July-September 1996, pp.13–14.

Gazelle ZA765 from 25 Flight AAC on the HLS at Cadenas OP on the south-west border between Belize and Guatemala in 1984. The Gazelle is fitted with flotation devices on the skids. (Army Flying Museum)

> We had a casualty call from the south of the country near the Honduras [*sic*] border and we had to go and we landed on this beach down south and went into the village. It was a little fishing village, and when I say 'village' it was straw huts, it was really pretty basic. And there was one of the largest ladies I had ever seen in my life, lying there, who'd been attacked by some mad bloke with a machete. Said mad bloke was tied to a tree and they were waiting for the police who would be about two days because it was one of the islands. This large lady, she was so large in fact that he'd hit her at the top of the shoulder and the machete had cut through about three inches of fat and just nicked the muscle, so actually her injury, although spectacular and very bloody, actually wasn't in any way life threatening. So, we got some burly men to help us put her on a stretcher and take her to the aircraft. As we put her into the aircraft it actually sank the skids on that side into the sand so I wasn't quite sure if we'd get off but, thankfully we were fairly light on fuel so as I picked it up we just crept out of the sand and off we went back to Belize [City] to drop her off…
>
> We were the only helicopters in the country, we were the only people who could go and get people and so you did some very, very odd things. I flew right out to the Blue Hole [50 miles out to sea on Lighthouse Reef] to pick up a US diver with the bends who'd come up from deep in the Blue Hole. There was no heli-pad on the ship so I had to hover with one skid on the ship and one skid not on the ship whilst they put him on into the stretcher and then flew him to San Pedro to the compression chamber.[37]

As well as providing military aid to the civil community through casevac and other means, the British Government also provided support to the Belize Defence Forces (BDF) to ensure they were

37 AFM Interview Recording: I. Curry, 20 May 2019.

robust enough to deter any Guatemalan resurgence. This included providing loan service military personnel to bolster and advise the BDF. As part of that agreement both Captain Leslie Henderson AAC and Major James Hazou AAC spent two years with BDF Air Wing. As well as being a fixed wing pilot Major Hazou also acted as the Air Wing's Operations Officer and aviation adviser to the Belizean CO. The Air Wing operated the twin piston engine Britten-Norman Defender. The aircraft's roles of communication, liaison and surveillance were similar to those conducted by the AAC's Britten-Norman Islanders in Northern Ireland. In addition, the BDF Air Wing's aircraft had a CAS role. Major Hazou described the limitations of the Defender in the CAS role and other tasks he helped to develop for the Air Wing.

> In the CAS role the Defender carries four General Purpose Machine Guns (GPMG) in two underwing pods. Using a single point sight the pilot can, with practice, put down quite accurate fire. However, the aircraft is placed in a very vulnerable position during an attack. Carrying out a dive attack from 500 feet in a relatively slow moving and unprotected aircraft is really a last-ditch manoeuvre.
>
> I decided earlier this year to go back to basics and see if the Defender would make a good Air Observation Platform (AOP) for the adjustment of artillery and mortar fire. A 'porpoise' technique was developed using a low-level four minute race track circuit, a zoom climb (this is a slight exaggeration in the case of the Defender) in order to observe the fall of shot, while trying not to expose the aircraft for more than nine seconds. In order to prove the concept, the Royal Artillery 105mm Light Gun Troop had its fire adjusted by the BDF Defender along with the BDF's own 81mm mortars. The concept worked very well and it is hoped in the New Year Naval Gunfire Support (NGS) will be attempted with the help of the Royal Navy's West Indian Guard ship. Even Forward Air Control has been successfully attempted (during the last visit of Harrier to Belize). These roles have offered an offensive capability to the Air Wing and allow a more flexible use of the aircraft.
>
> Low level parachute resupply is another role being developed. A trial has taken place using three different designs of locally made parachute (for the princely sum of approximately £15 each), allowing a simulated load of 40lb of rations to be dropped accurately from 150 ft. The trial was very successful and a cross shaped parachute design has been chosen, to be made locally. This rations paradrop will allow the BDF infantry patrols to be extended to a two week patrol cycle.[38]

In October 1995 AAC Lynx visited Belize for the first time, with 657 Squadron AAC deploying on Exercise Condor Guard.[39] Eighteen months later in May 1997 the Lynx Mk 7 joined the Gazelles to improve 25 Flight's capability.[40] Lynx operated in Belize for a relatively short period however, and 25 Flight was re-equipped with the highly capable Bell 212 in 2003. In 2011, during a further drawdown of BATSUB, 25 Flight was disbanded, ending Army aviation's association with Belize, Central America and South America that had lasted nearly 50 years. The final Bell 212 sortie, flown by 25 Flight's OC Major Dave Roberts AAC, took place on 15 July 2011.[41]

38 J. Hazou, 'Loan Service in Belize', *Army Air Corps Newsletter,* No.166, January-March 1995, pp.10–12; L. Henderson, 'Defender in Belize', *Army Air Corps Newsletter,* No.160, July-September 1993, pp.10–13.

39 AFM 190: 25 Flight AAC – The Review of the Year, October 1996.

40 AFM 190: Monthly log of Matters of Historical Importance, May 1997.

41 D. Roberts, account to the authors, 3 September 2020.

Chapter 9

Northern Ireland

See Map 14

Operation Harvest – The Border Campaign

England and Great Britain's fractious relationship with the island of Ireland is centuries old and its full history is outside the scope of this work.[1] The AAC's operations across the Irish Sea have their roots in the partition of Ireland, which was enacted on 3 May 1921 following the Anglo-Irish War. Partition led to the Irish Free State being created with 26 of the 32 counties in the south and west of Ireland. The remaining six counties in the north-east opted out of becoming part of the Free State and instead became Northern Ireland, part of the United Kingdom with Great Britain.[2] The Irish Free State gained independence from the United Kingdom on 6 December 1922, adopted a new constitution and was renamed Ireland in 1937 and, finally, was declared the Republic of Ireland in 1949.[3] For Irish nationalists and republicans the continued existence of Northern Ireland within the United Kingdom was the cause of discontent and protest. The issue was complicated by religious sectarianism, with the Republic being a Catholic state while the population of Northern Ireland had a Protestant majority. The Irish Republican Army (IRA) had formed in 1917 and fought against the British during the Anglo-Irish War of 1919–1921. The IRA fought and lost the Irish Civil War within the Free State in 1922–23 but survived and became a focus for violent attempts to remove British influence from the island of Ireland and to bring Northern Ireland into the Republic.

After ineffectual campaigns in England in 1939 and Northern Ireland from 1942 to 1944 the IRA regrouped and focused on the idea of launching armed action against the British Army in Northern Ireland from the Republic. During the early 1950s the IRA steadily rearmed by raiding British Army armouries in Northern Ireland and England. At the end of 1956 the IRA launched Operation Harvest, also known as the Border Campaign, which envisaged guerrilla units of approximately 50 men known as flying columns crossing the border from the Republic to attack targets in Northern Ireland.[4] In the early hours of 12 December 1956 the campaign was launched with simultaneous

1 For a broader history of Ireland and its relationship with England and Britain see R. Bourke and I. McBride (eds), *The Princeton History of Modern Ireland* (Princeton: Princeton University Press, 2016), S. J. Connolly (ed.) *The Oxford Companion to Irish History* (Oxford: Oxford University Press, 2000) and T. W. Moody and F. X. Martin, *The Course of Irish History, Fourth Edition* (Lanham: Roberts Rinehart, 2001).

2 Northern Ireland is also often referred to as Ulster, although the six counties only made up part of the traditional Irish province. British servicemen have also often referred to Northern Ireland colloquially as 'The Province'.

3 The Republic of Ireland is also referred to as *Poblacht na hÉireann* in the Irish language, often abbreviated to Éire.

4 For a full chronology of violent Irish resistance to the British from the seventeenth century and including the Border Campaign see T. Geraghty, *The Irish War; The Military History of a Domestic Conflict* (London: Harper Collins, 1998).

attacks by around 150 IRA members on targets across the border. These targets included a BBC relay transmitter that was bombed in Londonderry; a courthouse in Magherafelt and a police post near Newry, both set on fire; a half-built Army barracks at Enniskillen that was blown up; and a failed raid on Gough Barracks in Armagh.[5] The campaign continued into 1957 and by the end of the year 341 incidents involving the IRA had been recorded. In response the British Army reinforced its presence in Northern Ireland. In early 1957, 1913 Light Liaison Flight RAF's Austers were sent to RAF Aldergrove, 12 miles north-west of Northern Ireland's capital of Belfast. On 1 September 1957 the Flight, commanded by Captain Peter Wilson of the Royal East Kent Regiment (Buffs), became 13 Liaison Flight AAC and continued its operations to counter the ongoing border campaign. Along with the Malayan and Cyprus Emergencies, Northern Ireland was the third ongoing operational campaign that the AAC found itself in the midst of at the moment of its establishment.

> The Flight continued to support the Army and Royal Ulster Constabulary on operations against the IRA. Towards the end of the period under review [July to September 1957] the IRA changed their tactics in favour of ambushes. This has given the Flight a new job, which is to check roads prior to and during any appreciable military movement.
>
> As an example, on 5th September two large convoys moved between BALLYKINLER and the port of LARNE carrying important stores. The Flight patrolled the roads in front of the vehicles between 0600 hrs and 1900 hrs. Both convoys maintained a 62 set wireless link with the aircraft.
>
> Some interesting information came from the police chief at LARNE, District Inspector SPEARS. He has undertaken a number of air reconnaissances over the more remote parts of his area and on one of those flights spotted a group of buildings that had not been marked on his large scale map. There was an immediate follow up on the ground, and as a result some arms and ammunition were recovered.[6]

Flying with police officers as observers became a common mode of operating and communication with the ground was improved through use of the officer's VHF radio that he carried in the Auster during the sortie. Photographic and search sorties were frequently flown to find and disrupt the IRA's activities close to the border with the Republic. Security force operations on the ground were proving effective and IRA guerrillas found it increasingly difficult to safely lodge with sympathisers in Northern Ireland. They were forced to resort, therefore, to living in carefully concealed dugouts constructed all over the six counties. Locating the dugouts became a task for 13 Flight.

> On Monday 28th [October 1957], Police HQ in BELFAST requested a flight for County Inspector WOLSLEY in order to investigate some dug-outs used as hiding places for members of the IRA from EIRE. The aircraft landed him a mile from the first of these to be investigated and a police car then took him and the pilot to the spot, a police guard having been mounted on the aircraft. The hide-out was well located, spacious and excellently camouflaged, being under a hedge on a hill overlooking the landing area. Later when viewed from the air, it was found to be difficult to locate, but aerial photography may well be the answer in future. The others were of a similar pattern and all three were within a radius of six miles.[7]

---

5 Both the city and county are known alternatively as Derry or Londonderry by the two sides of the Unionist/Republican divide. In this history both will be referred to as Londonderry as that is how they appear in the majority of contemporary accounts and documents being used for reference.

6 AFM 182: '13 Liaison Flight AAC, Quarterly Newsletter for July, August and September 1957', 24 October 1957.

7 AFM 182: '13 Liaison Flight, Army Air Corps – Operations Report for the Week Ending 3rd November 1957', 8 November 1957.

In February 1958 13 Flight AAC returned to England to join 651 Light Aircraft Squadron AAC at RAF Feltwell in Norfolk. It left a detachment of three Austers at RAF Aldergrove which came under the direct control of HQ 39 Inf Bde. The search and convoy escort sorties that the detachment continued to conduct required prolonged low flying which could prove hazardous. On 8 July 1958 Auster T7 WJ407 hit high-tension power cables and crashed into a field near Castledawson, northeast of Magherafelt in County Londonderry, killing its pilot Staff Sergeant Ralph Hall RA. Just four months later, on 13 November 1958, Auster AOP6 VF661 hit trees in a low level turn and crashed near Ballypitmave, 5 miles northwest of Lisburn, killing the pilot Captain Michael Cracknell RA and his passenger, Flt Lt L. R. Readman RAF.[8]

In July 1959 13 Liaison Flight AAC returned to RAF Aldergrove to reinforce the detachment and take up the affiliation with HQ 39 Inf Bde. The Flight, under the command of Major F. R. I. 'Dai' Gotaas RA, immediately began an intensive programme of training and operational flying. In November it established a permanent detachment in a tented camp close to the border with the Republic at St Angelo, just north of Enniskillen in County Fermanagh. This detachment carried out daily flying along the border in conjunction with units from 39 Inf Bde and the Royal Ulster Constabulary (RUC). Many of those patrols were carried out in extremely poor flying weather; however a remarkably large number of flying hours were completed during the winter of 1959. In February 1960 three Saunders-Roe Skeeter helicopters under the command of Captain Pat Speedy RE joined 13 Flight. These greatly enhanced the capability of the Flight as there were very few suitable landing sites in the rugged border counties of Northern Ireland.[9] Unfortunately, due to engine and undercarriage issues, all Skeeters were grounded in June 1960 and did not fly again until September that year after modifications. After the teething problems, however, the Skeeters proved a success. Work with the RUC went from strength to strength with friendly and effective liaison between aircrew and police officers. Two successful operations that resulted in the capture of IRA guerrillas were enabled by effective reconnaissance from the air. Despite successes, however, there were still minor frustrations, particularly for the Auster pilots, as recalled by Major David Manktelow KORBR.

> During the period of my time in Northern Ireland [1961–62] we saw the IRA campaign slowly fizzle out. Our Skeeter work was very much in support of HQ Northern Ireland but on the fixed wing side the majority of my work was in support of the police. At that time we had one Auster 7 mounting an F24 camera and its main role was to provide aerial photographs for police searches of farm buildings in the border areas, chiefly in Fermanagh. St Angelo was a deserted airfield in those days and frequently an Auster went down for three or four days to carry out extensive border patrols. In the event of anything suspicious being seen one then had the frustration of being unable to communicate with anyone for we had no police radio and military patrolling was virtually non-existent. One also did a number of aerial arms searches, carrying large, heavy footed Irish policemen, although the value of an Auster in such a role was extremely limited.[10]

As Major Manktelow suggested, by the end of 1961 Operation Harvest, the IRA's Border Campaign had practically run its course. It had cost the lives of eight IRA men, four republican supporters and six members of the RUC. An additional 32 RUC members had been wounded. More significantly,

8 Anon, '651 Light Aircraft Squadron, Army Air Corps—1958', *Army Air Corps Association Journal*, vol.1, no.1, Spring 1959, p.6.

9 Anon, '651 Light Aircraft Squadron (1st March 1959 – 1st January 1960)', *Army Air Corps Association Journal*, vol.1, no.2, Spring 1960, p.12.

10 AFM 182: Letter, 'Major D. L. Manktelow to Brigadier M. W. Sutcliffe', 28 May 1974.

the internment of known IRA members and sympathisers had been enacted on both sides of the border during the campaign. A total of 256 Republicans were interned in Northern Ireland and approximately 150 in the Republic. Of those in Northern Ireland, 89 signed a pledge to renounce violence in return for their freedom. The Border Campaign officially ended on 26 February 1962 with a statement from the IRA leadership in the Sinn Féin newspaper *The United Irishman*. In it they blamed the failure of the campaign on, 'the attitude of the general public whose minds have been deliberately distracted from the supreme issue facing the Irish people – the unity and freedom of Ireland. The Irish resistance movement renews its pledge of eternal hostility to the British Forces of Occupation in Ireland.'[11]

A Brief Period of Peace

With the end of the Border Campaign there was some relaxation of the British military effort in Northern Ireland. Following the success of the Unit Light Aircraft concept trial the introduction of the Integrated Flight Scheme brought changes to the AAC's establishment in the Province. 13 Liaison Flight left Northern Ireland to join 653 Light Aircraft Squadron in Aden, leaving behind a small nucleus from the Flight workshops. Aircrew and ground crew from 2 RTR Air Squadron came into Aldergrove and joined the remaining REME technicians to form 2 Recce Flight, under the command of Captain Richard Eccles RTR. The Flight operated three Auster AOP9s and three Skeeter AOP12s and had an establishment of approximately 30 personnel forming an integral part of the Northern Ireland armoured car regiment. The winter in early 1963 was particularly bitter with exceptional snow fall making life extremely difficult for rural communities across Northern Ireland. 2 Recce Flight was called on to assist, carrying farmers and fodder to stranded farm animals, moving officials around the countryside, and lifting civilian casualties, the critically ill and pregnant women to hospitals and medical facilities. On 7 February Lieutenant Martin Tweed RTR had already flown one woman in labour to Lagan Valley hospital and a casualty to Carrickfergus hospital in Skeeter XN339 when he was tasked again to fly to a farm some three miles away and lift Mrs Madge McIlroy, who had gone into labour earlier that day.

> It was by now late afternoon and what little there was of the light was fading fast. Snow was falling again, the cloud base had lowered further and the attempt to reach Mrs McIlroy had to be abandoned. He began the slow and difficult return to base. The terrain over which Martin Tweed was flying was a flat plain of farmland, close by the rocky outcrop of Knockagh Hill, with its war memorial monument on the summit. It was crossed by the pylons carrying the main electricity cable from Kilroot Power Station. Just outside the little village of Mossley the helicopter struck the high-tension wires. The local farmers still remember the mighty flash which lit up the sky and the colossal bang. They had been cut off for four days, with snow piled up to the top of the hedgerows. Farmer Barry Garrett rushed out to help. The tail rotor had been severed from the helicopter which was lying on its side. Fortunately the pilot had escaped serious hurt. He sheltered in the farmhouse until the police arrived and the road was cleared sufficiently for an Army lorry to get through and take him back to Aldergrove. As for the locals, they now also had a power cut to contend with. Even worse, one old farmer had run out of tobacco and was reduced to smoking tea.[12]

11 Geraghty, *The Irish War*, p.4.
12 G. Warner, 'Ulster's "Big Snow" of 1963', *Ulster Airmail,* February 2008.

Towards the end of 1964 the QDG moved from BAOR to Omagh in County Tyrone to take over from 2 RTR. The QDG Air Squadron was established at Aldergrove but in January 1965 had to detach one Auster for operations in Borneo leaving a single Auster and four Skeeters. The Air Squadron was initially commanded by Major Jeremy Railton QDG who was succeeded by Major Francis Chamberlain AAC. The aircrew comprised Sergeants Ken Kendall QDG and Martin Forde QDG and Lieutenant Mike Richards QDG, who recalled the flying they carried out from Aldergrove.

> The facilities were good if somewhat bleak. The airfield was famous for its huge army of hares which grazed happily accompanied by vast flocks of seagulls… Tasking continued in support of the Regiment where liaison sorties were interspersed with Squadron and Regimental training deployments. Aircraft were periodically detached to support recruiting tours in Ulster, Wales and the Borders. We were tasked frequently by HQ Northern Ireland at nearby Lisburn. This led to work with the UK Strategic Reserve airlifted to the Province and the Ulster TA Brigade on deployment to Otterburn. We flew in the beautiful Mourne Mountains and ferried the Squadron out to Rathlin Island where we helped a Belfast University team chart the forest of wrecks in the tortuous surrounding waters. QDG were the only permanent Army Aviation resource in the theatre and we were kept busy hosting visiting RAF and AAC aircraft as well as supporting the police and emergency services .... Martin Forde provided a bit of drama with a nasty wire strike from which he thankfully emerged unscathed while Ken Kendall had an engine failure near Carlisle. Not to be outdone, I had a tree strike so the repair and recovery teams were not idle.[13]

1966 saw the fiftieth anniversary of the Easter Rising in Dublin and the authorities in Northern Ireland were expecting trouble from the republican community to mark the event. In January of that year the Skeeters were modified to ensure they could work effectively with the RUC.

> During the month our aircraft were fitted with police 'Bantam' radio sets, enabling us to operate on the RUC network. This was mainly in anticipation of possible riots and troubles later in the year, but the pilots and their passengers derived hours of interesting and often amusing in flight entertainment from listening in on those sets .... April brought the 50th anniversary of the Easter Uprising ('so it did') but, in fact, nothing came of it as far as we were concerned. We flew various Ulster Constabulary officers and observers over the parades and processions and at one point a Skeeter was detailed to hover above [the Reverend Ian] Paisley's rostrum, the bellow of its exhaust effectively drowning his seditious ranting.[14]

Towards the end of 1966 the QDG Air Squadron had its Skeeters replaced by the Westland Sioux before they left Northern Ireland to move to Aden that November. They were replaced by the Air Squadron of 4/7 RDG, who made the reciprocal move from Aden. They arrived in the province during a relatively benign period, a lull before the storm. 4/7 RDG's tour in Northern Ireland was a long round of exercises across the UK. Meanwhile the IRA was regrouping and rethinking its future strategy.[15]

13 M. Richards, account to the authors, 25 October 2020.

14 Anon, 'Air Squadron', *The Journal of the Queen's Dragoon Guards*, 1966, pp.626–627.

15 Anon, 'Air Squadron 4th/7th Royal Dragoon Guards', *Journal of Army Aviation*, 1968, No.3, pp.82–83 and Anon, 'Air Squadron 4th/7th Royal Dragoon Guards', *Journal of Army Aviation*, 1969, No.4, p.90.

Following the failure of the Border Campaign there was recognition that it could not succeed unless it was representing the genuine grievances of the Republicans and Catholics in Northern Ireland. The leadership came up with a strategy of agitation within the Catholic population that could then be supported by the Republican movement and protected by the IRA. On 13 August 1966 an umbrella organisation known as the Northern Ireland Civil Rights Association (NICRA) was formed, which informally encompassed a number of shady republican, socialist and Marxist groups. On the surface NICRA was representing the legitimate grievances of the Catholic minority, which included discrimination by the Protestant dominated authorities, particularly in areas such as housing and employment. NICRA began to organise a number of peaceful protests across Northern Ireland based on the example of the civil rights movement in the USA. On 5 October 1968 a peaceful protest in Londonderry was baton-charged by the RUC and the resulting television footage was broadcast around the world. On 1 January 1969 NICRA and other allied organisations began a deliberately provocative march from Belfast to Londonderry. As the marchers approached their destination, they were ambushed near Burntollet Bridge by an angry Protestant mob, while the RUC failed to stop the ensuing violence. There were calls for the IRA to rearm and protect NICRA, but they were denied by the leadership in the Republic. The violence across the sectarian divide began to build its own momentum and reached a crescendo during the Protestant Orange Order's marching season in July 1969. On 12 August a march by the Apprentice Boys of Derry resulted in sustained violence that became known as the Battle of the Bogside. At the same time RUC stations in Belfast were attacked and set alight. Shots were fired by the RUC in both locations and innocent Catholics were killed. On 14 August, with the RUC fully committed, the British Prime Minister, Harold Wilson gave consent for troops to be deployed to attempt to restore law and order. Later that day a company of the 1st Battalion Prince of Wales's Own Regiment of Yorkshire (1 PWO) moved into Londonderry; the first unit to deploy onto the streets of Northern Ireland in a campaign, known as Operation Banner that would last nearly 38 years.[16]

'We are in Great Britain, aren't we?'

In March 1969 4/7 RDG Air Squadron left Northern Ireland for BAOR and were disbanded under the AAC recentralisation programme. They were replaced in Aldergrove by the Air Squadron of the 17th/21st Lancers (17/21 L) commanded by Captain Reggie Purbrick 17/21 L. Soon after the Lancers' arrival six bomb attacks were carried out to sabotage electricity and water installations, causing considerable disruption. It was judged that the actions of 17/21 L Air Squadron helped to curtail further attempt. 'Through speed, surprise and its ability to cover large areas in a short space of time the troublemakers could never be certain of success when an aircraft might arrive at any time of the day or night; there is little doubt that the use of helicopters in this particular role prevented any further serious sabotage.'[17]

The deterrent effect of having a helicopter in the air would become an enduring feature throughout Operation Banner. At the time it was presumed that the 'troublemakers' were members of the IRA, but it later emerged that the bombings had been carried out by the loyalist UPV (Ulster Protestant Volunteers) and UVF (Ulster Volunteer Force) in an attempt to incriminate the republican movement. The late spring was relatively quiet for 17/21 L, but when the marching

16 Geraghty, *The Irish War*, pp.3–21. For a full assessment of Operation Banner see Anon, *Operation Banner; An Analysis of Military Operations in Northern Ireland* (London: MOD, 2006).

17 Anon, 'Army Aviation in Northern Ireland', *Army Air Corps Newsletter*, No.5, June 1970, p.4.

season began in July their aircraft were constantly employed ferrying Army commanders between Belfast and Londonderry.

1 PWO's Air Platoon had also moved to Ulster in the spring of 1969, based initially in Ballykinler but moving to Ballykelly when the Battalion deployed to Londonderry.[18] The Air Platoon returned to the mainland in August, however, leaving the Lancers as the only Army aviation asset in Northern Ireland as Operation Banner began. They were soon joined by other units from the mainland as, on 16 August 1969, units from 24 Inf Bde were sent across the Irish Sea. 8 Flight AAC, HQ 24 Inf Bde's affiliated flight, was put on standby and moved its aircraft from its home in Plymouth to Prestwick. On the same evening they flew their Scouts and Sioux over to base themselves at RAF Ballykelly, on the southern bank of Lough Foyle in County Londonderry. RAF Ballykelly had the strange distinction of the main Londonderry to Belfast train-line cutting across one of the runways, which caused more issues for the RAF's Avro Shackletons than it would for the AAC's helicopters. The Flight CP vehicle was flown to Ballykelly by RAF C130 Hercules, arriving in time to provide immediate control and briefing facilities for the aircrew. The rest of the vehicles moved by Landing Ship Logistic from Southampton to Londonderry. The aircraft were immediately employed flying commanders for reconnaissance of the border area, conducting border patrols and liaison sorties. To make more efficient use of border flying hours, one Sioux and one Scout were deployed forward to support the 17/21 L from Lisanelly Barracks in Omagh along with a CP vehicle.

Soon after arrival, in October 1969 8 Flight was reorganised as part of the AAC centralisation programme and became 666 Aviation Squadron under the command of Major Adrian Woodford RA. The change from flight to squadron, along with the move to Ballykelly caused considerable confusion as recalled by Corporal Bob Harrod RTR.

> My regiment, 3 RTR, moved from BAOR to Tidworth towards the end of 1968. During the summer of 1969 I was informed that I would be going to 8 Flight, Army Air Corps as their Squadron clerk, promoted to full Corporal and the icing on the cake, they were apparently based in Coypool, Marsh Mills, Plymouth. My home town! Imagine my surprise when I arrived at Coypool in November, 1969 to find their accommodation empty because they had been in Northern Ireland since August when the Troubles first blew up. Furthermore, they were no longer 8 Flight AAC, but were now known as 666 Aviation Squadron. This was my first experience of what, in those days, was apparently 'situation normal' in the Army Air Corps at the time, the left hand not knowing what the right hand was doing…
>
> It also turned out that as the former Flight was now a Squadron, the establishment had changed in that they required a Sergeant clerk rather than a Corporal. Having been recently promoted to full Corporal, further promotion was of course out of the question. Also out of the question seemed to be admitting to the mistake, sending me back and asking for a Sergeant. The solution was to promote me to 'acting Sergeant as a temporary measure for the purposes of pay and allowances'. This meant I was paid as a Sergeant but continued to wear the rank of Corporal. Problem solved, at least in the short term, more about the long-term solution later! The change from a Flight to a Squadron seemed to confuse people at all levels. I remember receiving a signal from the MOD replying to a previous signal sent to them. It seems that they, the MOD, had asked who 666 Squadron was, and the person replying had misread the signal as asking where 666 Squadron was. The signal read, "Not where, what is 666 Sqn?" This signal clearly tickled a few people as it took up permanent residence on the Squadron notice board![19]

18 Anon, 'The Air Platoon, The Prince of Wales' Own Regiment of Yorkshire', *Journal of Army Aviation*, 1970, p.89.

19 R. Harrod, account to the authors, 20 February 2021.

The Squadron was reinforced with men and equipment, and it split into two mixed flights of Scout and Sioux. One flight was made responsible for supporting operational requirements while the other was left free to carry out training, as long as it was also able to support operations with one hour's notice. The Squadron was tasked by G3 Air at the Bde HQ, although the Omagh detachment was under the operational control of 17/21 L. 666 Squadron quickly developed standard operating procedures, along with 72 Squadron RAF based in Aldergrove, to create an airborne QRF capable of delivering troops anywhere in the Province at short notice. A Sioux acted as the command and control aircraft with a Scout carrying an infantry platoon or company HQ. 72 Squadron's Wessex acted as troop carriers. The infantry commander in the Sioux, with a bird's eye view of the situation, could give orders and direct his troops over the UHF radio which was being monitored by his subordinate commanders in the Scout and Wessex. This again proved to be an effective tactic that would be developed over the following years.

To ease command and control Northern Ireland was essentially split in two for aviation purposes, roughly on the line of the River Bann. 666 Squadron took responsibility for operations west of the Bann, including the counties of Londonderry, Armagh, Tyrone and Fermanagh. 17/21 L Air Squadron operated east of the line in Antrim and Down, including Belfast. As Operation Banner began 17/21 L cancelled all leave and the Air Squadron doubled its flying effort.

> In August life began to hot up quite considerably. The troubles in Londonderry saw the flying of VIPs and two Brigade Commanders up to Derry at all times of the day and night, and then Belfast "blew up," almost literally, as well. The Squadron had a very hectic few months, running a permanent detachment of 1 hel[icopter] at the Bde HQ in Belfast, the aircraft sitting on the flat roof of a police station, and on occasions a second detachment of 1 hel[icopter] at Main Bde HQ at Lisburn. From these two places patrolling was carried out over Belfast by day and night to spot crowds, barricades, fires, snipers and traffic jams caused by the troubles.
>
> In the late summer we were given Sky Shout, an airborne loud speaker system, to play with and on occasions this has been used to address crowds from the air. Sky Shout was the first of several pieces of kit we have been given to try. We were asked to do a lot of photography, and we have now got three different cameras...and an M10 Polaroid which have all proved very good and the Polaroid great fun as well! Our latest piece of kit has been Night Sun, an extremely powerful searchlight with which we hope to illuminate any crowds or any suspicious movement at night.[20]

To support and maintain flying operations across Northern Ireland 666 Squadron's Light Aid Detachment moved to a shift system to provide servicing and repair 24 hours a day, with technicians working in shifts for at least 18 hours in every 36. Additionally, 701 Mobile Servicing and Repair Detachment deployed from 70 Aircraft Workshops REME at Middle Wallop to reinforce the first line and provide second line servicing and repair. As a result, a very high level of availability was maintained across the Scout and Sioux fleets.[21]

666 Aviation Squadron departed Ulster at the end of January 1970 to be replaced by 660 Aviation Squadron from Topcliffe. The initial civil unrest of the summer of 1969 had ebbed to sporadic rioting and public disorder in the cities with rural areas remaining reasonably quiet. Many members of the IRA in Northern Ireland, however, unhappy with the support they had received from Dublin the previous summer, created a splinter group, the Provisional IRA (PIRA) with a far more militant agenda. At the end of March widespread violence returned to Belfast with

20 Anon, 'Air Squadron 17th/21st Lancers', *Journal of Army Aviation*, 1970, p.98.
21 Anon, 'Army Aviation in Northern Ireland', pp.2–5.

666 Squadron Sioux XW187 on patrol over Belfast docks being flown by Sergeant Slade and Corporal McLachlan. (Army Flying Museum)

the Easter marches. Barricades were erected around the Shankhill and Ballymurphy estates that were cleared by infantry units and then removed by the RE. Further violence erupted in June as the Orange Order began its marches. On the night of 27/28 June six people died and 61, including three soldiers, were injured by paramilitary gunfire. Riots were used as cover for gunmen to shoot across the sectarian divide or at the security forces. The Army estimated that nearly 2,000 rounds were fired during the night; the security forces responded with over 1,600 CS gas canisters.

At the height of the violence 663 Aviation Squadron was given 24 hours' notice to move to Northern Ireland. The order was received by the squadron commander at home in Netheravon who then had the difficult task of corralling his troops. Sioux Flight was close by at Perham Down, but Scout Flight had recently returned from Exercises Hardfall and Arctic Express in Norway and was at that time on exercise in the Orkneys. Nevertheless, the entire Squadron was on its way across the Irish Sea on 29 June 1970 and, having relieved 660 Squadron, were operational by 1 July. 663 Squadron was based at Long Kesh, a former RAF airfield approximately 10 miles southwest of Belfast city centre. The conditions at the base were austere, particularly as the Squadron's four-week emergency tour was extended to four months. The aircraft were engaged on border patrols and supporting cordon and roadblock operations on the ground with aircrew flying an average of 60 hours per month.[22]

22 Anon, '663 Aviation Squadron', *Journal of Army Aviation,* 1971, pp.82–84.

The billets and hangars on the eastern edge of the former RAF Long Kesh seen from the air in 1972. The Nissen huts of the internment camp are towards the top of the image. (Army Flying Museum)

The 17/21 L Air Squadron was also heavily involved in operations to counter unrest in July 1971. In April Captain Purbrick had handed over command to Major Francis Gradidge 17/21 L and the Squadron acquired a further four Sioux giving them a total of ten. The original six aircraft remained at Aldergrove with the extra four forming a detachment at Ballykelly to assist 8th Infantry Brigade (8 Inf Bde) who had by then taken over responsibility for the Province west of the River Bann, with 39 Inf Bde looking after the east. The extra aircraft were soon involved in the thick of operations.

> July 5th saw one of the biggest ever operations in Belfast with the Battle of the Lower Falls. We also had our first taste of shooting when Capt Simon Arthur and 39 Bde Commander, Brig [Peter] Hudson, made a precautionary landing after hearing bullets rattling past the aircraft. The former was rather disappointed in being unable to actually find any bullet holes...
>
> As part of the build-up for July 12th and the celebrations to commemorate the Battle of the Boyne, we were joined by 663 Aviation Sqn who were stationed at Long Kesh. They busied themselves with most of the county areas and assisted the [armoured car] troop in Omagh, whilst those of us at Aldergrove concentrated on the Belfast area.[23]

It was a period of trial and flux. Commando air troops also operated in Northern Ireland including Montforterbeek Flight with 45 Cdo RM and Salerno Flight with 41 Cdo RM. A solitary Beaver, flown by WO2 Clifford Taylor RHA, was sent to Aldergrove for a short trial that included border patrols and flare dropping to assist patrols at night. In October 653 Aviation Squadron took over from 663 Squadron and four of its Scouts were detached to 17/21 L to give its Air Squadron more utility.

The sortie conducted by Captain Arthur with Brigadier Hudson had far reaching consequences. From the air the Brigade Commander had seen barricades being erected in the Lower Falls and ordered a large-scale cordon and search operation of the area involving four infantry battalions. It was later suggested that the Brigadier's decision making had also been influenced by the shots fired at his Sioux. The cordon and search operation effectively imposed an overnight curfew and produced 107 weapons, over 25lbs of explosives and 21,000 rounds of ammunition. The Army, however, fired over 1,400 rounds of live ammunition resulting in the deaths of four civilians with a further 68 being injured. The Army was widely perceived to have been indiscriminate in the manner it executed the operation and what was later referred to as 'the Rape of the Falls' handed a propaganda victory to the republicans as the sympathy of many Catholics shifted from the security forces to the PIRA. The situation deteriorated as PIRA began to target the Army while able to rely on support and shelter from many of sympathisers within the Catholic community. On 6 February 1971 Gunner Robert Curtis RA was the first British soldier killed in what became known as 'the Troubles'. PIRA employed incendiary attacks and directional anti-personnel devices (improvised Claymore mines) to target British Army patrols. The Troubles had moved from civil unrest to open insurgency and helicopters became a prime target for attack.

Although no rounds had hit the Sioux, the shooting incident involving Captain Arthur and Brigadier Hudson resulting in a precautionary landing had demonstrated how vulnerable aircraft and their crews and passengers could be if deliberately targeted. Evidence of the intent of PIRA to bring down a helicopter was found in their publicly available 'manual' *The Irish Guerrilla*, entitled 'Helicopters and their Destruction'.

23 Anon, 'Air Squadron, 17th/21st Lancers', *Journal of Army Aviation,* 1971, pp.94–95.

> The "border" of Ireland is being more and more brought under the surveillance of the occupation army using helicopters. These vehicles are vulnerable to the shots of a skilled marksman. In the Vietnamese liberation over 800 helicopters have been disposed of by guerrillas since 1967. If we can achieve a rate of destruction equal to only one quarter of this, the presence of these pirate spy ships over our country will be removed, in a few months. The point of attack for your bullets on a helicopter is the rotor arm. One strike on this will disable the propellers. Firing through the Perspex visor can kill the pilot, but shots at the belly of the craft are likely to be deflected by sheet steel fitted for that purpose. So make your mark the rotor arm. Shots aimed at this will, if they miss, go unheard by the pilot. The one he does hear will be the one that will force his craft back to earth. When his craft does crash a small PE charge set in the engine space will put it out of action for good.[24]

The aircrew were given some degree of protection when the cabin floor and seats of the Sioux were 'armoured' using Makrolon, reinforced polycarbonate sheets, widely reported in the press as 'moon armour' due to its use in the Apollo programme.[25] 17/21 L also trialled fitting a crew operated GPMG to the Sioux. It was mounted on the starboard side of the aircraft on a rig secured to the skid and cabin floor. It was demonstrated to the press on the sand dunes of Magilligan Point in an aircraft flown by Captain 'Dickie' Whidborne AAC with Lance Corporal Dillon acting as the gunner. The purpose of publicising the demonstration was to act as a deterrent to further PIRA attacks on helicopters, nevertheless the Squadron Commander, Major Gradidge, was at pains to explain to the assembled press that, 'As a rule we would fire only at someone who was shooting at us first, and only if we had a clearly defined target.'[26] Despite the reassurance there was a great deal of political reluctance to arm helicopters in Northern Ireland and the parallels the republican movement might make with American gunships in Vietnam, hence it would be another two decades before AAC aircraft routinely carried door gunners in Northern Ireland.

While 17/21 L Air Squadron provided a degree of continuity from their new home in Lisanelly Barracks, Omagh the AAC aviation squadrons began to settle into a regular rotation, or roulement, to reinforce the military effort. Initially, through 1970 and 1971 all the squadrons involved came from the UK and were based in Aldergrove for a roughly four-month tour. 663 Squadron was relieved in October 1970 by 653 Aviation Squadron. They were followed in February 1971 by 664 Parachute Aviation Squadron who were relieved in turn by 665 Aviation Squadron. It was during 665 Squadron's tour that the British and Northern Ireland governments took decisive action in an attempt to quell the republican insurgency. On 9 August 1971 the security forces launched Operation Demetrius, an effort to prevent known and suspected PIRA insurgents from being able to operate. 342 people were arrested and, once they had been put through a sorting process, 238 were interned. The operation backfired, however. Faulty intelligence meant that many of those arrested were innocent Catholic civilians with no links to the republican movement while several senior PIRA members were tipped off and escaped the net. There were also complaints that no members of loyalist militant groups, such as the UVF were arrested despite the organisation beginning its own bombing and shooting campaign against the Catholic population. Demetrius provoked widespread civil unrest and rioting, vehicles were hijacked, factories and homes were set on fire, burning barricades were erected in Belfast and Londonderry and PIRA took control of areas of the cities. Twenty-four people were killed (11 of them civilians in the Ballymurphy estate in Belfast) including two members of the British Army and two PIRA members. The operation

24 L. A. Palmer, 'Six Months in the Six Counties', *Journal of Army Aviation*, 1972, p.24.
25 AFM ALB 16: various press cuttings, 5 August 1971.
26 *The Telegraph*, 7 April 1971.

and the resulting internment policy, which endured until the end of 1975, proved effective in bringing recruits to the republican cause.

PIRA stepped up its campaign against the security forces including attempts to bring down a helicopter with three aircraft from 17/21 L Air Squadron engaged inside a month. On 15 September 1971 a journalist was flying in a Sioux being piloted by Captain Whidborne on sortie to check a burnt-out post office van that had been hijacked near the border at Clogher, midway between Enniskillen and Dungannon, earlier in the day. During the sortie Captain Whidbourne was called on the radio by an armoured car from 17/21 L to inform him he had been shot at by automatic fire from hidden gunmen on the other side of the border. Neither Captain Whidborne nor the journalist heard or saw any sign of the engagement and the aircraft was not hit.[27] A few weeks later, on 9 October a team of four PIRA gunmen approached a customs post near Mullan in County Monaghan in the Republic. The gunmen fired a shot, forced the customs officers outside, wired the building with explosives and blew it up. They then stole one of the officers' cars to make their getaway. Sergeant Philip Tonks 14th/20th King's Hussars was soon on the scene in his Sioux XT240 and decided to follow the gunmen's most likely escape route. He found the car, which appeared to have been abandoned in the mountains. As he circled the car Sergeant Tonks reported, 'There was a pinging noise as we moved near the car and we pulled out of the area fast. We were told from the ground that we had been shot at.' Sergeant Tonks flew back to Omagh where, on inspection it was found that one bullet had entered the Sioux's cabin and hit the radio.[28] Less than a week later, on 14 October Captain Tim Holmes QDG was flying a patrol along the County Tyrone Donegal border near Castlederg, 13 miles northwest of Omagh, when he was engaged by three gunmen in the Republic.

> I heard a clattering noise – three bursts – and at first I thought it was the noise of the rotor blades. I then became suspicious and decided to land and have a look. I climbed and flew five miles and landed. When I was flying off I was told over my radio that gunfire had been reported in the area and this confirmed my suspicions. When I landed I inspected the aircraft and found a bullet lodged in the block.

It was thought that about 20 shots had been fired at Captain Holmes with two rounds hitting the aircraft, one lodging in the engine casing just behind his seat. Following a REME inspection in the field the Sioux was cleared to fly back to Omagh.[29]

17/21 L no doubt breathed a collective sigh of relief when they moved to Germany shortly after the spate of helicopter engagements to be replaced by the 16th/5th Queen's Royal Lancers (16/5 L). With the escalating violence in Northern Ireland a second armoured regiment was deployed on Operation Banner and the recently amalgamated Scots Dragoon Guards (Scots DG) moved to Ulster in October 1971 along with its Air Squadron. The Squadron's six Sioux, commanded by Major Nick Muers-Raby Royal Hussars (RH) flew from their base at Harewood Barracks, Herford in Germany on 15 October 1971, arriving at Long Kesh five days later as the first BAOR aviation unit to serve in Northern Ireland.

Long Kesh then housed one of the internment camps and as such some improvements had been made to the accommodation and facilities since 663 Squadron had been there the previous year. The internees were accommodated in Nissen huts within a double, high wire fence. For those with more freedom of movement a village of new portacabins had been installed. Regimental

27 *The Guardian*, 15 September 1971.
28 AFM ALB 16: 'Accident message signal', 12 October 1971 and unattributed press cutting, undated.
29 *The Tyrone Constitution*, 22 October 1971.

A Sioux from the Air Troop 14th/20th King's Hussars working with ground troops during rural operations in 1974. (Army Flying Museum)

Headquarters Scots DG was located in Bessbrook, County Armagh, a location to soon become familiar to generations of military aircrew and ground crew, while two of its armoured squadrons were at Gosford Castle, 6 miles southeast of the city of Armagh. Sergeant Howard Gater-Smith RTR deployed as a pilot with the Air Squadron and recalled their new area of responsibility and the hazards it presented to those on the ground and in the air.

> We were operating in South Armagh, roughly below the line Caledon – Armagh – Castlewellan. This area of real estate was sometimes referred to by the British press as 'bandit country'. The landscape is of outstanding natural beauty and dotted with lush green gently rolling meadows, mostly surrounded by stone walls or hedgerows. This patch-work is interspersed with narrow lanes, a haven for IRA terrorists. Some of the border crossings had been cratered to prevent 'inter-state' movement. Others were merely marked by a white cross painted in the middle of the road...Many incidents involved cross-border engagements and culvert bombs (IEDs). The latter were often constructed in milk churns and activated through command wires of several hundred yards in length. The Regiment's ground element spent much time patrolling in the reconnaissance vehicles of the day, Saladin armoured cars and Ferret scout cars. Mobile vehicle check points were operated and control and inspection of cratered border road crossing points took place daily. Where ground activity was on-going the Regiment's aircraft were always near to hand...
>
> The IRA perfected the use of so-called 'culvert bombs'. Numerous small road bridges existed throughout the rural areas, some being constructed over minute tributaries or drainage ditches. They became favourite places to site these explosive devices, most of which were

> activated by long command wires, lengths of telephone cable through which an electrical charge was transmitted. On approaching a bridge, a patrol would halt whilst a couple of dismounted soldiers inspected the object before the commander allowed further movement. For whatever the reason, somewhere near to Poyntz Pass one morning an armoured pig, a four wheeled armoured van, drove over a small bridge without the crew carrying out the checking procedure. A resounding boom blasted the vehicle into the air, turning it through one hundred and eighty degrees. One soldier received a vicious cut to his upper lip, splitting it up to the base of his nose. It was caused by his face connecting with a rifle barrel, the foresight smacking him full in the chops. He must have suffered excruciating pain and a 'medevac' was requested. As the nearest helicopter to the incident I quickly picked him up and started off in the direction of Belfast hospital. I must say he looked in a sorry state with an enormous wound dressing wound around the lower half of his face. I contacted our forward base at Gosford Castle to inform them of my intentions. Our Regimental Medical Officer, Dr Marcus Smith, quickly came over the radio and suggested that a trip to Belfast would be an overkill and that I should ferry the man to the medical staff at Gosford. "I'll sew him up again", was his cheery remark!
>
> The Sioux's engine was fitted with a turbo-charger. This device rarely gave any trouble but chose to malfunction on an early morning flight over the Mourne Mountains near to Castlewellan [15 miles northeast of Newry]. We had not long been airborne and therefore close to the aircraft's maximum permitted All-Up Weight. Without the turbo charger's assistance there was insufficient power available to keep us afloat and therefore we committed to a forced landing. I selected a flattish looking meadow, bordered by a road a short distance from a farm house and carried out a limited powered approach and landing. There we sat for a quarter of an hour or so awaiting help. I have to say, we were pretty poorly placed for a hostile engagement. Although the field was surrounded by a stone wall our defensive armament consisted of two Browning pistols and twenty-four rounds of ammunition between us. Anxiety was soon to be dispelled, because on our way down my observer, Jim McGregor, had transmitted a distress call to another aircraft in the vicinity. Within a short space of time, and in true fashion, the cavalry came to the rescue. A Scots DG Sabre Troop charged down the lane and surrounded the field in all-round defence, Browning machine guns pointing outwards. Saved! Sometime later a REME recovery team appeared on the scene, removed the rotor blades, and our injured Sioux was airlifted out by Wessex helicopter and transported to the aircraft workshop at Aldergrove.[30]

As 1971 drew to an end, aviation squadrons began to return for their second Operation Banner tour in the Province. In October 665 Aviation Squadron was replaced by 666 Squadron on its second tour, who were then relieved in turn by 663 Aviation Squadron also on its second tour. In February 1972 more AAC aviation squadrons from BAOR were deployed to Northern Ireland to reinforce the effort. 651 Aviation Squadron provided a scout flight of five aircraft to Aldergrove to work alongside a Sioux detachment at Long Kesh from 657 Aviation Squadron. Major L. A. 'Arnold' Palmer R Anglian, OC 651 Squadron, summed up some of the tactical steps that had taken place on the ground and in the air up until his Squadron's arrival.

> Under present arrangements it is not possible thoroughly to check each of the 10,000 vehicles and 50,000 people who daily cross the border. There are 20 approved crossings, with customs facilities (frequently bombed or burnt) and some 130 unapproved crossings. It was decided in

30 H. Gater-Smith, previously written unpublished account to the authors, 20 November 2020.

> October to close these unapproved crossings. This was a political decision but a military task, and one which involved an extremely heavy flying effort for reconnaissance, photographic and passenger sorties, and for positioning of RE stores and personnel. Although all have been blocked at least once the operation continues. In some places local people from both the South and the North are kept busy unblocking them and the Sappers kept busy re-blocking the road. One Sioux pilot reported that 60 minutes before one road was first cratered a lorry load of labourers and a republican County Council bulldozer were positioned just South of the border ready to fill in the crater when it was blown. It was while investigating one of the blown crossings that one Sioux pilot appreciated that tactical flying is not necessarily low flying. He observed from behind low cover and his aircraft was hit by a low velocity bullet as he turned away. He has now agreed that in Ulster high is good and low is bad. On numerous occasions helicopters have been shot at. The defensive action is to fly as high as possible compatible with the task, and to make judicious use of the wind. Varying ground speeds and drift make accurate shooting by fairly poor shots extremely difficult. There have been a total of three hits on helicopters in Ulster in the last six months. On no occasion was the aircraft forced to land nor was serious damage caused…
>
> Flying intensity has been high. In the six months up to the end of January 1972 Army aircraft flew almost 7000 hours in Ulster. Serviceability has been steady at about 90%. It is interesting to note that pilot fatigue has not been a real problem. This intensity achieved in the winter months, is less of a flight safety hazard than the same figure would be in the summer months. In the winter the hours of darkness permit a full night's sleep each night and so allow a sustained flying effort each day. Perhaps we should look again at the possible benefits of that maligned expression "crew duty time."[31]

651 and 657 Squadrons arrived in Northern Ireland at a particularly volatile period of the Troubles. On 13 January 1972 a Sioux from Kangaw Flight of 3 CBAS was flying above Londonderry when a major firefight broke out between Catholic gunmen and 1st Battalion Coldstream Guards in the Creggan district in the southwest of the city. Shots were also fired at the aircraft and when it returned to Ballykelly at the end of the sortie a subsequent inspection revealed a single round had gone through the synchronised elevator on the tail boom.[32] On 30 January 1972 NICRA organised a demonstration in Londonderry to protest against internment. As the protestors marched through the city they were forced to re-route due to British Army barricades. Some of the marchers broke away from the main body of the protest and began to throw stones at the soldiers from 1 Para who were manning the barricades. The paratroopers responded with small arms fire and the protest broke down in confusion with some rioters responding to the actions of 1 Para and peaceful marchers trying to escape the violence. The paratroopers continued to answer the rioters with live rounds and in a short time 26 people were shot. Thirteen of those hit were killed at the scene and a fourteenth died in hospital four months later. The incident, one of the most notorious of the Troubles, became known as 'Bloody Sunday' and cast a long shadow that still endures to this day. Three days later when the majority of those killed during Bloody Sunday were buried there was a general strike in the Republic of Ireland and irate crowds burnt down the British embassy in Dublin. Anglo-Irish relations reached a nadir as the Irish Minister for Foreign Affairs went to the United Nations Security Council in New York to demand the involvement of a UN peacekeeping force in the Northern Ireland conflict. Bloody Sunday fuelled Catholic and Republican hostility towards the British Army in the north and support for PIRA, with a corresponding surge

31 Palmer, 'Six Months in the Six Counties', pp.22–24.
32 S. Taylor, *Air War Northern Ireland*, p.36.

of recruitment, particularly in County Londonderry. Exactly a week after Bloody Sunday Sergeant Gater-Smith was on a sortie above another protest march which had attracted some prominent support.

> Trying to obtain photographic evidence of the actress Vanessa Redgrave from a height of a thousand feet proved nigh on impossible. On 6 February 1972 she, together with other 'dignitaries' was leading the 'Newry March' which we had been tasked to monitor. The demonstration had been deemed illegal and, to make the point, the RUC police officer flying as a passenger in a second helicopter tried in vain to put his message across with the aid of 'Sky Shout'. This specialist piece of equipment consisted of a large loudspeaker fitted to the exterior of the aircraft and wired to the intercom system. His efforts to get the demonstrators to disperse failed, mainly due to the fact that his instructions could not be understood. Eventually the pilot from another aviation squadron, David Morley, received an order from the ground to put someone on who could speak clear English. Alas, his own cultured tones, although understood, were completely ignored and the march continued into the centre of Newry.[33]

Following Bloody Sunday, the Republican insurgency intensified, and the security situation deteriorated further. In May, June and July 1972 there were over 5,000 shooting incidents in the Province. During that period 664 Aviation Squadron and the wider aviation community experienced the full effect of PIRA's campaign. The incident was described by Second Lieutenant David Marfleet AAC of 664 Squadron at the time.

> On 24 June 1972, Sergeant Dave Whitney Para, (ex-Red Devils free-fall team) was flying a Sioux back to Aldergrove from Londonderry when he experienced a turbocharger failure in the Glenshane Pass, a major route cutting through the Sperrin Mountains. Having put out a 'Mayday' call, he landed in a field alongside the A6 road. He then spent a very cold and lonely hour hiding in a ditch with his 9mm pistol drawn, while awaiting a platoon of infantry to arrive to secure the scene. In the meantime, HQ Northern Ireland made the fateful mistake of organising a road recovery of the helicopter. A flat-bed Bedford RL and two Landrovers [*sic*] with a REME FRT and an escort of Paratroopers from 664 Squadron arrived at the scene late in the afternoon. The Bedford truck got stuck in the mud but eventually the Sioux was loaded and the convoy set off towards Aldergrove. However, the recovery had not gone unnoticed by the local IRA unit and half a mile down the road a massive bomb consisting of 120lbs of Semtex in two milk churns destroyed the leading Landrover.
>
> Sgt Whitney had been asleep in the back of the vehicle with his flak jacket over his head. He woke up in a field, his flak jacket in a tree, and his left foot badly mangled. The other soldiers in the vehicle were all killed .... Dave Whitney recovered from his injuries and returned to flying duties with a metal plate in his foot. He used to joke about having to compass swing the aircraft just for him.[34]

Sergeant Stuart Reid REME, Lance Corporal David Moon REME and Private Christopher Stevenson Para were all killed in the attack.[35] The deaths of the three members of 664 Squadron were part of an increasingly grim total. In July 1972 alone there were 95 terrorist-related deaths

---

33 H. Gater-Smith, account to the authors 20 November 2020.

34 D. Marfleet, account to the authors, 6 June 2022.

35 Anon, 'Obituaries', *Journal of Army Aviation*, 1973, p.43.

culminating in what became known as 'Bloody Friday' on 21 July. Twenty bombs exploded across Belfast in the space of 80 minutes. Nine people were killed: five civilians, two British soldiers, a RUC reservist and a member of the loyalist paramilitary UDA (Ulster Defence Association). A further 130 were injured including 77 women and children. Bloody Friday proved an own goal on several levels for PIRA. The degree of public revulsion caused many Catholics to turn away from supporting the Republican movement both in Northern Ireland and the Republic and it gave the security forces the impetus to clamp down on the insurgency and the loyalist response across Ulster.

The Army was reinforced by a brigade HQ and seven infantry battalions under Operation Glasscutter. On 28 July civil communities were warned of the security forces' intentions and given an opportunity to dismantle the barricades that had been erected by Republican and Loyalist groups in Belfast and Londonderry. At 0400 hours on Monday 31 July HQNI launched Operation Motorman as an outer cordon was set up around both cities. The Bogside and Creggan estates in Londonderry, and Andersonstown and the Ballymurphy in Belfast were sealed off. Thousands of troops with armoured vehicles moved in to clear the barricades. All areas had been secured and cleared by 0700 hours with no security force casualties. It was a turning point in Operation Banner. PIRA was in disarray following Motorman as the Army kept up the pressure and the high concentration of troop numbers reached operational peak at over 28,000 during the year. With the increase in infantry on the ground came calls for additional aviation support. By the end of 1972 651 Aviation Squadron was on a tour at Lisanelly Barracks in Omagh, 654 Squadron was at Long Kesh, 658 Squadron at Ballykelly and 665 Squadron was at Aldergrove having recently taken over from 653 Squadron. In addition, 16/5 L and the Scots DG's Air Squadrons operated from Omagh and Long Kesh respectively and there was a detachment from 3 CBAS at Ballykelly. There was often little time for training, preparation or familiarisation before a squadron was plunged into internal security operations. Lieutenant Roger Sheldrake RA described the strange feeling of dislocation experienced by thousands of soldiers during the Troubles.

> We are in Great Britain, aren't we? So why do all the roads have funny lumps on them, particularly outside police stations, and all the PCs carry SMGs [submachine guns]. The vehicles are covered in atrophied cardboard and some of them have very odd looking dents. 'We are in Great Britain aren't we'? 'You certainly are my lad, and you're here to fly, so draw a flak jacket, a combat dressing, a 9mm pistol and ten rounds and go and have a look around.' 'But surely this country hasn't killed all these people'? It's about the size of Yorkshire and a beautiful land. From the Sperrin Mountains to the Giant's Causeway, from Mountains of Mourne to the lakes of Enniskillen, it is the Emerald Isle. Small bright green fields with the farmer waving at you from beside his farmhouse. The names are wonderful too – the market town of Tondragee, the cathedral at Armagh, the tongue twisters of Mullaghclogha, Donaghcloney, Donaghadee. This is a country steeped in history – from the Gaelic Celts to the Rebellion of '98, from Wolfe Tone to the Anglo Irish Treaty. Its religious and political history is a fascinating tapestry which has many threads, all too many of which are so easy to loosen. It is a land of gentle fairy tales and violent tempers, of machine guns and the Blarney Stone, of Pope and Orange Lodge.[36]

At the end of 1972 the brunt of the roulement effort was being taken by squadrons from BAOR, often with reduced notice before an Operation Banner tour. With little time for specific

36 Anon, 'A Squadron in Ulster', *Journal of Army Aviation*, 1972, p.48. The article came from 661 Aviation Squadron who were in Long Kesh, June-October 1972.

656 Squadron Sioux grounded by fog at Long Kesh in March 1972. (Army Flying Museum)

pre-deployment training aircrew and ground crew alike had to adapt existing Cold War tactics to the situation in Northern Ireland, as explained by one squadron commander.

> Three roles are prominent and it is satisfying to note how accurately they reflect the role of army aviation worldwide. About half the hours flown are on routine patrols generally in conjunction with ground troops. Routine in a security situation is dangerous from the point of view of ambush. Latterly, patrols have tended to be conducted as mini operations in themselves. There is a certain deterrent value in this and terrorists are more inhibited in their use of country. Another value is in showing the flag. High intensity air operations help to reassure the local population and indicate that security forces are active.
>
> Special operations .... are planned as a result of intelligence reports. They involve searching and arresting or 'lifting', which is the colloquial equivalent. In this context air support has two functions. It provides general surveillance of the area under search particularly for people leaving or avoiding the area. It also facilitates command of the operation. Many search operations have been effectively controlled by the commander from the air using the good communications available to him and a valuable view of progress on the ground. Care must be taken in the early stages not to compromise the operation by the early arrival of aircraft.
>
> Liaison: Operational priorities and the need to conserve flying hours are powerful arguments to deter would be 'swanners'. That still leaves a long queue of potential passengers with cogent reasons for air travel. One problem is that ground movement in the IS situation becomes expensive in terms of manpower. Where minimum escorts in vehicles are mandatory the demand for an air taxi service increases. In many cases liaison can be as effectively conducted in staff cars, private cars or civilian army transport such as mini buses. In Northern Ireland staff cars, mini buses and pool minis are coloured every colour except khaki (they are even orange and green). Liaison tasks are really a question of speed. The Sioux and Austin 1800 cruise at comparable speeds. Priorities are necessary which seem to be: commanders

> to urgent briefings or reconnaissances, MO's to accidents and incidents, (with subsequent casevac if conditions permit) and ammunition technicians to bombs or explosions. Darkness should impose no limits on these tasks but training of the supported unit in setting up a night LS [landing site] is vital and the pilot must have confidence in their ability.
>
> The Small Arms Threat: The army aviation pamphlet I referred to earlier [Pamphlet No.5 of Army Aviation, Volume II, (Part 2) Aircrew Techniques and Tactics] claims that flight at 2000 feet AGL or more, will remove the threat from random small arms fire. Experience has shown that this is not always so as one cannot discount the chance of a lucky shot. Apart from this the observation task cannot always be effectively conducted from 2000 feet, even with stabilised binoculars. Here are some typical questions asked of the recce air crew: "Is that car abandoned or is there someone in it? What is its registration number and make? Is there anyone in that apparently deserted farm house? Can you see anyone in the customs post? Has that cratered border crossing been filled in? Do you think there are any claymore devices along this road?" In most cases these questions can only be satisfactorily answered by taking a fairly close look at the target. Some risk is inevitable since good intelligence is the key to successful IS operations…
>
> Since we now have Macralon [*sic*] and fibreglass protection the flak jacket has been retrieved from its position in the seat pan where it offered vital protection during Aden days. Further progress in aircrew protection is desirable but must be viewed against the weight problem. All pilots are briefed that speed and height are safety techniques. Prudent flying and a recognition of potentially dangerous situations such as recent outrages, bombs, hijacked vehicles and such like are survival factors.[37]

It was noted that the experience being gained by squadrons in Ulster was extremely valuable, particularly in improving surveillance and reconnaissance techniques and the ability to operate at night. It was also proving that the two-man crew was essential for all but the simplest liaison tasks. New tactics were also being developed, such as the 'Eagle Patrol', which would become tried and tested and familiar to a generation of aircrew. An Eagle sortie essentially provided a flexible ARF capable of putting an infantry patrol on the ground to conduct a variety of tasks at short notice. 661 Aviation Squadron reported that Eagle sorties accounted for around 50 percent of their effort, in terms of hours flown, in 1972.

> The sortie involved both Sioux and Scout, the former having a patrol leader as a passenger while the latter carried a four-man patrol. With some units the Sioux was dispensed with and the operation was done by one Scout. The patrol leader was briefed to stop and search road traffic, search farmhouses, carry out a general surveillance of the border or simply to advertise the Army presence. So, he acted as the spotter and would indicate to the Sioux pilot a particular target; the pilot would then relay the information to the Scout and if necessary talk it onto the target. The type of flying on these sorties was mainly low level, as height was a disadvantage when the target was a moving vehicle. Once the target was identified, the requirement was for a fast deployment of the patrol, hoping that the intentions of the driver had been guessed correctly. In the first few weeks many vehicles were missed at the first attempt, so we either lost them altogether, if close to the border, or a fast re-deployment was necessary. However, having got used to the area, we rarely deployed twice for one vehicle.[38]

37 Anon, 'Army Aviation in Northern Ireland, a Squadron Commander's Impressions', *Army Air Corps Newsletter*, No.26, April 1972, pp.2–3.

38 Anon, 'A Squadron in Ulster', *Journal of Army Aviation*, p.46.

See Map 15

1972 had been a challenging year for the Army Air Corps with nine aviation squadrons deployed at various times during the course of the year. It was a tough year for the British Army as a whole, with 102 soldiers being killed or dying in Northern Ireland, the highest annual loss since the Korean War. It had, however, been equally difficult for PIRA following Motorman and their campaign switched from the previous urban insurgency to a more limited rural insurgency mainly centred in South Armagh and a campaign of terrorism in the Province, the British mainland and occasionally elsewhere in Europe. With the shift to more rural operations Army aviation came into its own. Eagle patrols could be particularly effective but were not without risk, as described by Second Lieutenant Marfleet.

> Once while flying a Scout on Eagle patrol, I spotted a stolen car and gave chase. It was eventually abandoned in a quiet lane and I dropped troops from 3RRF [3rd Battalion, Royal Regiment of Fusiliers] in a field away from the car with instructions to search the area but not to approach the car. As I took off again, the car blew up – leaving my troops unharmed but very shaken. The command wire for initiating the explosion was traced back to the hedge just a few metres from where I had landed. A very cool operator – but he detonated it too early. He got away undetected.[39]

As 1973 progressed the aviation effort increased with more squadrons being committed to roulement tours as part of Operation Banner. For those returning to Northern Ireland for a second tour the developments after just three years were significant and not entirely positive. Having first deployed with 663 Squadron in 1970, Major William Duthoit of the Staffordshire Regiment returned to the Province in 1973 and described the changes as he saw them.

> The first contrast is the sheer scale of aviation involvement. In June 1970 the presence, up to 663 arrival, consisted of a reinforced RAC air sqn totalling 10 Sioux. With our assistance that rose to 16 Sioux and 5 Scouts. The "work rate" was, as far as my memory recalls, about 50 hours per month on Sioux and 70 hours per month on Scout. Nowadays the work rate is about 80 hours per month on each type and the fleet is 15 Scout and 23 Sioux – a sizeable increase. From this fact I deduce that customers are making wider use of our services and are more aviation minded but I freely admit that that is open to argument…
>
> The next thing that springs to mind is equipment. In 1970 the ultimate in equipment sophistication aviationwise was to have Pye radio, Skyshout and Nightsun [sic]. In those days it was only the local air sqn that had those – nowadays everyone has them and Kenlab stabilised binoculars, DECTRAC [navigation aid], macralon [*sic*] armour, and up to recently one heli-telly! [*sic*] Pilots wear flak vests and the new improved model shortly to be in issue is said to be able to stop an Armalite round at short range…
>
> In 1970, with the exception of some very small areas, the Army could and did have relatively normal relations with the civil populous. One could put on plain clothes and go out, to the pictures, to the pub, museums and concert halls as your recreational fancies lay. Time off was at a premium but when available could be used. The Army of 1973 wears a different image …. Army locations are now forts. The wire and sandbag sangars of 1970 have been extended; in 1973 sangars are now much more resilient and normally incorporate more than sandbags – anti-rocket screens 21 foot high, anti-vision screens, security flood lighting, sleeping policemen, 40 gallon drums full of concrete and linked together, etc, etc all play their part in reducing the target that is offered to the enemy. These outward defences also to me

39 D. Marfleet, account to the authors, 6 June 2022.

Sergeant Bailey and Corporal Parsons of 666 Squadron in Scout XT640 being flown over the Harland and Wolff dockyard buildings in Belfast. (Army Flying Museum)

> symbolise the change of attitude in the soldiers. In 1970 their attitude to the civil population was relaxed and friendly and, mentally they found it difficult to associate the public at large with the psychopathic killers of the IRA. Some 200 odd deaths later they are watchful, suspicious and ready to act at all times…
>
> In 1970 General Sir Ian Freeland briefed the Sqn in terms of the Army's role being to maintain and enforce a peaceful pause so that the politicians could mend the wounds. In 1973 we are fighting a war – a fitful one maybe, but still a war. General Sir Frank King's statement on the death of the 200th regular British soldier killed in Northern Ireland says all that I feel needs to be said on this subject. Before anything else here can be finally achieved the men of violence be they dark green or deep orange must be beaten. To win the war of today the Army must be prepared to give all that it has got… if the sum total of the aviation effort here in 1973 has only saved one soldier's life (and I believe it did considerably better than that) then it was all worthwhile.[40]

Much of the new equipment described by Major Duthoit was welcomed by the aircrew and increased the capability of the AAC's aircraft. The DECTRAC navigation equipment was simple to use. It had just one dial allowing the pilot to select any one of three coloured hyperbolic networks. These were powered by radio beams from one master and three slave masts located around Ulster.

40 W. P. Duthoit, 'Ulster Revisited', *Journal of Army Aviation,* 1974, pp.7–9.

The interlocking hyperbolic patterns could give a very accurate reading of the aircraft's position, particularly when operating near one of the masts where the beams were closer together. The accuracy of the system was demonstrated by Second Lieutenant Marfleet during a night sortie.

> One night a UDR patrol ran into a fire-fight in South Armagh taking several casualties, two of them serious. A Puma was tasked to go and pick them up, but the RAF at that time were not allowed to land at night unless they had their own Mobile Air Operations Team (MAOT) on the ground. So I offered to fly in a MAOT consisting of a Squadron Leader and a Sergeant in a Scout, ahead of the Puma, since we were under no such restriction.
>
> WO2 Dave Rowat flew the Scout and I operated the DECTRAC and followed the plot on my Decca overlaid 50,000 chart from the left seat. The Sqn Ldr was most impressed that we went straight to a six-figure grid reference with no lights and no landing aids. Twenty seconds out I called 'HLS should be directly ahead!' and Rowat called the UDR patrol to illuminate the HLS with their torches. The most dangerous part of the operation was indeed the moment of landing. As we approached the site I heard the tell-tale 'click' as the Squadron Leader inserted his magazine into his SMG [submachine gun]! My yell at him to remove it at once shook him up a bit. We did not need a cocking handle catching on seat straps and loosing off a burst of 9mm in the cab!
>
> Having landed, he ran off to set up his Bardic [night landing aid] and Rowat turned to me and said 'The crabs would be really pissed off if we just took the casualties back with us now, wouldn't they, Boss?' The Puma did land safely and quickly picked up the casualties but didn't hang around to pick up the MAOT so the Squadron Leader and his Sergeant had to spend the night with the UDR![41]

Heli-Tele was the first of a number of camera systems to be fitted to AAC aircraft operating in Northern Ireland. It was a large, cumbersome piece of equipment that was mounted in the back of the Scout and protruded out of the rear left door. Zoom and limited azimuth were controlled by a Royal Signals operator in the front left seat, who had benefit of a small black and white monitor that was set at 90 degrees to the camera angle. The equipment could beam live video to commanders on the ground. This led to the aircrew then receiving instructions from those watching the live feed that were not easy to respond to, as also described by David Marfleet.

> The problem was that the operations team back at Brigade HQ were following the proceedings on a colour monitor, so we would receive instructions such as 'zoom in on man in red jacket behind blue van!' I had to swivel in my seat and look down over my left shoulder along the line of the camera lens to see what they were talking about, and then point to the target on the monitor. It did produce some useful intelligence though and was used over Belfast and Londonderry during Op Motorman.[42]

When the Troubles first began the threat to off-duty servicemen was relatively low and soldiers could take risks that would become unacceptably dangerous in just a few years' time. On Christmas Eve 1969 Corporal Bob Harrod of 666 Squadron at Ballykelly was invited back to a local resident's house following an evening in the pub. Feeling slightly worse for wear Harrod went outside for some air.

41 D. Marfleet, account to the authors, 6 June 2022.
42 D. Marfleet, account to the authors, 6 June 2022.

> I started to feel considerably worse after the air hit me and took the decision to make my way back to the airfield. I knew it wasn't far away, I just didn't know in what direction it lay. Clearly I chose the wrong one because two hours later I was still stumbling along a country road in the dark. A small van eventually pulled up beside me, and out got a guy who seemed a good deal more under the influence than I was. By now I'd pretty well sobered up but was completely lost. He asked me where I was trying to get to, and when I told him Ballykelly Airfield he announced that I was walking in completely the wrong direction. It was however, my lucky night apparently because he decided that he would turn his van around and deliver me to the guard room. I was a bit worried about his ability to drive but quite apart from the fact I was completely knackered, I thought it would be rude to refuse so I gratefully accepted and within around fifteen minutes was deposited as promised at said guardroom. As I was getting out of the van and thanking him for his generosity, he suddenly stopped me and said, "Oh well, if you can't help a body at Christmas when the bloody hell can you? Oh, and by the way, if it hadn't been Christmas I'd have blown your f…..g head off!" With that, he pulled the door shut and shot off like a missile. To this day I have no idea if he was serious or not. If he was, it was a lucky escape![43]

Just a few years later the threat to off-duty members of the security forces was very real and brought into sharp focus for the aviation community in 1973. On 18 May four pilots were enjoying a night out at a regular Thursday dance at the Knock Na Moe Castle Hotel in Omagh. At the end of the night they returned to their car parked at the hotel as a bomb was detonated. Staff Sergeant Arthur Place PWO, Corporal of Horse Brian Cox of the Blues and Royals (RHD/G) and Sergeant Sheridan Young of the Royal Military Police, all from 652 Squadron AAC and Sergeant Derek Reed RM attached to 16/5 L Air Squadron were all killed in the blast.[44] Security force bases were frequently subjected to heavy and sustained attacks and action by AAC aircraft was often instrumental in preventing potential heavy loss of life. During 660 Squadron AAC's tour in the summer of 1973 the RUC stations at Crossmaglen and Forkhill in South Armagh were attacked by heavily armed PIRA active service units. At Crossmaglen a large band of PIRA insurgents attacked at night firing rockets and rifles at the station. The results were limited to minor structural damage, thanks, in part, to Scouts from the Squadron landing troops at pre-recced sites close to the station while the battle was raging. At Forkhill 16 mortar rounds and 26 RPG-7 rockets were fired at the station during a night attack, but the arrival of a Scout overhead flown by Sergeant Alan Wiles RM was enough to persuade the attackers to desist and withdraw having caused only one minor injury to the occupants.[45]

660 Squadron's pre-planned night sortie at Crossmaglen was part of wider developments in the aviation community to expand operations to a 24-hour footing and to ensure PIRA and other militant groups would not feel secure even in darkness. 2 RTR Air Squadron took over from 16/5 L at Lisanelly Barracks and developed a system for night flying they dubbed 'Night Eagle'. For the system to be safe and effective a number of conditions had to be met; meteorological conditions had to be within limits, landing sites required a recce in daylight, illumination had to be available and there had to be sufficient time for aircrew and troop briefing and planning. With those conditions in place planning and conduct of a night sortie could be relatively flexible.

43 R. Harrod, account to the authors, 20 February 2021.

44 Anon, 'Obituaries', *Journal of Army Aviation,* 1974, pp.45–46.

45 Anon, '660 Squadron Keep the Peace', *Journal of Army* Aviation, 1974, p.105.

A Sioux prepares to lift from the HLS at Bessbrook Mill in South Armagh. (Army Flying Museum)

The ideal is for the commander on the ground to select his target based on intelligence received. For example a typical target could be a farmhouse occupied by a suspect. The grid reference of the target is fed into the air operations room where the Sioux pilot who is to fly the night sortie processes it. Taking the Scout pilot with him he flies a covert, daylight recce of the target, i.e. transits across it at 1,500 feet without paying obvious attention to it. A suitable landing site is selected and Polaroid photos are taken to back up the night briefing. With experienced air crew one pass over the target is sufficient for night patrols to be successfully planned and flown. Once the HLS has been selected and the type of illumination has been decided upon the commander on the ground must be carefully briefed. Should it be decided that the HLS is to be illuminated by ground troops, they must be briefed as to the type and layout of the illuminant. The troops themselves can get to the HLS either by vehicle as part of an existing operation in the area, or on foot. Scout landings in the light provided by Nitesun have proved simple and most effective. Emphasis in this case is put on the daylight recce and to positive re-identification of the HLS by night. Only in extreme cases of emergency should this method be employed without a daylight recce. An example of such an emergency was the successful casevac of Lt Anderson from the Deanery (6 UDR) on the night of 2 May. Lt Anderson was lifted out very quickly to Omagh hospital by Scout and Nitesun Sioux, helping to save his life. The success of this operation is attributed to the high degree of training of both Sioux and Scout pilots.

Remembering that the aim is to retain an air presence by night, sorties are often flown by Sioux and Scout carrying no troops at all and effecting no landings. It is hoped that this activity by night in the right place is enough to deter the terrorist. It is important that the enemy knows we have a night capability.

Night Eagle has been criticised in the past as taking unnecessary risks and endangering aircraft and lives for no positive return. With the right training, however, and by obeying the principles, the risk is acceptable and the operation becomes a very valuable addition to the armoury. Positive results can be attributed to Night Eagle in that innocent people speak with awe having witnessed the arrival of two Scouts in the middle of the night or the sudden blinding light of the Nitesun, so those with a conscience are undoubtedly, it is hoped, "suitably impressed". Within the limitations of our present equipment and training, Night Eagle is extremely worthwhile and undoubtedly improves our war capability, be it in Ulster or anywhere else.[46]

'The Land of the Pilot'

As 1973 progressed the aviation roulement effort increased. During the year 13 AAC squadrons deployed to Northern Ireland; 10 from BAOR and 3 from the mainland UK. In addition, 2 RAC air squadrons and a RM detachment continued to serve in Ulster. With the increase of units and aircraft in the Province, an increase in the number of hours being flown per airframe per month and increasing night operations, the pressure on aircrew, ground crew and the REME rose and flying incidents and accidents became almost inevitable. On 18 April 666 Squadron Sioux XT174 hit telephone lines while low flying near the border at Caledon, 7 miles west of Armagh. The aircraft crashed and caught fire but the two crew managed to escape safely. On 20 November 663 Squadron Scout XP896 suffered an engine failure on take off from Armagh. The pilot attempted to make a forced landing in a vehicle park but struck a recovery vehicle injuring the two aircrew and three passengers. The effort was not going unnoticed however, and at the end of the year the GOC Northern Ireland, Lieutenant General Sir Frank King praised the work of the AAC.

> As the old year ends and a new one starts in Ulster, I am anxious to pay tribute to the very fine record of the many Army Air Corps Squadrons which have rotated (or flown fixed wing) through this Province. The statistics are themselves formidable.
>
> Some 37,000 hours were flown in 1973 with a very high proportion at night or under abominable weather conditions. The monthly average has been:
>
> Scouts 79.4 hours per aircraft
> Sioux 80.6 hours per aircraft
> Beaver 61.6 hours per aircraft
>
> Despite this very heavy usage rate, aircraft serviceability has always remained well above the equipment management target of 88.5%. This reflects great credit on hardworking, dedicated and shift operating REME craftsmen of all grades and also on the excellent spares backing and foresight provided by Ordnance.
>
> But above all Ulster is the land of the pilot. With its operational setting, very wide variety of tasks and a virtually non-stop service it provides an ideal training background under the most realistic and warlike conditions, for both the young and the experienced flier. Few branches of the Army can have profited more from its tours in Northern Ireland and none has greater cause for pride.
>
> I have yet to meet a unit that has not been full of praise and affection for its helicopter and Beaver pilots. I have not met one less than extremely grateful. Pilots have been shot

46 Anon, 'Night Operations by 1 RTR Air Squadron in Northern Ireland', *Army Air Corps Newsletter*, No.45, June-August 1974, pp.5–6.

> at and bombed. They have helped to arrest bank robbers, bandits and have even rescued bogged-down cows. But above all they have flown, at all hours, in all weathers and whenever tasked. Like every soldier in the Province I have been delighted and grateful to have them here through 1973. We wish them especially well in 1974.[47]

1974 would prove to be the high-water mark for the AAC in terms of the number of squadrons that deployed on Operation Banner. Fourteen squadrons served in the Province at some point during the year, 10 from BAOR and four from the UK, generally for four-month tours. Sioux and Scout flights from different squadrons often operated together, forming composite squadrons for the duration of their roulement tour. Long Kesh, Aldergrove, Omagh, Sydenham and Ballykelly had all become familiar to aircrew, ground crew and REME after five years of steadily increasing roulement activity. It was becoming clear that Northern Ireland was going to remain an operational theatre for some years to come and action was taken to streamline the flying effort and regularise operations as far as possible. There were by then three brigades operating in Northern Ireland. 3rd Infantry Brigade (3 Inf Bde) with its HQ at Lurgan was responsible for the southern half of the province including volatile South Armagh, often referred to as 'Bandit Country'. 8 Inf Bde based in Londonderry covered the city and the generally quieter northern half of the province. The Belfast area was the responsibility of 39 Inf Bde with its HQ co-located with HQNI at Thiepval Barracks in Lisburn. The AAC squadrons and detachments generally worked to the brigade within whose area of responsibility they were based, with tasking coming direct from the brigade HQ to the squadron or detachment HQ. The squadron based at Aldergrove acted as a reserve to be tasked directly by HQNI. The flying orders that governed operations were considered to be 'simple, practical and unambiguous and a refreshing change'. Standard tasking was also introduced to further simplify day-to-day operations as described by Captain David Skinner Staffordshire Regiment from 657 Squadron, deployed to Ballykelly from Soltau for a tour covering the first months of 1974.

> The daily tasks for the squadron included one aircraft deployed to the roulement regiment based in Londonderry on Piggery Ridge. This was the temporary camp constructed to overlook the Creggan estate, it was constructed entirely of wriggly tin and was positioned to overlook and dominate the estate. It had a landing site within the perimeter on the side furthest away from the estate. This aircraft was deployed at first light and remained until it was released by the regiment.
>
> Next was the daily aircraft to support the company detachment of the resident battalion [1st Battalion, Duke of Wellington's Regiment] at Magherafelt. Here the company was based in the police station and the HLS was divorced by about 500 metres further down the hill. The HLS was secure but unguarded, it had fuel in jerry cans which was the standard means of refuelling on this duty, Avgas only. The Magherafelt duty was first light until released by the company commander, basically this had to be by last light as the route over the Sperrins at night would mean that the pilot would have to be IMC to get home safely.
>
> There was always an immediate readiness aircraft ready to go and the remainder of the tasks were those passed down on a daily basis to support the brigade headquarters in Londonderry at Eglington barracks. A lot of time was spent looking for suspicious activities and generally dominating the area as the terrorist had little idea of what we could see from the air. The other tasks would be moving key personnel around the province, but in a Sioux a round trip to Belfast was about 2 hours flying and although there was no shortage of flying hours, you

47 F. King, 'Foreword', *Army Air Corps Newsletter,* No.42, January-February 1974, pp.1–2.

> lost the use of the resource in [the operational area] when flying long trips .... The routine was pretty much non-stop and there were no days off. However there was the running total of flying hours with 60 hours in 30 days as a maximum that was not to be exceeded, if you were about to exceed the total you were grounded until you were back within limits. Needless to say this immediately became a challenge to see if you could exceed the total.[48]

The Piggery Ridge detachment presented particular risks to the aircraft and aircrew due to its close proximity to the notorious Creggan estate. 661 Squadron, who followed 657 Squadron at Ballykelly, recorded several incidents in April and May 1974 where aircraft were fired at from unseen gunmen in the estate. The aircraft were particularly vulnerable as they transitioned to the hover before landing and as they lifted into the hover before flying away. Piggery Ridge was frequently targeted by RPG-7 fire from the Creggan and therefore the HLS was located on the opposite side of the camp from the estate to offer some protection.[49] Despite the risks, as Captain Skinner suggested, one of the benefits of deploying to Northern Ireland was that the vast majority of the flying was in support of operations, therefore had clear and obvious purpose. While the workload was heavy and potential hazards were abundant, all AAC and supporting REME personnel remained motivated and focused on the task at hand.

> Most HLSs away from base locations tend to be in confined areas. In the rural parts of the country they are usually the back yard of the local RUC station surrounded by a high anti-rocket fence, whilst in the urban areas they are normally small barrack squares or areas of grass once again surrounded by the anti-rocket fence or high buildings. Almost inevitably the HLSs are congested with piles of stores, fuel drums or parked vehicles and further confined by the ever-present wires. Extreme caution must be exercised when landing or taking off anywhere in the province except in known clear areas, particularly as most sorties are flown at or near maximum AUW and the usually strong westerly wind is not always blowing in a convenient direction…
>
> The number of radios to be monitored is a potential flight safety hazard especially with talkative passengers. All three radios in the aircraft are in use. VHF(AM) to Aldergrove or Sydenham ATC, VHF(FM) to the troops on the ground and UHF for the important, province wide, air safety net. If additional nets are required for city operations matters will get worse and confusing to say the least when there is other civil and military traffic in the area and everyone is trying to speak to Hawkeye [the aircraft]. Practical combinations of nets can be used to get around this difficulty but initially it can be extremely confusing…
>
> In many ways the flying is easier than elsewhere. Pilots fly long hours regularly and after the initial week or ten days of bedding in, fatigue ceases to be a problem within the specified limitation hours. Everyone is involved in operational flying – their *raison d'être* – happy in the thought that a well administered rear party will deal with all but the most pressing personal problems. Daily requirements are simplified; a wash, a meal and occasionally a bed, with plenty of books and coffee to fill in the gaps between work. One is able to work for long periods at a quicker pace and seeing a job well done, gain the satisfaction to sustain one for even longer. The importance of the AAC and its direct relevance to the soldier on the ground keeps the system motivated and the support of the 24 hour ground crew and LAD shift systems ensure that the helicopters are at maximum availability…

48 D. Skinner, account to the authors, 9 November 2020.

49 AFM 141: 'The Canterbury Tales', 661 Squadron AAC Newsletter, April and May 1974.

An Airtrooper armed with an SLR and wearing a flak jacket stands guard beside a Scout on its HLS platform at an unidentified security forces' base in 1975. (Army Flying Museum)

> One of the greatest joys of serving in Ulster is to see the aircraft being used operationally. Liaison flying is infrequent and the machines are being applied well to the tactical situation. Roulement units, especially those in the infantry role are delighted with "real" AAC support which they have not been used to and soon learn to make use of it and the various items of special role equipment.[50]

The high flying rate, intensive aircrew workload and the operational risks inevitably increased the chances of accidents. On 12 April 1974 Scout XR636 from 664 Squadron flew into rising ground near Richill, between Portadown and Armagh. The incident and its aftermath were recalled by Lieutenant David Marfleet:

> Sergeant Major Dave Rowat was on ARF out of Armagh, working with 3 RRF. He was flying alone from Armagh to Portadown, in good weather when he crashed into a rising slope while executing a 180° turn. Up until this time, we had always transited at 1,000ft or above because of small arms fire, but we had recently received intelligence that the IRA had taken delivery of a number of SAM-7 missiles and were intent on shooting down a helicopter. So our strategy was changed to ultra-low flying except in urban areas.
>
> At the time of impact, Dave had initiated a turn back towards Armagh, raising the question of whether he had seen something suspicious and wanted to take another look. Additionally,

50 Anon, 'Impressions of Northern Ireland', *Army Air Corps Newsletter,* No.52, June-July 1975, pp.3–8.

> the FM radio was neither on the 3 RRF frequency nor the unit at Portadown, with whom he had already made contact, but was midway between frequencies, suggesting that he was tuning the radio when he flew into the hill…
>
> I was Scout Flight OC at the time of the crash, and I flew directly down to the site as soon as we heard the news. Several hours later, I was authorised to remove his body from the wreckage and I flew it to Belfast in my cab, landing at the Royal Victoria Hospital. I then had to accompany the body to the city morgue to formally identify him. It was a very emotional and somewhat bizarre journey, travelling in the back of an Army land rover ambulance with Dave, and just two drivers up front – no escort – and I had a one-sided conversation with a pilot who had been my mentor as a young officer and one whom I greatly respected. He had joined the squadron on transfer from the Royal Artillery, completed P Company as a Staff Sergeant and was greatly loved by all ranks.[51]

A composite squadron of 662 Squadron's Scouts and 669 Squadron's Sioux, based at Long Kesh hosted a team from *Soldier Magazine* to give them a taste of the work of the AAC in Northern Ireland. Captain Mike Bell AAC flew the journalist on a sortie over South Armagh where he experienced patrolling this wild border area around the flashpoint villages of Forkhill, Crossmaglen, Newtonhamilton and Newry:

> Their job – observation and reconnaissance, supplies, night and day support for the company of 1st Battalion, Welsh Guards, whose task was to control a hardline Republican district with friendly faces few and far between.
>
> As the Scout skimmed over the ground at less than 200 feet, Captain Bell said it was better to risk the erratic aim of a shotgun-wielding IRA sympathiser than a possible rocket attack. No missile had yet materialised but it was safer to take the Provo blarney at face value. Fortunately there are few overhead lines to present problems to low-flying helicopters…
>
> All seemed well and before returning to Long Kesh we were taken on a guided tour of the area – a country post office robbed 18 times in three months, a derelict farmhouse blown up after a booby trap left for the Army had just been ignored for a week or two, debris on the side of the main railway line blown up last year. Clearly a hot area …. and one which the pilots of 662 Squadron now know as well as their own gardens.[52]

In the autumn of 1974 Forkhill certainly proved to be a 'flashpoint village' for former Parachute Regiment pilot WO2 Bill Scarratt AAC from 655 Squadron. On 4 September 1974 WO2 Scarratt was due to fly Scout XV132 to deploy a patrol from 45 Cdo RM in two lifts from Forkhill, about 2.5 miles to the west to Ballsmill. In the mid-afternoon as WO2 Scarratt approached Ballsmill at low level and the village came into sight, his infantryman's instincts began to suggest that all was not well. There were no people in sight or animals in the surrounding fields. Although it appeared to be washing-day in the village, two mid-terrace houses had no washing hanging out and the upstairs windows of both were wide open. The agreed landing point was in an open field in full view of the terrace so, after a brief conversation with the patrol commander, WO2 Scarratt selected an alternative landing site nearby in a dip in the ground behind the cover of trees. The four-man patrol deployed and took up fire positions. As the Scout lifted to head back to Forkhill it came under heavy small arms fire. Bill Scarratt could hear rounds hitting the airframe and elected to get

51 D. Marfleet, account to the authors, 6 June 2022. Anon, 'Army Air Corps Roll of Honour', HQ AAC, 10 November 2018; Anon, 'Aircraft Losses since 1957', HQ AAC, 13 June 2019.

52 J. Walton, 'Bird's Eye on the Border', *Soldier Magazine*, April 1974, pp.22–23.

clear of the area as fast as possible. Attempting to make a contact report he discovered that he had lost all communications and there was a strong smell of aviation fuel. Scarratt applied full power, spun the aircraft round and gained height, returning to Forkhill to pick up the second half of the patrol. To save time he landed the Scout downwind and got out of the aircraft to find fuel pouring from the tanks. XV132 was going nowhere. A subsequent examination found that six rounds had hit the Scout, two in the fuel tank, one clean through a main rotor blade and three in the cabin. One of the rounds had entered the cockpit and severed Scarratt's helmet communication (mic/tels) lead, missing him by inches, hence his loss of communication. Had WO2 Scarratt made different decisions during the incident it could have had far reaching consequences for aviation operations in South Armagh as his subsequent citation made clear.[53]

> At 1515 hrs on 3 Sep 74 WO2 SCARRATT was piloting a Scout helicopter of 655 Sqn AAC tasked with inserting a 4 man patrol of 45 Cdo RM in the area of South ARMAGH close to the border with Eire. Having landed the patrol, WO2 SCARRATT was just taking off and was in a low hover when his aircraft came under fire from 4 gunmen. A total of 88 shots were fired at the helicopter and the patrol, 6 of which hit the aircraft including 2 in the main fuel tanks.
>
> Despite the weight of fire from the terrorists and the fact that he had felt the bullets strike through the airframe WO2 SCARRATT continued with the take off procedure though he was aware that he was rapidly losing fuel. Displaying airmanship of a very high order he flew his damaged helicopter to the SF base some two miles away. Although the aircraft was badly holed through the main rotor blade, which adversely affected the handling characteristics, WO2 SCARRATT landed the helicopter safely.
>
> The alternative to this action, and probably less dangerous from a personal point of view would have been to land and abandon the aircraft immediately. In this case the helicopter would undoubtedly have been destroyed by hostile ground fire, a noticeable success for the terrorists. This would have resulted in a re-examination of helicopter tactics in border areas, and could have caused a reduction in the overall operational efficiency of close helicopter support. WO2 SCARRATT's gallant action was a blow to the terrorists' avowed aim to destroy a helicopter. It enabled a balanced appreciation of the situation to be taken, as a result of which normal helicopter operations have been allowed to continue.
>
> Throughout this incident WO2 SCARRATT's professionalism and coolness both under fire and afterwards were of the highest order.[54]

For his actions on 4 September 1974 WO2 Bill Scarratt was awarded the DFC on 25 March 1975.

In an attempt to give pilots a warning of when they were being engaged by ground fire, Scouts were fitted with the Hostile Fire Indicator. The system comprised a small dish fitted below the cockpit pointing towards the ground which, in theory, detected small arms fire aimed at the aircraft. It then gave the pilot a visual warning on a small dial on the instrument panel. Aircrew doubted the effectiveness of the system and there were no reports of it having produced positive results.[55]

PIRA announced a ceasefire over the Christmas period of 1974 and extended it in January 1975 until, on 8 February, the leadership issued a statement suspending offensive military action. A series of clandestine meetings took place between PIRA's commanders and British Government

53 M. Ashcroft, *Heroes of the* Skies (London: Headline, 2012, e-book), pp.338–342.

54 AFM 78: 'Distinguished Flying Cross, Warrant Officer 2 William Thomas Scarratt Army Air Corps', undated.

55 R. Petrie, account to the authors, 7 October 2020.

representatives which PIRA mistakenly believed signalled the start of a process of British withdrawal. The level of PIRA activity against the security forces significantly reduced although tit-for-tat sectarian killings of Protestant civilians and paramilitary groups continued. Attacks on members of the security forces fell to an average of 10 per month compared to nearly four times that number throughout 1974.[56] The lull in operations allowed the AAC to step back, survey its establishment, posture and *modus operandi*, and work towards a more efficient and effective approach to its support of Operation Banner, which was clearly going to be a long-term commitment. Over the course of 1975 the roulement plot was rationalised until only two squadrons were required in Northern Ireland at any given time, each rotating through on an approximately four-month cycle. One of the squadrons had its HQ in Aldergrove (call sign Albatross on the UHF aviation communications net) with detachments in Omagh (call sign Sandpiper) and Ballykelly (call sign Sparrowhawk). The second squadron had its HQ in Long Kesh (call sign Skylark). Additionally, a detachment from either the AAC or 3 CBAS was located at Sydenham (call sign Screech Owl), and the resident cavalry regiment retained its air squadron at Omagh. A fixed wing Beaver detachment also operated from Aldergrove. The entire effort was coordinated by a small team of staff officers and SNCOs based at HQNI in Lisburn (call sign Eagle).[57] This consolidation resulted in the AAC's commitment being 18 Sioux, 12 Scouts and three Beavers. Flying rates remained high, however, with aircraft flying 50–60 hours a month and pilots around 80 hours a month. Flying rates across the rest of the AAC had to be reduced to allow for the raised effort on Operation Banner.[58]

The ceasefire also gave the AAC a relatively quiet period in which to begin replacing the Sioux with the Gazelle in Northern Ireland. The additional equipment placed on aircraft to assist operations and protect the crew meant the Sioux was beginning to struggle, as described by Captain Skinner.

> The Sioux was the bog standard AH1 with no modifications pertinent to NI, but for night tasks it could be equipped with the Night Sun [*sic*], a powerful search light operated by the crewman, however the combined weight of the Night Sun [*sic*] and crewman put the Sioux very close to the edge of its safe operating limit, depending on the size of the pilot and crewman… In a Sioux you needed to be aware that a round trip to Belfast [from Londonderry] was about 2 hours flying and although there was no shortage of flying hours, you lost the use of the resource in the TAOR [tactical area of responsibility] when flying long trips.[59]

The Sioux could still provide good service and when XT214, flown by Sergeant Kenton Bryars R Sigs from 651 Squadron came under fire near Forkhill on 21 April 1975 he was commended by the infantry on the ground who were impressed that the aircraft continued to give support during the contact.[60] The Sioux was, however, restricted in what it could achieve in poor weather and at night, both of which were prevalent during flying operations in the Province. 660 Squadron became the first operational AAC unit equipped with the Gazelle which came into operational service on 6 July 1974. A year later in July 1975 660 Squadron deployed to Long Kesh for a roulement tour.

The PIRA ceasefire effectively ended on 22 September 1975 when 22 bombs were detonated across the six counties. Many of PIRAs IEDs were detonated by either a command wire or radio control. The security forces attempted to counter them by developing equipment that could

56 AFM 137: D. T. Young, 'Operational Review', *Visor Service News*, No.90, 13 November 1975.
57 AFM ALB 51: 'Pilot Callsigns Northern Ireland', 1 November 1975.
58 R. L. C. Dixon, 'Progress Report 1975', *Army Air Corps Journal*, 1975, p.8.
59 D. Skinner, account to the authors, 9 November 2020.
60 AFM ALB 51: 'Hostile Incident Report', 5 May 1975.

A 656 Squadron Sioux shuts down at Ballykelly having completed a sortie in a snowstorm at night. A member of the ground crew is carrying ground handling wheels to the aircraft. (Army Flying Museum)

remotely initiate a hidden IED from a safe distance before it could be triggered by a PIRA terrorist. One such system was known as 'Vetch' which could be carried in a vehicle or in a helicopter. For Vetch tasks an additional 8-foot whip antenna was attached to the right rear skid post of the aircraft which stuck out horizontally at a right angle to the line of flight. The equipment in the aircraft cabin was operated by members of the Royal Signals.[61] On 16 October 1975 Lieutenant David Waldock AAC was tasked to fly Scout XT663 on a Vetch task. He arrived at Bessbrook in the early morning for briefing and then headed for his allocated search area at one of the nearby border crossing points with the Vetch equipment, its operator and an RE soldier. Having conducted a recce of the area from high level he descended to between 150 and 200 feet AGL and conducted a systematic sweep of the area, flying a criss-cross pattern.

> Vetching had continued for approx. 2 minutes without incident, when suddenly the aircraft shook violently from the percussion caused by an explosion. The actual blast was not observed as it occurred in the aircraft's 6 o'clock, but the aircraft was banked immediately and the smoke and flying debris were clearly visible to the pilot and passengers. At the time of the blast, the aircraft was approx. 400 metres to the West. An immediate check of the control functions showed no apparent damage to the aircraft.
>
> Lt Waldock then flew North to the Glasdrumman area and climbed, to be well clear of any resultant small arms fire from the border area and passed a contact [report] to the Bessbrook Ops Room. The bomb, on subsequent inspection, appears to have been a device of 20–30lbs

61 P. Courtenay, account to the authors, 9 December 2020.

> hidden in a stone wall adjacent to the roadway; the fence was demolished and a small crater was blown on the side of the road.[62]

As Lieutenant Waldock was transmitting his contact report he observed a dark blue Fiat 125 drive up and stop at the site of the explosion. The driver got out, observed the destruction and then drove to Glasdrumman on the coast, 18 miles east of Newry, where the driver was observed speaking to several people in the village. The suspicious car then began to speed towards the border. Lieutenant Waldock had been updating Bessbrook on the car's activity and a Puma with a QRF on board was despatched to stop the vehicle. Lieutenant Waldock, however, could see the Puma would not arrive in time to stop the car escaping across the border and so he landed his two passengers on the road ahead of the Fiat and brought it to a halt. The Puma arrived a couple of minutes later; the QRF took charge of the suspect and his car and Lieutenant Waldock's passengers re-embarked. Search teams were also sent to the area, but the Vetch operator in the Scout insisted they were held back until another Vetch sweep had been conducted.

> Another systematic sweep of the area commenced and after approx. 30 seconds of operation another bomb was detonated at a T junction leading North from [the border] about 250 metres from the initial explosion. This bomb consisted of a series of devices detonated by a timed relay device, possibly to improve its capability against moving vehicles. Subsequent aerial inspection revealed that four craters had been formed about 25 feet in length, 10 feet wide and approx. 4–5 feet deep. The device was thought to have consisted of approx. 100lbs of explosive. It was further thought that an aerial-type device could be detected from the air in a nearby tree, one of the RE passengers believing that this was used to detonate the device.[63]

Later in the day a further Vetch sweep of the area was conducted without incident and so an Ammunition Technical Officer (a bomb disposal officer) was flown to the site to examine it. His opinion was that the first, smaller device was a 'come on' and that when search teams arrived they were likely to have parked in the area of the T junction when the second larger device would have been detonated. At the time it was believed that it was the first instance of Heli-Vetch being used successfully to prematurely explode IEDs where the initiator was a radio signal.

The return to full PIRA operations led to a proportional increase in aviation tasking and, heading into winter, there was a concern that the risk of accidents would also increase. A series of accidents over the Christmas and New Year period appeared to confirm everyone's worst fears. On 18 December 1975 Sioux XW195 flew into wires near Crossmaglen and was written off, although the pilot escaped alive. On the same day Major David Hicks AAC, the OC 662 Squadron was the standby pilot flying Sioux XT818. At 1620 hours he departed Bessbrook with an observer on board heading for Long Kesh. Major Hicks was aware that there was poor weather on the planned route and on reaching the Banbridge area, 6 miles south of Lurgan, he ran into fog forcing him to hover taxi and search for gaps. He knew there was a pylon line ahead in the Hillsborough area, 3 miles south of Lisburn, and having identified it, lifted to hover taxi above the cables. The Sioux ran into fog again, however, and on attempting to turn back XT818 struck the cables and crashed. Major Hicks was killed and his observer suffered a broken pelvis.[64] Three weeks later on 5 January 1976

62 AFM ALB 51: 'Account of Events Which Occurred in H24 Area on 16 Oct 75 Involving Scout XT 663', undated. The same account was also sent to the authors by J. Waldock and was published almost word for word in the 1976 *Army Air Corps Journal* but with the references to Vetch and Lieutenant Waldock removed.

63 AFM ALB 51: 'Account of Events Which Occurred in H24 Area on 16 Oct 75'.

64 Anon, 'Wirestrikes', *Army Air Corps Newsletter*, No.63, September 1976, pp.21–23. Anon, 'Army Air Corps Roll of Honour', HQ AAC,10 November 2018; Anon, 'Aircraft Losses since 1957', HQ AAC, 13 June 2019.

WO2 Brian Jackson AAC, an experienced Scout pilot, was the overnight standby Nitesun pilot at Bessbrook in XV133 with Corporal Ken Ford R Sigs of 652 Squadron as his Nitesun operator. At 0550 hours the Scout, along with a Wessex, flew to Crossmaglen to pick up a patrol for an insertion in the darkness. The patrol was dropped without incident, the Scout's Nitesun illuminating the landing point for the Wessex. The weather soon began to deteriorate, however, and the Wessex crew, in discussion with WO2 Jackson decided to return the patrol to Crossmaglen. Arriving over the pick-up point the Scout illuminated the landing site while flying tight circuits. In doing so, XV133 entered a patch of low cloud and flew into a hillside bursting into flames. Both WO2 Jackson and Corporal Ford were killed in the accident.[65]

The loss of life was a severe blow to the AAC, but its members were consoled by the fact that they were having a positive effect in Northern Ireland, not only in terms of supporting troops on the ground and acting as a deterrent to terrorist activity, but also as an important casevac asset, saving the lives of soldiers and civilians wounded in the Troubles. While on duty at Dungannon, County Tyrone supporting the Royal Hussars, Staff Sergeant Ian Harrison REME of 651 Squadron AAC was tasked to take a doctor to Aughnacloy, 10 miles southwest of Dungannon, to treat a wounded soldier. On route the aircraft entered cloud and Staff Sergeant Harrison had to adopt instrument flight rules and make an approach to Aldergrove. Having broken cloud, he landed, refuelled and then made another attempt to reach Aughnacloy flying low level. On arrival it was clear the soldier required evacuation to a hospital immediately. Despite failing light and atrocious weather conditions, Staff Sergeant Harrison set off with the casualty to fly the 45 miles to Musgrave Park Hospital in Belfast. Once again, the weather closed in and he had to divert and make an approach to Aldergrove and, on breaking cloud, immediately tried to make it to Belfast low level. It proved impossible to make it through the Divis Mountains and, calling ahead, flew back to Aldergrove where an ambulance was waiting to transfer the casualty. Staff Sergeant Harrison flew nearly two hours during the casevac, an hour and a half of which was on instruments. In a separate incident WO2 Vivian Dicks AAC was flying in the area of Silverbridge in County Armagh when he received instructions to pick up a critically wounded civilian. Despite a cloud base as low as 50 feet AGL and visibility down to a couple of hundred metres, WO2 Dicks located the site of the incident but found it was impossible to land due to boggy ground. He therefore sat in a low hover while the casualty was embarked and then had to use a combination of low flying and the hover taxi with the landing lamp on to make the 10 miles to Bessbrook. The conditions were so poor that the team waiting for him only spotted the aircraft as he hovered across the perimeter fence. Without the difficult casevac it was assessed that the civilian 'would have died at the scene of the incident had it not been for the courage, determination and flying skill displayed by Warrant Officer Dicks.'[66]

While flying operations continued, in the background the AAC continued to make organisational changes to its establishment in Northern Ireland. 15th/19th The King's Royal Hussars (15/19 KRH) became the last cavalry unit to provide an air squadron at Omagh. As part of the steady demise of the Integrated Flight Scheme it became 3 Flight AAC on 1 May 1976 and Major Tony Collis 15/19 KRH continued in command. Although the change helped rationalise Army aviation in the Province, the chain of command of 3 Flight remained rather convoluted. Nominally the Flight was part of 7 Regiment AAC at Netheravon. Operationally it was under command 3 Inf Bde, however day to day it was answerable to the CO of the cavalry regiment resident in Lisanelly

65 Anon, 'Accident', *Army Air Corps Newsletter*, No.64, October-November 1976, pp.25–26. Anon, 'Army Air Corps Roll of Honour', HQ AAC, 10 November 2018; Anon, 'Aircraft Losses since 1957', HQ AAC, 13 June 2019.

66 AFM 136: *Visor Service News*, 29 October 1976.

Barracks and was treated much like a regimental sub-unit. For special-to-arm, aviation specific matters the Flight came under the remit of the SO1 Aviation at HQNI. Most of the Flight's sorties took place in Counties Tyrone, Fermanagh and Londonderry with occasional tasks in Armagh. Two aircraft per day were tasked by the resident cavalry regiment, two aircraft were available to be tasked by HQ 3 Inf Bde and 8 Inf Bde and a fifth aircraft remained on standby in Omagh. At night there was one aircraft on standby at Omagh and one at St Angelo where Nitesun was often required.[67] While life was varied, and often exciting and testing for aircrew, for the ground crew, confined within the wire of their respective bases for much of the time, life could become monotonous and frustrating.

> Being ground crew we were mostly kept in the dark when it came down to operations. For many at the time we were regarded as just bubble washers. With six Sioux and with each due a 'cab wash' every six weeks we knew with some certainty that each week contained at least one cab wash. Each day was much the same as the last, to a degree we were no more than a car wash. As a soldier I personally found it disheartening.... I can understand as to why some people later regarded it as a punishment posting.[68]

In October 1977 Major Colin Sibun AAC replaced Major Collis as Flight Commander and the Flight's Sioux were replaced by six Gazelles. They were the last Sioux to operate in Northern Ireland.

The progress that had been made in combined infantry and aviation cooperation and tactics was ably demonstrated by an incident on 23 June 1976. In the late evening two covert OPs from 3 Para observed a previously hijacked car arrived in Drumiliy, 3 miles southeast of Bessbrook. Four masked and armed men got out of the car and loitered around nearby houses. The sighting was communicated to Bessbrook as a firefight broke out between the gunmen and one of the OPs. Immediately, a four-man ARF was scrambled from Bessbrook in a Scout along with 3 Para's OC Patrols Company in a Gazelle. A further eight-man QRF sat in a Puma with the engines running, ready to deploy. As the ARF flew the short distance to the incident the gunmen began to disperse into nearby houses. The QRF joined the ARF in the incident area and, directed by OC Patrols in the Gazelle they landed, disembarked and surrounded one of the houses. After shots were fired into the house, two of the gunmen emerged without their weapons and surrendered. The Scout and Puma moved back to Bessbrook, picked up additional troops and appeared back over the incident as the OC Patrols spotted a further gunman moving towards another house. As darkness closed in the Gazelle used its Nitesun to illuminate landing points for the Scout and Puma, but the disembarked troops failed to find the gunman. The area was cordoned off and at first light two search teams were deployed. The subsequent search discovered a third gunman asleep in an outbuilding clutching an Armalite automatic rifle. Several other weapons and a large amount of small arms ammunition were also recovered. It was noted in the incident report that the entire operation was controlled by OC Patrols in the Gazelle, including briefing the ARF, QRF and follow-up troops as they were *en route* in the air.

> The immediate operation mounted as a reaction to the sighting of the gunmen was an entirely airmobile one in that all troop lift, observation and command and control were carried out by helicopter. The Op[eration] would not have succeeded without the excellent helicopter

67 C. Sibun, account to the authors, 9 February 2021; P. Courtenay, account to the authors, 6 December 2020; AFM 171/172: '3 Flight AAC Standing Orders', undated.

68 M. Keating, account to the authors, 9 November 2020.

Aircrew return to Ballykelly following a Gazelle and Scout sortie in the late 1970s. The Gazelle is equipped with Nitesun on its port boom. (Army Flying Museum)

> support and first rate flying provided at the time. Reaction by vehicle would have been altogether too slow.[69]

At the end of the year Army aviation in Northern Ireland received a significant boost to its capability in the shape of Passive Night Vision goggles (PNG). The previous year 660 Squadron had experimented with inserting patrols by night using the Scout. Selected infantry patrol commanders were trained to sit in the left-hand seat of the aircraft wearing PNG. Following a daylight recce pilots, flying without goggles would be talked down onto the landing site by the patrol commander wearing PNG in the left-hand seat. Once on the ground the patrol would disembark leaving the PNG in the cockpit for the next patrol to fly.[70] Although the technique had some success it was by no means the most effective use of the equipment. Through September and October 1976 658 Squadron trained and qualified 11 pilots and eight aircrew to fly using PNG and deployed on Operation Banner in the November. Although the aircrew were proficient in the mechanics of flying on PNG and the general utility was obvious, it was up to the Squadron to determine how the advantages of night vision could be used to best effect during operations. An example of the range of scenarios in which PNG proved invaluable was illustrated by following a single night duty conducted by two Scouts forward based at Bessbrook.

69 AFM 139: 'Incident Report', 2 July 1976.
70 J. Hathaway, account to the authors, 26 February 2021; B. Baynes, account to the authors, 1 March 2021.

Both night Scouts arrived well before dark. This allowed time for briefing and discreet HLS reconnaissance. An early changeover also ensured that day standby aircraft returned to Long Kesh before last light when, during the winter, the weather often deteriorates quickly. Although fog was forming further north, the South Armagh terrain was once again creating its own weather pattern. On this occasion the area was generally clear, though the forecast indicated some cloud and possible snow showers later.

Soon after dusk a Scout took off, equipped with white-light Nitesun. The requirement was to find and illuminate a series of eight figure grid references for use as support helicopter landing sites. Several of these were within a few hundred metres of the border. Last year the crew might have used the Nitesun for several minutes to map read between landmarks, avoid a border incursion and guarantee the accuracy of the destination grid reference. Now, with PNG and some moonlight, they could see the ground clearly to identify landing sites. At each the light was only illuminated for about twenty seconds as a Puma SH made its final approach.

After returning to Bessbrook to refuel, the Nitesun aircraft moved to its next task. This involved providing illumination for foot patrols moving cautiously across waste ground, and searching for mortar base plate positions which could threaten the nearby company base. By now thin mist had begun to form and moisture in the air caused the Nitesun beam to be reflected back, dazzling the crew. By using a dispersed beam and PNG, the aircrewman was able to see through the mist and into the pool of light. He could then detect movement which might otherwise have been obscured.

Later in the evening battalion headquarters received reports from the police of illegal road blocks established near border crossing points. A helicopter was tasked to check all routes in the area. Much of the reconnaissance was done with PNG. The need to show lights was reduced and the risk of being fired at minimised.

At 0230 hours a patrol was to be landed by two Scouts. The moon was now obscured by thickening cloud cover and was in any case about to set. There was doubt about whether the light level would be suitable. An accurate assessment at the base HLS is normally impossible, because of false light levels created by security lighting. The extent of low mist patches was increasing and the light level in remoter areas was likely to be very low. At first sight such conditions would preclude PNG work but practical experience had shown that covert landings might well still be possible. The area of the drop was in flat country, with prominent line features. There were no wires above 50 feet.

We had decided that, with the onset of winter, low-level transit on PNG from the base HLS could bring unexpected problems. All the routes between Bessbrook and its outstations cross saddles linking higher ground. The saddles are 500 ft or more above sea level. An aircraft moving low-level could often be climbing steadily towards a ridgeline. Landmarks beyond the feature would be in dead ground, and the crew could not be sure that the way ahead was clear. The ability of PNG to penetrate thin mist means that a pilot can find himself in thickening hill fog, low cloud or mist rising from the low ground beyond a saddle. Outside reference can suddenly be lost at low-level surrounded by high ground.

Normal navigation was therefore used in transit, with the two Scouts well separated and showing beacons. On arrival in the PNG area, the light level was found to be acceptable. Aircraft lights were extinguished and pilots' goggles lowered. The rear aircraft closed in to the "seven o'clock position" of the lead aircraft, keeping station by its shielded tail light and skid-mounted Beta lights. As the aircraft descended low-level, the aircrewmen were mainly concerned with checking for bad weather not visible to the pilot through PNG. They used goggles intermittently to check landmarks against the map.

'An average sort of day at Bessbrook.' RAF Pumas and Wessex compete for space with an AAC Sioux as seen from the cockpit of a Scout. Refuelling is carried out from the drums and jerry cans. (Army Flying Museum)

> The aircrewman in the leading aircraft map-read to an easily distinguished initial point. The last two miles from the initial point had been memorised by all crew members and could be flown at 100 ft. On arrival, short run-on landings were made and, within seconds, the patrol had deployed and melted into the night. The aircraft moved low-level to three diversionary landing sites and then climbed to return to Bessbrook, using normal night flying methods.
>
> The final task of the night was to support the landing of a much larger patrol by a Puma, shortly before first light. A Scout equipped with infra-red Nitesun loitered near the landing site, ready to give an artificial boost to the ambient light level if asked. In the event, the first glow of dawn and the frost-covered ground generated sufficient light, without the infra-red Nitesun.[71]

At the end of 1976 the AAC operated 30 helicopters in support of Operation Banner. Three Scouts and Six Sioux flew from Omagh in support of 3 Inf Bde and 8 Inf Bde. At Aldergrove three Scouts and three Gazelles operated in direct support of HQNI. Three Gazelles at Ballykelly supported 8 Inf Bde, principally over Londonderry. The largest contingent of six Scouts and six Gazelles flew from Long Kesh in support of 3 Inf Bde. In addition, Beaver Flight operated from Aldergrove

71 Anon, '"How Green is Your Valley?" A Look at the South Armagh Winter, Through PNG', *Army Air Corps Journal,* 1977, pp.74–77.

in support of HQNI and the Reconnaissance and Intelligence Centre and Dieppe Flight from 3 CBAS operated three Gazelles from Sydenham in support of 39 Inf Bde.[72]

See Map 16 1977 began with a disheartening repeat of the previous year's events. On 2 January Sioux XT241 of 3 Flight AAC was being flown on a sortie from Aldergrove to Kilrea, 25 miles to the northwest, carrying an NCO of the Intelligence Corps. At the request of his passenger the pilot decided to approach Kilrea by flying low level up the River Bann from the south. Not far from its destination, however, XT241 struck wires and crashed into the river. Both men managed to extricate themselves from the aircraft but, while the pilot managed to swim for the shore, the passenger was swept downstream and drowned.[73]

In 1976 the British Government announced a new strategy of 'police primacy' in Northern Ireland, with responsibility for security and counter terrorist operations being handed to the RUC. Violence began to reduce markedly during 1977 with total deaths attributable to the Troubles falling by almost two-thirds from 295 in 1976 to 111 in 1977. The trend was so pronounced that it led the Northern Ireland Secretary, Roy Mason MP, to declare that the tide had turned against PIRA's campaign. Aircraft tasking, however, remained at a high rate and during 1977 the AAC flew over 22,000 hours in support of Operation Banner.[74] Despite Mason's assertion PIRA continued to evolve and develop new tactics and the security forces, including the AAC had to adapt to respond.

Although the security forces had a clear dominance in terms of aviation assets in Northern Ireland, PIRA also occasionally attempted to exploit the air with varying levels of success. On 31 October 1973 a PIRA gang hijacked an Alouette helicopter and forced the pilot to fly it into the top security Mountjoy prison in Dublin. In a carefully planned operation, the aircraft picked up three senior members of PIRA, including chief of staff Seamus Twomey, before making its escape. In the most notorious incident, on 24 January 1974 a MBB Bo 105 helicopter was hijacked by a PIRA team that included English heiress Rose Dugdale. The hijackers dropped two milk churns full of explosives onto Strabane RUC station, but the improvised bombs failed to explode. The incident led to flying restrictions being imposed on civilian aircraft in areas of the Province. On 28 September of the same year a Socata Rallye training plane was hijacked from the Dundalk Aero Club. The PIRA team intended to launch an air attack on an Army permanent OP at Drumackavall near Crossmaglen, coordinated with a ground attack using mortars and small arms. The air element did not go according to plan however, as the PIRA hijackers could not locate the OP from the air and dropped their bombs approximately 10 miles away where they failed to explode.[75]

On 10 August 1977 the Queen was due to begin a two-day visit to Northern Ireland as part of her silver jubilee celebrations. There was genuine anxiety that PIRA would attempt a spectacular attack on the royal visit and they might try to exploit the air as a means of assault. Lieutenant Iain Adams-Cairns Black Watch had recently returned to Detmold following a roulement Operation Banner tour with 655 Squadron AAC when he was ordered to redeploy at short notice to join 662 Squadron AAC along with two further pilots and three aircrew. The aircrew flew to Northern Ireland on 5 August and spent a few days re-familiarising themselves with the ground before contributing to the plans to protect the royal visit. To Lieutenant Adams-Cairns' surprise he was tasked to provide a single aircraft CAP above the royal party when they visited Coleraine in County Londonderry:

72 AFM 139: 'Outline Deployment and Tasking of AAC/RM Helicopters in Northern Ireland as at 1 September 1976', 14 September 1976.
73 Anon, 'Accident', *Army Air Corps Newsletter*, No.69, April 1977, pp.17–18.
74 AFM 139: 'AAC Northern Ireland – Script for DAAC Conference 1977', undated.
75 Taylor, *Air War Northern Ireland*, pp.44–47.

> Temporarily prohibited airspace had been established above Coleraine and we were to intercept any aircraft which entered 'her' airspace and, if the intruding aircraft didn't obey our interception signals, Cabinet approval had apparently been given for us to open fire on it. This simple plan presented a number of interesting technical difficulties. First, interception was a subject that I and most others had not taken seriously during the pilots' course at Middle Wallop on the grounds that it was highly unlikely we'd ever have to intercept anything, especially in a helicopter .... Second, the Gazelle was not armed. This issue was solved by someone much higher up the food chain who decided that if the Army Air Corps was going to do RAF-like tasks such as 'CAPs', their aircraft should be armed. Unfortunately, this did not mean that the Gazelle would be fitted with an awesome, tank-busting cannon but with a hand-held, 7.62mm SLR in the grip of a trusty aircrewman. He would sit in his normal front seat on the port side with his door removed to allow him to bring his mighty weapon to bear...
>
> The last problem was that the trusty aircrewman in my case was Airtrooper W. When awake, he was a very pleasant and amusing man but wakefulness was not his watchword; in the Squadron he had an unrivalled reputation for being able to sleep uninvited, anywhere and at any time. I doubt if the Cabinet were told that the defence of the Realm would be in the hands of Airtrooper W and that all he could reasonably be expected to do would be to wave his rifle in a suitably discouraging manner. I was more worried that he would drop it from 4,000 feet onto the Queen and that I'd end up as the forgotten man in The Tower, or worse. In order to provide a continuous CAP over the Queen, we were to be based at Ballykelly airfield with Robin Tait and his aircrewman in another Gazelle. This arrangement would allow the first aircraft to fly in dizzying circles over the Queen for about an hour before he was relieved by the second aircraft in time to nip back to Ballykelly for a quick refuel before heading back in turn to relieve the second aircraft. As far as I remember, there was no spare aircraft and no 'Plan B' if either of the 2 Gazelles became unflyable for any reason...
>
> To start with all went well and our change-overs worked seamlessly. However, flying round in circles was dull and it was difficult, especially for Airtrooper W, to remain alert. Imagine our consternation, therefore, when a little light aircraft suddenly bumbled out of nowhere into the airspace overhead the Queen. Luckily, he wasn't going very fast and by dint of pulling more power than I care to remember, we managed to draw alongside him in the approved manner. Initially, the pilot didn't see us as he was hunched over the controls. In the days before the twin towers, it dawned on me with a rising sense of horror that I could be looking at a kamikaze wannabee about to dive onto the assembled company below in one of the biggest terrorist outrages in history – no prizes for guessing who'd get it in the neck if he didn't prevent it. At this point, the pilot suddenly looked up, perhaps because he'd been studying his map, saw this (hopefully) threatening military machine very close to him, probably got the fright of his life and luckily banked away sharply to starboard. Airtrooper W remained as calm as ever. For my part, I pretended that my heart hadn't stopped and we did our best to resume our CAP as if nothing had ever happened.[76]

The temporary drama of the Queen's visit aside, the general downturn in violence across Northern Ireland provided another opportunity to review structures and methods of operation and take further steps towards formalising the AAC's establishment on Operation Banner. Rather than HQ AAC centrally designating squadrons to fill the roulement plot, a BAOR based AAC regiment would be given responsibility for manning the Operation Banner commitment for a set period of

76 I. Adams-Cairns, account to the authors, 29 June 2020. The authors thank Ian Adams-Cairns and the late Ross Mallock for permission to reproduce this account.

time. During that time the regiment would stagger the deployment of one squadron to Long Kesh and one to Aldergrove with a two month overlap in the middle. The responsibility would then be passed to a different regiment. This system, however, led to 'one of the more bizarre command and control arrangements seen in Northern Ireland.'[77] The new system allowed the CO and two further officers from Regimental Headquarters to deploy for up to two months when 'the bulk of the regiment' was deployed during the overlap period. The CO, however, would have no operational control over his regiment while he was deployed. In effect, the CO would '...continue to exercise the normal 'A' but will have limited 'G' and 'Q' functions...' In other words, a deployed CO would find himself in the unsatisfactory position of being responsible for the training, welfare and efficiency of his personnel while having no control over either operational planning and execution or the logistics of maintaining the aircraft to meet the requirement. In operational matters the CO had to defer to the SO1 Aviation in HQNI.[78]

Early in 1978 the first AAC Gazelle was lost during Operation Banner. It was also the first aircraft loss directly attributable to terrorist action and resulted in the death of the most senior officer to be killed during the Troubles in Northern Ireland. On 17 February 1978, near Jonesborough in County Armagh, a covert OP from 2nd Battalion, Royal Green Jackets came under heavy small arms fire from a PIRA squad operating close to the border. When the contact report reached Bessbrook the CO of 2nd Battalion, Royal Green Jackets, Lieutenant Colonel Iain Corden-Lloyd, decided he needed to be at the scene as quickly as possible. The standby Gazelle pilot, Sergeant Brian Ives AAC was notified and at 1620 hours he was airborne in XX404 with Corden-Lloyd and his adjutant, Captain Schofield. The ARF Scout also launched but held off until it received more information. Sergeant Ives had not been in the Province long and, on approaching Jonesborough and conscious of the threat, he made a fast and low pass of the OP site, travelling from north to south close to the border at 120 knots and 50 feet AGL. Having seen nothing, Corden-Lloyd asked Ives to make another pass. As Sergeant Ives prepared to make another run, more small arms fire came from the border, but it was unclear if it was aimed at the OP or at XX404. Sergeant Ives initiated a sharp cyclic climb and when at 200 feet AGL made a tight turn with a large angle of bank to bring him back on a reciprocal heading. During the turn the Gazelle's nose dropped and the aircraft began a rapidly descending spiral. The aircraft did not recover and struck the ground heavily before cartwheeling and impacting the ground a second and final time. The ARF was landed nearby and rushed to the crash site. Corden-Lloyd, Ives and Schofield were pulled clear of the wreckage and into the shelter of a nearby ditch. More troops arrived by Wessex to cordon the area while the PIRA gun team slipped away over the border. When a medic reached the casualties he found Ives and Schofield severely injured but Lieutenant Colonel Corden-Lloyd was dead at the scene.[79]

The incident, and the death of Corden-Lloyd, was overshadowed in the British press by a PIRA bomb on the same day at the La Mon House Hotel near Belfast that killed 12 civilians in a hideous fireball. PIRA stated that the Gazelle had been brought down by a special 'magnetic' bullet, a claim the British Army unsurprisingly denied, pointing out that no such weapon existed.[80] In fact the subsequent accident investigation found no evidence that XX404 had been hit by any type of small arms fire. Nevertheless, it concluded that PIRA's actions on the day were a substantial contributing factor in the incident, stating that, '...the pilot was under operational stress and the threat of hostile

77 Anon, 'The Army Air Corps in Northern Ireland: Past, Present and Future', *Army Air Corps Newsletter,* No.105, May-June 1982, p.3.

78 AFM 137: 'Directive for the Commanding Officer of the AAC Roulement Regiment in Northern Ireland', 30 November 1976.

79 Anon, 'Accident', *Army Air Corps Newsletter,* No.82, September 1978, pp.18–19; Taylor, *Air War Northern Ireland*, pp.54–57.

80 AFM ALB 205: unreferenced newspaper cutting, 'Army silent on helicopter death crash', undated.

action, to which he was not accustomed, and therefore his actions were not regarded as having been blameworthy. Indeed, the accident was largely attributable to the terrorist activity in the area.'[81]

Aside from the terrorist threat aircraft continued to be lost to accidents. In the late morning of 10 April 1978 Captain Michael Kett RA of 655 Squadron was tasked with taking an army cadet on an air experience flight from Aldergrove. The planned departure time had to be delayed due to heavy snow showers and when the sortie eventually got under way in Scout XV132 there were still snow showers in the local area. The aircraft departed to the south and soon after ATC lost contact with the sortie and reported it missing. Late that night wreckage was found off the shore of Loch Neagh and the following day XV132 was located by divers under the water immediately south of Langford Lodge, 4 miles southwest of Aldergrove. Captain Kett and his passenger were both killed.[82] In the early hours of 2 December 1978 Captain Alan Stirling AAC of 659 Squadron was tasked with flying a Nitesun sortie to assist RAF Wessex support helicopter operations in South Armagh. Captain Stirling along with aircrewman Corporal Roger Adcock Para took off from Bessbrook in poor weather in Scout XW614. Engine problems delayed the take off of the Wessex and when it arrived in the planned area of operations close to Crossmaglen there was no sign of the Scout. The Wessex returned to Bessbrook and initiated overdue action. Later in the day debris was spotted on the surface of Lough Ross and XW614 was subsequently found below the surface. Captain Stirling and Corporal Adcock were killed in the accident and Andy Simkins, a Lieutenant RA at the time, recalled the after effect:[83]

> Six months after I arrived in 669 Squadron AAC, 9 Regiment AAC in Detmold, we were warned off for a tour in Northern Ireland. Our sister squadron, 659 Squadron AAC would deploy two months prior to us, so they started their training in earnest. Sensing I was an apprehensive, newly qualified Gazelle pilot, the 2IC 659, Captain Alan Stirling took it upon himself to tutor me in the geography, base locations and threats and hazards of the Province. This went a significant way in giving me confidence and assurance as 659 deployed. It therefore came as a massive shock to hear that Alan and his aircrewman had died in a Scout accident north of Crossmaglen, in Lough Ross only days before we flew out to the Province. It was a tragic and salutary lesson for us all, as we started our tour.[84]

XW614 was the last of five Scouts to be lost during Operation Banner although there were further lucky escapes.

On 3 March 1979 Sergeant Bob Petrie RE from 655 Squadron was on duty at Bessbrook. The South Armagh roulement infantry battalion was in the process of handing over and Petrie was tasked to fly a familiarisation sortie of the battalion's area of responsibility. Normally an aircrewman would have sat in the front left-hand seat, but on this occasion it was occupied by a company commander from the outgoing infantry battalion, Major Charles Woodrow Grenadier Guards. Four passengers from the 3rd Battalion Queen's Regiment, the incoming infantry

81 Anon, 'Accident', *Army Air Corps Newsletter,* No.82, September 1978, pp.18–19. Brian Ives, promoted to WO2, was subsequently killed in a Lynx crash in France on 17 September 1986.

82 Anon, 'Accident', *Army Air Corps Newsletter,* No.83, October-November 1978, pp.22–25; Anon, 'Obituaries', *Army Air Corps Journal,* 1979, p.65; Anon, 'Army Air Corps Roll of Honour', HQ AAC, 10 November 2018; Anon, 'Aircraft Losses since 1957', HQ AAC, 13 June 2019.

83 Anon, 'Accident', *Army Air Corps Newsletter,* No.88, July-August 1979, pp.29–31; Anon, 'Obituaries', *The Army Air Corps Journal,* 1979, pp.65–66; Anon, 'Army Air Corps Roll of Honour', HQ AAC, 10 November 2018; Anon, 'Aircraft Losses since 1957', HQ AAC, 13 June 2019.

84 A. Simkins, account to the authors, 22 February 2021.

battalion, were loaded into the back of Scout XV126 at Crossmaglen and Sergeant Petrie set off to fly a routine familiarisation sortie of South Armagh.

> Towards the end of the flight we were transiting south along the border at approximately 1,000 feet with the border crossing point H24 out on the left-hand side. As the outgoing company commander was briefing the rear passengers I was fortunately staring ahead when I observed a barrage of tracer fire approximately 200 metres directly in front of the aircraft, the idea being for us to fly through the anti-aircraft fire. I immediately turned hard to starboard and dived to low level as I assumed the IRA gunners would not be able to quickly depress their weapons. At this stage my passengers were aware that we were in contact and as we descended we were hit several times as rounds came through both the cockpit and rear cabin. Maj Woodrow was hit in the leg by rounds which came up through the floor between the tail rotor pedals. I had been slightly injured by flying glass from the round which came through the co-pilots windscreen, but apart from that no other passengers were hit. That in itself was amazing, if you look at the entry and exit points through the cabin, how no one in the back was hit was incredible! Within a short time we had descended to virtually ground level and the firing had ceased. I therefore headed straight to the helipad at XMG [Crossmaglen] and shut down.[85]

On arrival at Crossmaglen Major Woodrow was cross-loaded to a Wessex for a casevac to Musgrave Park Hospital in Belfast. Over 100 rounds had been fired at XV126 from a M60 machine gun and a Garand rifle, and several rounds had hit the Scout in the cockpit, cabin and through one of the main rotor blades. Sergeant Petrie was understated in his recall of the incident as Major Woodrow later reported that the pilot was quite badly injured with glass from the cockpit embedded in his face. The fact that no one was killed in the attack is testament not only to the robust Scout but also to Petrie's quick reactions, flying skill and determination.[86]

After 10 years continuous flying operations in support of Operation Banner the AAC had much to be proud of. Gazelles, Scouts and Sioux had all contributed to deterring, suppressing and countering terrorist activity in Northern Ireland. New tactics had been developed and new equipment had been successfully brought into service. The REME had done prodigious work in maintaining the exceptionally high flying rate. Nonetheless, the attrition rate had been high, with 12 aircraft written off resulting in 10 aircrew and passengers killed, with seven more members of the aviation community murdered by PIRA in IED attacks. More serious in the long term however, the relentless turnover of squadrons in and out of Northern Ireland to fill the roulement plot was having a clear negative effect on morale across the AAC with aircrew in particular leaving the Army in worrying numbers. With no sign of an end to the Troubles, more changes would be required as Operation Banner and the AAC moved into the 1980s. As the AAC moved into a new phase of its development, stability and oversight would be watchwords in Northern Ireland.

85 R. Petrie, account to the authors, 7 October 2020.

86 Taylor, *Air War Northern Ireland*, pp.57–58 and R. Petrie, account to the authors, 7 October 2020.

Scout XV126 that was being flown by Sergeant Petrie when it was hit by small arms fire in South Armagh on 3 March 1979. Bullet holes are clearly visible in the windscreen, port window and cockpit chin. (B. Petrie)

Chapter 10

Coming of Age

The AAC experienced a turbulent infancy. From 1957, and throughout the 1960s and 1970s, it had to constantly manage change and cope with multiple challenges to its growth and progress while trying to establish itself as a new and effective arm. Those challenges included the recruitment of soldiers and officers into the new Corps and creating a recognisable, unified identity from those that answered the call. During its first two decades the AAC was subject to a bewildering array of organisational changes, and units and headquarters seemed to be in an almost constant state of flux as titles, locations and functions appeared, evolved and disappeared. Issues with equipment and its procurement, as ever in the British military, were a constant source of concern. New aircraft types were introduced while others were rendered obsolete as the Corps attempted to match form to function and sometimes *vice versa*. Underpinning all these challenges was the need to contribute to multiple operational theatres across the globe. The AAC's experience in some theatres, such as British Guiana, was relatively short-lived. Others – including Cyprus, Belize, Hong Kong and Northern Ireland – created legacies that would occupy aircrew, ground crew and technicians for decades to come.

By 1979, the initially controversial light blue beret no longer attracted derisory glances and comments and the cap badge was recognised throughout the Army. These were important factors in creating a unique identity within the Corps and in presenting a unified image to those without. This was essential during a period when there were few 'home-grown' AAC soldiers or officers. In 1961 the permanent cadre of the AAC, that is those officers and soldiers whose parent regiment was the AAC, numbered just 61 individuals: 41 officers and 20 warrant officers and SNCOs.[1] The first soldiers were not recruited directly into the AAC until October 1973 when the aim was to recruit 150 individuals per year. The first ground crew course began at the Army Air Corps Centre at Middle Wallop in February 1974, with the first 'thoroughbred' AAC ground crew arriving in their squadrons during May of the same year. Airtrooper Anthony Power was singled out for the privilege of being considered the first airtrooper. Alongside their ground crew training, Airtrooper Power and his contemporaries were also trained as driver/signallers, HGV drivers or clerks. In the longer term a number of them might have expected to be trained and later employed as observers, air gunners and pilots.[2] The AAC resisted the trend towards employing exclusively officer pilots adopted by the RAF and RN. The prospect of becoming a pilot, following the observer or air gunner route, therefore remained realistic, although challenging, for many soldiers who initially joined the Corps as ground crew. Although the AAC had accepted direct entry officers on short service commissions from the early 1960s, it continued to recruit the majority of its officers through transfer from other arms in order to ensure that the Corps would, 'never lose sight of our primary

1 Anon., 'Permanent Cadre AAC', *Army Air Corps Association Journal*, 1961, pp.10–11.
2 A. C. Jerrard, 'Training the Army Air Corps Soldier', *Journal of Army Aviation*, 1974, pp.71–72.

aim, which is to fly in support of the ground troops.' The first Regular Commissioned AAC officers passed out of the Royal Military Academy Sandhurst, alongside officers from the rest of the Army, in 1974 and began their flying training in March the following year. The pool of 'true blue' officers remained initially small, however, as the AAC took just two officers per year through this route.[3]

The AAC's structure and organisation went through dramatic change during the first two decades of its existence. In 1962, just five years after the Corps' establishment, trials were conducted to assess the value of integrating aircraft into the RAC, RA, infantry, RE and R Sigs at unit level. The trial was considered a success and the following year the Sioux was selected as the unit light aircraft. In 1964 the final paper on the reorganisation of Army aviation was signed off and the Integrated Flight Scheme was launched with regiments and battalions across the Army and around the world receiving their very own helicopters.[4] The AAC appeared in danger of losing its own unique purpose and identity before it had a chance to gain a real foothold. Even the AAC's in-house publication changed its title in 1966 from the *Army Air Corps Association Journal* to the *Journal of Army Aviation*, which seemed to confirm the dilution of its own reason for existence. The Integrated Flight Scheme was considered a great success by those regiments and battalions lucky enough to operate their own air troops, platoons or squadrons. That was particularly the case in operational theatres such as Borneo where the austere conditions, poor infrastructure and dispersed operations meant that the Sioux provided a commanding officer with an invaluable asset with which to conduct reconnaissance, liaison, communication, resupply and casevac. The Integrated Flight Scheme, however, bore the seeds of its own demise. To make the scheme work the AAC had been forced to introduce compromises, particularly concerning pilot training. To cope with the sudden expansion in aviation, a new pilots' course began once every five weeks and the course itself was reduced to 31 weeks.[5] The aircraft were spread so far and thinly that their maintenance stretched the provision of spares, the supply chain and the REME manpower required for servicing and repair almost to breaking point.

To avoid a catastrophic collapse of the maintenance system, in 1972 the reorganisation necessary for centralisation was approved. All air troops and platoons were absorbed into the AAC leaving just four RAC regiments with their own squadrons. The following year those squadrons were also transferred to the Corps. With centralisation came a change in titles and terminology.

> ...it has been decided to drop the title "Army Aviation", which is not a corps designation and no longer serves any useful purpose, and substitute "Army Air Corps" in all titles such as my own and in that of the Journal and Association. Similarly, the title "wing" (always an anomaly in Army phraseology) has been discontinued .... It may still sometimes be difficult to avoid using the word "aviation" (with a small "a")in the same general sense as one would use "armour" or "artillery"; but whenever the flying arm of the Army is being referred to, "Army Air Corps" is the correct term.[6]

Fifteen years after its establishment, the AAC was finally back on a firm footing. During the period of the Integrated Flight Scheme, army aviation headquarters had been established within divisional HQs with the dual function of providing aviation advice to the divisional staff and exercising special-to-arm command over the independent air troops, platoons and squadrons. As the integrated flights were merged into squadrons, the divisional aviation headquarters became

3 R. L. C. Dixon, 'Progress Report 1975', *Army Air Corps Journal*, 1975, pp.6–7.
4 Anon., 'Editorial', *Army Air Corps Association Journal*, 1964, p.5.
5 Anon., 'Editorial', *Army Air Corps Association Journal*, 1964, p.5.
6 Dixon, 'Progress Report 1975', p.6.

regimental headquarters. Centralising into squadrons quickly paid dividends. Standards were raised, manpower overheads were reduced, maintenance was streamlined, and helicopter support was apportioned fairly. By the end of the 1970s the centralised AAC squadron and regimental system was considered a success.

> AAC regiments have truly arrived and the Corps has at last a viable, regimental entry into almost every aspect of Army life. It has taken just twenty years, from September 1957, to achieve this and the dividends which the regimental system is now paying are there for all to see. Command is more clear cut, loyalties are less divided and AAC regiments are slowly but surely achieving recognition throughout the Army. [7]

The AAC's equipment programme, although generally successful overall, did suffer setbacks requiring improvisation and stop-gap measures. The 4,000lb AUW limit imposed on the Corps during its establishment had not proved to be a serious hindrance to development. The AAC had perhaps been slow to see the value of and adopt the helicopter as its primary aviation asset. At times it appeared that the RAF might steal a march in that field, but the AAC did steadily develop and gain experience in rotary flight. Having overcome most of its early problems with ground resonance the Skeeter began deliveries in 1958 and was being widely employed across the AAC by the following year. The Skeeter's size, however, meant that it was not suitable or robust enough for any duties apart from AOP and liaison, although it continued to see service through to the mid–1960s. In 1959 the Corps was anticipating the introduction into service of the Saunders-Roe P.531. Due to development and industrial issues, however, it did not enter service until 1963 as the Westland Scout. Although the Scout suffered from a chronic shortage of life in its Nimbus engines during its early years of service, once the teething problems had been resolved it proved to be a reliable and robust aircraft. Popular with the aircrew that flew in it, the Scout would go on to give decades of valuable operational service around the world. Perhaps more significantly, the Scout, particularly in Aden, had proved that helicopters had a role directly contributing to offensive operations and, as such, it helped pave the way for the anti-tank Lynx and the Augusta Westland Apache AH1.

With the delays to the Scout's arrival, in 1961 the Alouette II was brought into the AAC as a stop-gap measure. The Corps' first turbine powered aircraft proved to be much more than a simple expediency and, although its use was never widespread, the few aircraft that were procured gave sterling service on operations, particularly in Cyprus. Along with the Integrated Flight Scheme the Sioux entered service in 1964. It proved an excellent aircraft for observation and could provide some wider utility through limited resupply and casevac. It was, however, slow, had limited capacity and was withdrawn from service at the end of the 1970s.

1974 saw the introduction of the Gazelle which represented a significant increase in performance over the Sioux and the Scout. So much so, in fact, that the test pilots who conducted the initial trials with the aircraft saw fit to warn the rest of the AAC of the potential hazards:

> Gazelle can be a "Hot-rod". Its limits are beyond the personal limits of most helicopter pilots and conversion training will have to raise the pilots' limits to match those of the aircraft.... It is one of the facts of life that aircraft which are capable of sudden, high-speed turns are also capable of sudden collision. Aircraft which can fly faster can hit things harder. And it's no good saying, "Well, I'll just take it easy then – I'm no speed merchant!" because by the time you have one or two hundred Gazelle hours under your belt you find yourself secretly

7 R. Canning, 'The Growth of the Regimental System in the Army Air Corps', *Army Air Corps Journal*, 1978, pp.17–19.

> wishing that it could fly just a little bit faster – another thirty knots, say.... A generation of faster helicopters is just beginning to arrive. Start thinking about how to use the speed – not how to dodge it.... If you are a Sioux pilot, start worrying about how to navigate at twice the speed you do at present.[8]

Although not as robust as the Scout, nor suitable for all theatres of operation, the Gazelle inspired great loyalty from those who flew it and would go on to give nearly 50 years' service to the AAC. In 1978 the AAC took another step forward with the introduction of the Lynx into squadron service. Lynx would prove to be one of the mainstays of operational capability as the AAC entered its next period of development.

Despite embracing rotary flight, the AAC continued to see the advantages of operating fixed wing aircraft. The Auster had been inherited from the RAF AOP flights and squadrons and on establishment in 1957 was the only aircraft the AAC initially operated. It continued to provide sterling service in all operational theatres until it was finally retired in 1966. The Beaver was introduced in 1961 and the robust, utilitarian aircraft was immediately useful and popular across operational theatres. Friendly rivalry between fixed wing and rotary pilots was a feature that persisted in the Corps. 'No matter how the fortunes of the Army Air Corps may wax or wane, the Beaver remains indestructible. In a Gentlemen v Players match, the Beaver would be crewed by gentlemen almost as old and discoloured as their aircraft. The rotary Players would be younger, fitter, disappointed and determined to do better next time...'[9]

A plethora of role equipment was brought into service to improve the capability of those aircraft, some of it very *ad hoc* and temporary but with many examples that were carefully developed and withstood the test of time. The former category included the Mudlark modifications, locally developed and fitted to the Scouts in Hong Kong specifically to assist with the rigours of Operation Culex. More enduring examples included Nitesun, Heli-Tele and PNG which had utility on multiple operations and became familiar to generations of AAC personnel. Nevertheless, many equipment issues appeared cyclical and would resurface many years after apparently being recognised and solved. A young Captain Robin Grist had observed the vital contribution played by sand filters in protecting the Scout's gas turbine engine in Aden, only to be dismayed when, 25 years later as Director Army Air Corps, Brigadier Grist discovered the same equipment had been dropped from the Lynx equipment table when it was required during the Gulf War of 1991.[10]

Despite the RAF in 1957 insisting that the Army's aircraft should remain unarmed, the GPMG equipped Scout had proved useful in Aden. Nevertheless, the AAC continued to struggle with the idea of routinely arming aircraft in that fashion. In Borneo it was considered that an armed Scout firing into the jungle from above might cause more harm than good. In Northern Ireland, despite very public, early trials with fitting the GPMG to the Sioux it would take many more years before AAC aircraft would be routinely fitted with door guns. Meanwhile, throughout its first two decades, in the UK and Germany (and outside the scope of this history) the AAC grappled with the development of the helicopter fitted with anti-tank missiles.

Every new aircraft brought into service and each modification made to those aircraft required hours of dedicated work by the REME's aircraft engineers to ensure they could be employed as efficiently and effectively as possible. Alongside them the AAC's ground crew assisted in developing and integrating a raft of ground support equipment including logistic vehicles, refuelling bowsers and equipment, communications suites and weapons for defence of the HLS. They also

8 Anon, 'Gazelle Intensive Flying Trials', *Army Air Corps Newsletter*, No.42, January-February 1974, p.13.
9 Anon., 'Flight Test No.7 – The De Havilland Beaver AL1', *Army Air Corps Journal*, 1987, pp.66–67.
10 R. Grist, personal account to the authors, 4 July 2020.

helped introduce and develop enduring aviation tactics and procedures that would become second nature to generations of AAC soldiers and officers. In Malaya and Borneo AAC aircraft had flown to and from infantry FOBs in the jungle. It was in Aden, however, that dedicated aviation FOBs were first developed. By centralising the planning, refuelling, rearming and maintenance facilities in one forward and often austere location, aircraft reaction times to an objective were reduced and loiter times once in position increased. The aviation FOB also allowed the development of a greater understanding and synergy between the aircraft and the troops on the ground they supported. The introduction of centralised tasking via an air and aviation coordination cell led to more efficient and effective allocation and use of aircraft.

That the AAC managed to cope with most of the manpower, organisation and equipment challenges that it faced during its first two decades is commendable. The fact that it did so while maintaining operational effectiveness across multiple theatres around the world is nothing short of remarkable. On the day of its establishment the Corps inherited units on operations in Malaya, Cyprus and Northern Ireland. Each presented its own unique challenges, and each required a tailored approach from the officers and soldiers of the fledgling AAC. The lessons learned, and experience gained, were valuable and transferable to concurrent and future operations. The skills required to operate in and above the jungle, learned in Malaya, were efficiently carried forward to Borneo and then Brunei and transferred to the other side of the world for operations in British Guiana and Belize. The peculiarities of operating in the hot and high environment of Cyprus were equally applicable in Aden. Many of the skills required to find and stop illegal immigrants in Hong Kong were similarly valid to those required in the fight against the IRA in Northern Ireland. Fixed wing and helicopter TTPs were constantly evolving during operations and adapting across theatres. Many would become enduring and still be valid into the twenty-first century. Aviation force packaging, as practised during Operation Flash-Bulb Charlie in Aden, quickly became an established concept. AOP, ABFAC and NGFS would continue to provide vital support to troops on the ground. Casevac remained a critical lifeline and contributed to the maintenance of morale.

As a snapshot, in 1964 members of the AAC were involved in fighting Indonesian insurgents in the jungles of Borneo, coping with the partition of Cyprus and operating as part of a UN peacekeeping force, adapting from fighting in the mountains of the Radfan to countering terrorist actions in the Crater in Aden, contributing to internal security operations in the austere conditions in British Guiana, and maintaining a security presence in Hong Kong and Northern Ireland. The AAC remained a relatively small and tight-knit organisation and the lessons and experience from different operations could be spread around the Corps quickly and effectively by several means. First, soldiers and officers carried the expertise with them when they moved between theatres and the training establishment and HQ AAC. Second, the training establishment could ensure that aircrew and ground crew were taught the latest techniques and tactics during initial and subsequent training, utilising the direct experience of its staff. Third, information could be disseminated from HQ AAC through official doctrine and tactics publications, but also less formally through publications such as the *Army Aviation Newsletter* which was first distributed in 1970 and subsequently renamed the *Army Air Corps Newsletter* in 1975.

Wherever the AAC operated, danger was never far away. Thirty-nine members of the Army aviation family lost their lives during operations between 1957 and 1979. This included seven soldiers killed by IRA bombs in two separate incidents in Northern Ireland. In addition, seven passengers being carried in AAC aircraft lost their lives in crashes or through ground fire during the same period. The bravery displayed by members of the Army aviation family operating in these hazardous environments is reflected in nearly 200 awards for gallantry and distinguished service made to officers and soldiers as a result of their actions on operations during this period. These included: 10 DFCs, nine DFMs, 12 AFCs, 2 AFMs and 16 QCVSAs.[11]

By the end of the 1970s the AAC could reflect on over two decades of effective support to operations around the world. Looking forward to the 1980s and beyond there was no sign that the operational tempo might reduce. The Corps could, however approach the future with a confidence gained through hard-won experience. It was an attitude best summed up by the Director Army Air Corps at that point in time, Major General William 'Billy' Withall.

> I commend to you a look back to our illustrious predecessors and, at the same time, a look forward to what is in store. To conclude, I wish to report that today we have thriving Corps. It is fully manned for the first time in its recent history and the quality of our officers and men is second to none. Our aircraft are brand new and impressive and are expected to increase our capability and performance in a very short while. We can therefore face the future with confidence and the knowledge that as an Army Air Corps we are second to none.[12]

11 The authors extend their thanks to Jason Grover for the statistics on honours and awards provided on 20 November 2023. The authors acknowledge that there is an element of uncertainty in some award categories and therefore the overall figure but are confident that the numbers for the medals detailed are accurate.

12 W. N. J. Withall, 'Preface', *Army Air Corps Journal*, 1982, p.8.

Bibliography

Primary Sources (Archives)

The National Archives

War Office Papers:
WO 32: War Office and successors: Registered Files (General Series)
WO 288: Suez Campaign Headquarters: Papers and War Diaries
WO 373: Recommendations for Honours and Awards for Gallant and Distinguished Service (Army)
Cabinet Papers:
CAB 129: Cabinet Memoranda
Ministry of Civil Aviation Papers:
AVIA 101: Accidents Investigation Branch Case Files

Army Flying Museum (AFM)

AFM ALB 10: 656 Squadron AAC 1959–1962
AFM ALB 16: 658 Squadron AAC 1970–1973
AFM ALB 51: 652, 653, 661, 662 and 664 Squadrons AAC – Northern Ireland 1974/5
AFM ALB 55: 11 Flight AAC 1951–1959
AFM ALB 56: 11 Flight AAC 1959–1971
AFM ALB 205: 3 Flight AAC 1977–1991
AFM 123: Central America
AFM 136: Northern Ireland – Tasking, Reports and Correspondence
AFM 137: Northern Ireland – Annual Historical Records
AFM 139: Northern Ireland – Flying Statistics
AFM 141: Northern Ireland – Various Regimental Journals
AFM 171/172: 3 Flight AAC
AFM 174: 7 Flight AAC
AFM 179: 10 Flight AAC
AFM 182: 13 Flight AAC
AFM 183C: 14 Flight AAC
AFM 190: 25 Flight AAC
AFM 204C: 653 Squadron AAC
AFM 208: 656 Squadron AAC
AFM 209: 656 Squadron AAC
AFM 210: 656 Squadron AAC
AFM 215: 660 Squadron AAC
AFM 221: 665 Squadron AAC
AFM 221A: 660 Squadron AAC
AFM 259: UNFICYP

AFM 259A: UNFICYP
AFM Interview Recordings:
Curry, I., 20 May 2019

Liddell Hart Centre for Military Archives *(LHCMA)*
Papers of Brigadier Lunt
Papers of Major General Dunbar

Airborne Assault Museum (AAM)
Report on Operations in the Radfan by 3 PARA Group, 19–27 May 1964

Royal Anglian Museum *(RAM)*
HQ Middle East Land Forces, Operations in Radfan, 14 Apr – 30 Jun 64

656 Squadron Association Archive *(656 AA)*
656 AA 187: Hong Kong

Private Collections
Papers of F. Graham-Bell
Papers of J. Grover

Primary Sources (Published – Books)

de la Billiere, P., *Looking for Trouble: SAS to Gulf Command* (London: Harper Collins, 1994)
Hamilton, R., *The Uneven Road* (London: John Murray, 1955)
Harris, R., *Memories of a Happy Soldier* (Self-Published, 2017)
Kellock, D., *A Wonderful Aircraft* (Hull: Steve Taylor Design, 2016)
Mallock, R., *The Likes of Leicester* (Self-Published, 2013)
Mitchell, C., *Having Been a Soldier* (London: Hamish Hamilton, 1969)
Riley, J., *Oft in Danger* (Solihull: Helion Books,2015)

Primary Sources (Published – Articles)

Anon., '656 Light Aircraft Squadron', *Army Air Corps Association Journal*, 1959
Anon., 'Recovery in the Rain – Malaya', *Army Air Corps Association Journal*, 1959
Anon., '653 Light Aircraft Squadron, Army Air Corps', *Army Air Corps Association Journal*, 1959
Anon., 'Appreciation – Captain T.D.G. Mulady, Queen's Royal Regiment' *Army Air Corps Association Journal*, 1959
Anon., '651 Light Aircraft Squadron, Army Air Corps—1958', *Army Air Corps Association Journal*, 1959
Anon., '6 Liaison/Depot Flight', *Army Air Corps Association Journal*, 1960
Anon., '653 Light Aircraft Squadron', *Army Air Corps Association Journal*, 1960
Anon., '651 Light Aircraft Squadron (1st March 1959 – 1st January 1960)', *Army Air Corps Association Journal*, 1960
Anon., '653 Light Aircraft Squadron', *Army Air Corps Association Journal*, 1961
Anon., 'Permanent Cadre AAC', *Army Air Corps Association Journal*, 1961
Anon., '656 Light Aircraft Squadron, Army Air Corps 1961', *Army Air Corps Association Journal*, 1962
Anon., '10 (Independent) Recce Flight', *Army Air Corps Association Journal*, 1962

Anon., '656 Squadron, A.A.C.', *Army Air Corps Association Journal,* 1963
Anon., 'Army Air Corps Section (Cyprus)', *Army Air Corps Association*, 1963
Anon., 'Editorial', *Army Air Corps Association Journal*, 1964
Anon., '656 Light Aircraft Squadron', *Army Air Corps Association Journal*, 1964
Anon., '19 Liaison Flight', *Army Air Corps Association Journal*, 1964
Anon., '21 Recce Flight', *Army Air Corps Association Journal*, 1964
Anon., 'Stop Press – Cyprus', *Army Air Corps Association Journal*, 1964
Anon., 'Army Air Corps Section (Cyprus)', *Army Air Corps Association Journal*, 1964
Anon., '656 Squadron, A.A.C. Headquarters and Workshop', *Army Air Corps Association Journal*, 1965
Anon., Headhunting with 14 Flight in Sarawak', *Army Air Corps Association Journal*, 1965
Anon., '10 Flight in Borneo', *Army Air Corps Association Journal*, 1965
Anon., '"Pegasus Airways" – 21 Flight, Army Air Corps', *Army Air Corps Association Journal*, 1965
Anon., 'Cyprus – Aden', *Army Air Corps Association Journal*, 1965
Anon., '3 Flight AAAC, UNFICYP', *Army Air Corps Association Journal*, 1965
Anon., '24 Flight A.A.C. British Guiana', *Army Air Corps Association Journal*, 1965
Anon. (Alpha Zero Seven), 'Far Away Places', *Army Air Corps Association Journal*, 1965
Anon., 'The Air Platoon in Borneo', *Journal of Army Aviation*, 1966
Anon., '10 Flight A.A.C.', *Journal of Army Aviation*, 1966
Anon., 'The Guards' Sioux', *Journal of Army Aviation*, 1966
Anon. (Ironside), 'Denise and Delilah in Borneo', *Journal of Army Aviation*, 1966
Anon., 'Air Platoon 2nd Battalion Royal Anglian Regiment', *Journal of Army Aviation*, 1966
Anon., '6 Flight in Cyprus', *Journal of Army Aviation*, 1966
Anon., '25 Flight A.A.C. British Guiana Garrison', *Journal of Army Aviation*, 1966
Anon., '15 Flight A.A.C.', *Journal of Army Aviation*, 1966
Anon., 'Air O.P. Troop 1st Regiment Royal Horse Artillery', *Journal of Army Aviation*, 1966
Anon., 'Air Squadron', *Journal of the Queen's Dragoon Guards*, 1966
Anon., 'Headquarters 4 Wing Army Air Corps', *Journal of Army Aviation,* 1967
Anon., 'Air Squadron Life Guards', *Journal of Army Aviation*, 1967
Anon., '7 Flight Army Air Corps', *Journal of Army Aviation*, 1967
Anon., '10 Flight Army Air Corps', *Journal of Army Aviation*, 1967
Anon., '11 Flight Army Air Corps', *Journal of Army Aviation*, 1967
Anon., '4 Air OP Troop RA, UNFICYP', *Journal of Army Aviation*, 1967
Anon., 'UNFICYP Flight Army Air Corps', *Journal of Army Aviation*, 1967
Anon., 'HQ Army Aviation FARELF (4 Wing AAC)', *Journal of Army Aviation*, 1968
Anon., '14 Flight Army Air Corps', *Journal of Army Aviation*, 1968
Anon., 'Air Platoon 2nd Gurkha Rifles', *Journal of Army Aviation*, 1968
Anon., 'Infantry Air Platoon Cyprus', *Journal of Army Aviation*, 1968
Anon., 'Obituary of J. Baulcomb', *Journal of Army Aviation*, 1968
Anon., 'Air Squadron 4th/7th Royal Dragoon Guards', *Journal of Army Aviation*, 1968
Anon., 'Infantry Air Platoon Cyprus', *Journal of Army Aviation*, 1969
Anon., 'Air Squadron 4th/7th Royal Dragoon Guards', *Journal of Army Aviation*, 1969
Anon., 'HQ Army Aviation FARELF', *Journal of Army Aviation*, 1970
Anon., 'The Air Platoon, The Prince of Wales' Own Regiment of Yorkshire', *Journal of Army Aviation*, 1970
Anon., 'Air Squadron 17th/21st Lancers', *Journal of Army Aviation*, 1970
Anon., 'Army Aviation in Northern Ireland', *Army Air Corps Newsletter*, No.5, June 1970
Anon., 'Ferry Flight – The Return of the Beavers of 130 Flight from Singapore to Middle Wallop', *Army Air Corps Newsletter,* No.12, January 1971

Anon., 'UNFICYP Aviation Flight', *Journal of Army Aviation*, 1971
Anon., '663 Aviation Squadron', *Journal of Army Aviation*, 1971
Anon., 'A Squadron in Ulster', *Journal of Army Aviation*, 1972
Anon., 'Army Aviation in Northern Ireland, a Squadron Commander's Impressions', *Army Air Corps Newsletter*, No.26, April 1972
Anon., '28 ANZUK Aviation Squadron', *Journal of Army Aviation*, 1973
Anon., 'How to Land on a Submarine Without Really Trying', *Journal of Army Aviation*, 1973
Anon., '656 Aviation Squadron in the Far East', *Army Air Corps Newsletter*, No.32, December 1972-January 1973
Anon., 'UNFICYP Flight', *Journal of Army Aviation*, 1973
Anon., '28 ANZUK Squadron/11 Flight Army Air Corps', *Journal of Army Aviation*, 1974
Anon., 'An Attachment to the UNFICYP Flight', *Journal of Army Aviation*, 1974
Anon, 'Gazelle Intensive Flying Trials', *Army Air Corps Newsletter*, No.42, January-February 1974
Anon., 'Two Shades of Blue – The Army Air Corps and UNFICYP', *Army Air Corps Newsletter*, No.44, April-May 1974
Anon., '653 Sqn AAC – Belize', *Army Air Corps Newsletter*, No.43, March 1974
Anon., '660 Squadron Keep the Peace', *Journal of Army Aviation*, 1974
Anon., 'Night Operations by 1 RTR Air Squadron in Northern Ireland', *Army Air Corps Newsletter*, No.45, June-August 1974
Anon., '11 Flight AAC', *Army Air Corps Newsletter*, No.56, December 1975
Anon., 'Sioux Performance or Lack of it!', *Army Air Corps Newsletter*, No.48, January 1975
Anon., 'Impressions of Northern Ireland', *Army Air Corps Newsletter*, No.52, June-July 1975
Anon., 'Flight Safety – Accident (2)', *Army Air Corps Newsletter*, No.65, December 1976
Anon., 'Blue Water – White Rum', *Army Air Corps Journal*, 1976
Anon., 'Wirestrikes', *Army Air Corps Newsletter*, No.63, September 1976
Anon., 'Accident', *Army Air Corps Newsletter*, No.64, October-November 1976
Anon., '"How Green is Your Valley?" A Look at the South Armagh Winter, Through PNG', *Army Air Corps Journal*, 1977
Anon., 'Accident', *Army Air Corps Newsletter*, No.69, April 1977
Anon., 'The Return of the Alouette to Cyprus', *Army Air Corps Journal*, 1978
Anon., 'Comrade, is your Conscience Pricking?', *Army Air Corps Newsletter*, No.77, January-February 1978
Anon., 'Accident', *Army Air Corps Newsletter*, No.82, September 1978
Anon., 'Accident', *Army Air Corps Newsletter*, No.83, October-November 1978
Anon., 'Accident', *Army Air Corps Newsletter*, No.88, July-August 1979
Anon., 'UNFICYP Flight – March to May 1980', *Army Air Corps Newsletter*, No.95, September – October 1980
Anon., 'AAC Detachment Belize', *Army Air Corps Newsletter*, No.92, March-April 1980
Anon., 'Belize Flight AAC', *Army Air Corps Newsletter*, No.94, July-August 1980
Anon., 'AAC Flight UNFICYP', *Army Air Corps Newsletter*, No.99, May-June 1981
Anon., '16 Flight AAC – Apr-Jun 81', *Army Air Corps Newsletter*, No.101, September-October 1981
Anon., 'Flight Safety – an Unwelcome Passenger', *Army Air Corps Newsletter*, No.97, January-February 1981
Anon., 'The Army Air Corps in Northern Ireland: Past, Present and Future', *Army Air Corps Newsletter*, No.105, May-June 1982
Anon., 'Air Incident at Dhekalia Airstrip, 21 July 1982', *Army Air Corps Journal*, 1983
Anon. [A. J. Davies], 'Flight Test No.3 – The Alouette II', *Army Air Corps Journal*, 1983
Anon., 'AAC Detachment, Belize', *Army Air Corps Newsletter*, No.112, July-August 1983

Anon., 'A Little Goes A Long Way', *Army Air Corps Newsletter*, No.114, November-December 1983
Anon., '16 Flight AAC (Oct-Dec 83)', *Army Air Corps Newsletter*, No.116, March-April 1984
Anon., 'Maintaining a Ceasefire', *Army Air Corps Newsletter*, No.117, May-June 1984
Anon., 'Army Flying Profile – Capt A.J. Davies AAC', *Army Air Corps Journal*, 1987
Anon., 'Flight Test No.7 – The De Havilland Beaver AL1', *Army Air Corps Journal*, 1987
Anon., 'UNFICYP Flight AAC', *Army Air Corps Newsletter,* No.143, April-June 1989
Anon., untitled photographic article, *Army Air Corps Journal*, 2003
Begbie, B., 'Why the Light Blue Beret?', *Army Air Corps Journal*, 1982
Bennett, J., 'Long Somado ECU Change', *The Chinthe*, Iss.1 2011
Bushell, J., 'After the Wave', *Army Air Corps Journal*, 2005
Canning, R., 'The Growth of the Regimental System in the Army Air Corps', *Army Air Corps Journal,* 1978
Carey, C. (CJC), 'Hong Kong Memoir 1963', *The Chinthe*, Spring 2010
Chambers, C. H., 'Aviation Mauritius', *Journal of Army Aviation*, 1969
Chambers, G., 'Narrative of a National Serviceman', *The Chinthe*, Summer 2009
Crouch, C., 'Borneo et al', *The Chinthe*, Iss.1, 2011
Davies, B., 'Borneo 1963–1966; A Day to Remember', *The Chinthe,* Iss.2, 2011
Deane, T. M., 'Austers over Borneo', *Army Air Corps Association Journal*, 1964
Dixon, R. L. C., 'Progress Report 1975', *Army Air Corps Journal*, 1975
Duthoit, W. P., 'Ulster Revisited', *Journal of Army Aviation*, 1974.
Elwell, R. S., 'Jungle Survival in Belize', *Army Air Corps Newsletter,* No.95, September-October 1980
Emery, D., 'The Extraordinary Case of the Bounced Czechs', *Army Air Corps Association Journal*, 1964
Fleming, M., 'With the Auster Mk 9 in Malaya 1959–62', *The Chinthe*, Spring 2009
Fogden, S., 'Hong Kong December 1974 – July 1977', *The Chinthe*, Spring 2010
Gill, P. N., 'Island of Tragedy, Cyprus – Summer 1974', *Army Air Corps Journal*, 1975
Hall, R. C., 'The Army Air Corps in support of the Royal Hong Kong Police', *Army Air corps Newsletter*, No.149, January-March 1989
Hayhurst, A. J., 'The Cyprus Emergency', *Army Air Corps Journal*, 1975
Hazou, J., 'Loan Service in Belize', *Army Air Corps Newsletter*, No.166, January-March 1995
Henderson, L., 'Defender in Belize', *Army Air Corps Newsletter*, No.160, July-September 1993
Hill, B., 'The First Flight', *The Chinthe*, Spring 2008
Jerrard, A. C., 'Training the Army Air Corps Soldier', *Journal of Army Aviation*, 1974
King, F., 'Foreword', *Army Air Corps Newsletter*, No.42, January-February, 1974
Markham, T., 'Service with 656 Squadron AAC in FARELF', *The Chinthe*, Iss.2, 2012
Mitchell, R., 'Air Portable', *The Chinthe*, Iss.2, 2014
Morgan, B., 'Memories of Sarawak 1963–1964', *The Chinthe*, Summer, 2008
Palmer, L. A., 'Six Months in the Six Counties', *Journal of Army Aviation*, 1972
Penfold, F., 'Auster Mk9 Recovery in Thailand', *Army Air Corps Association Journal,* 1961
Pettyfer, P., 'Austers in Honkers', *Army Air Corps Journal*, 2003
Redshaw, B., 'A Near Miss – Auster AOP 9 XK375', *The Chinthe*, Iss.2, 2010
Richardson, P. R., 'Whirlwind Operations in Cyprus', *Army Air Corps Association Journal*, 1959
Robinson, R., 'Casevac – A Birth in Flight', *Army Air Corps Newsletter*, No.172, July-September 1996
Rooke, P. W. S., 'Oh! To be an aircrewman in Hong Kong', *Army Air Corps Newsletter*, No.114, November-December 1983

Rutherford, M., 'With 25 Flight Army Air Corps in Belize', *Army Air Corps Newsletter*, No.168, July-September 1995
Salter, S., 'Beer Bottle Bombing in Borneo', *The Chinthe*, Spring 2007
Smith, T., 'No Turning Back', *Army Air Corps Journal*, 2006
Swan, D., 'Hong Kong 1975 to 1977', *The Chinthe*, Spring 2010
Thomson, A.F., 'Some unique features of 660 Squadron Army Air Corps LAD REME in Hong Kong', *Army Air Corps Journal*, 1980
Thursby, N., 'Cyprus – the Divided Isle', *Army Air Corps Journal*, 1980
Walker, D., '656 Squadron Tour, Malaya, 1958–1961', *The Chinthe*, Iss.2, 2013
Waters, S. P., 'Alouette, Gentille Alouette', *Army Air Corps Journal*, 1978
Welsh, R., 'Memories of 656 Sqn in Hong Kong – Dec 1969 to Oct 1971', *The Chinthe*, Spring 2010
Whidborne, R., 'Hong Kong Revisited 2007', *Army Air Corps Journal*, 2007
Withall, W. N. J., 'Preface', *Army Air Corps Journal*, 1982
Whitcombe, M. J., 'Smugglers in the South China Sea', *Army Air Corps Newsletter*, No.152, October-November 1990
Williams, P., '20 Indep Recce Flt, 1962 to 1963, Hong Kong', *The Chinthe*, Iss 1. 2013

Primary Sources (Unpublished – Manuscripts)

Graham-Bell, F., *The Other Side of the Coin, Suez 1956* (Unpublished manuscript, 1967)
Graham-Bell, F., *Helicopter on the Heights* (Unpublished manuscript, 1968)

Secondary Sources (Books)

Anon., *Operation Banner; An Analysis of Military Operations in Northern Ireland* (London: MOD, 2006)
Arthur, M., *Men of the Red Beret* (London: Hutchinson, 1990)
Ashcroft, M., *Heroes of the Skies* (London: Headline, 2012, e-book)
Bourke, R. and McBride, I. (eds), *The Princeton History of Modern Ireland* (Princeton: Princeton University Press, 2016)
Connolly, S. J., (ed.) *The Oxford Companion to Irish History* (Oxford: Oxford University Press, 2000)
Deacon, R., *Tales from the Frontline: Middle East Hunters* (Barnsley: Pen & Sword, 2020)
Easter, D., *Britain and the Confrontation with Indonesia 1960–66* (London: Tauris Academic Studies, 2004)
Edwards, A., *Mad Mitch's Tribal Law: Aden and the End of Empire* (Edinburgh: Mainstream Publishing, 2015)
Eklöf, S., *Pirates in Paradise: A Modern History of Southeast Asia's Maritime* Marauders (Copenhagen: NIAS Press, 2006)
Everett-Heath, J., *British Military Helicopters* (London: Arms and Armour Press, 1986)
Farrar-Hockley, A., *The Army in the Air: The History of the Army Air Corps* (Stroud: Alan Sutton, 1994)
Felton, M., *China Station: The British Military in the Middle Kingdom, 1839–1997* (Barnsley: Pen & Sword, 2013)
French, D., *Fighting EOKA: The British Counter-Insurgency Campaign on Cyprus, 1955–1959* (Oxford: Oxford University Press, 2015)

Gavin, R. J., *Aden Under British Rule, 1839–1967* (London: C. Hurst & Company, 1975)
Geraghty, T., *The Irish War: The Military History of a Domestic Conflict* (London: Harper Collins, 1998)
Holland, R., *Britain and the Revolt in Cyprus, 1954–1959* (Oxford: Clarendon Press, 1998)
Jackson, R., *Suez, The Forgotten Invasion* (Shrewsbury: Airlife Publishing Ltd, 1996)
Jackson, W., *Withdrawal from Empire* (London: Batsford, 1986)
Kennett B.M. and Tatman J. A., *Craftsmen of the Army, REME Official History* (London: Leo Cooper, 1970)
Lucas, S., *Britain and Suez: The Lion's Last Roar* (Manchester: Manchester University Press, 1996)
Mallock, R., *The Likes of Leicester* (Self-Published, 2013)
Moody, T.W. and Martin, F. X., *The Course of Irish History, Fourth Edition* (Lanham: Roberts Rinehart, 2001)
Neillands, R., *A Fighting Retreat; The British Empire, 1947–97* (London: Hodder and Stoughton, 1997)
Postgate, M. R., *Operation Firedog; Air Support in the Malayan Emergency, 1948–1*960 (London: HMSO, 1992)
Robertson, B., *The Army and Aviation* (London: Robert Hale, 1981)
Riley, J., *Oft in Danger: The Life and Campaigns of General Sir Anthony Farrar-Hockley* (Stroud: Helion, 2015)
Spinner, T.J., *A Political and Social History of Guyana, 1945–1983* (Boulder: Westview Press, 1984)
van der Bijl, N., *The Cyprus Emergency; the Divided Island, 1955–1974* (Barnsley: Pen & Sword, 2010)
van der Bijl, N., *British Military Operations in Aden & Radfan* (Barnsley: Pen & Sword, 2014)
van Tonder, G., *Malayan Emergency; Triumph of the Running Dogs* (Barnsley: Pen & Sword, 2017)
Taylor, D., *The First Helicopter Boys* (Barnsley: Air World, 2019)
Taylor, S., *Air War Northern Ireland* (Barnsley: Pen & Sword, 2018)
Tsang, S., *A Modern History of Hong Kong* (London: Bloomsbury Academic, 2007)
Twigg, A., *Understanding Belize: A Historical Guide* (Madeira Park, BC: Harbour Publishing, 2006)
Walker, J., *Aden Insurgency: The Savage War in Yemen 1962–67* (Staplehurst: Spellmount, 2014)
Warner, G., *First in the Field; 651 Squadron AAC* (Barnsley: Pen & Sword, 2011)
Warner, G., *From Auster to Apache; The History of 656 Squadron RAF/AAC, 1942–2012* (Barnsley: Pen & Sword, 2013)

Secondary Sources (Articles)

Anon., 'Obituaries', *Army Air Corps Association Journal*, 1965
Anon., 'Obituaries', *Journal of Army Aviation*, 1967
Anon., 'Obituaries', *Journal of Army Aviation*, 1969
Anon., 'Obituaries', *Journal of Army Aviation*, 1973
Anon., 'Obituaries', *Journal of Army Aviation*, 1974
Anon., 'Obituaries', *Journal of Army Aviation*, 1975
Anon., 'Obituaries', *Army Air Corps Journal*, 1979
Anon., 'Survival Scout (Adaption of an Aircraft)', *Army Air Corps Newsletter*, No.119, September – October 1984
Fitzgerald, A.D., 'An Outline History of the Joint Experimental Helicopter Unit 1955–1959', *Army Air Corps Journal*, 1980
Graham-Bell, F., 'The Army Goes Rotary', *Journal of Army Aviation*, 1967

Norman-Walker, R. A., 'The Formation of the AAC – 1955 to 1957', *Army Air Corps Journal*, 1982
Parker, R. J., 'The Third Flank', *Army Air Corps Association Journal*, 1959
Warner, G., 'Ulster's "Big Snow" of 1963', *Ulster Airmail*, February 2008

Secondary Sources (Newspapers and Periodicals)

Air International
Aircraft Illustrated
Blue Beret
Daily Mail
Defence Helicopter World
Foundation News
Soldier Magazine
South China Morning Post
The Dhow
The Guardian
The Hotspur
The Lion; British Services Cyprus Weekly
The Telegraph
The Tyrone Constitution
Visor Service News

Secondary Sources (Unpublished)

Anon., 'Army Air Corps Roll of Honour', HQ AAC, 10 November 2018
Anon., 'Army Air Corps Aircraft Losses Since 1 September 1957, HQ AAC, 13 June 2019'
Grover, J., 'Honours & Awards to the Army Air Corps/RA AOP/GPR', undated

Index